Sustainable Development Goals Series

The **Sustainable Development Goals Series** is Springer Nature's inaugural cross-imprint book series that addresses and supports the United Nations' seventeen Sustainable Development Goals. The series fosters comprehensive research focused on these global targets and endeavours to address some of society's greatest grand challenges. The SDGs are inherently multidisciplinary, and they bring people working across different fields together and working towards a common goal. In this spirit, the Sustainable Development Goals series is the first at Springer Nature to publish books under both the Springer and Palgrave Macmillan imprints, bringing the strengths of our imprints together.

The Sustainable Development Goals Series is organized into eighteen subseries: one subseries based around each of the seventeen respective Sustainable Development Goals, and an eighteenth subseries, "Connecting the Goals", which serves as a home for volumes addressing multiple goals or studying the SDGs as a whole. Each subseries is guided by an expert Subseries Advisor with years or decades of experience studying and addressing core components of their respective Goal.

The SDG Series has a remit as broad as the SDGs themselves, and contributions are welcome from scientists, academics, policymakers, and researchers working in fields related to any of the seventeen goals. If you are interested in contributing a monograph or curated volume to the series, please contact the Publishers: Zachary Romano [Springer; zachary.romano@springer.com] and Rachael Ballard [Palgrave Macmillan; rachael.ballard@palgrave.com].

Sinan Küfeoğlu

Financing the Green Transition

Principles and Instruments

Springer

Sinan Küfeoğlu
University of Vaasa
Vaasa, Finland

ISSN 2523-3084 ISSN 2523-3092 (electronic)
Sustainable Development Goals Series
ISBN 978-3-032-10439-7 ISBN 978-3-032-10440-3 (eBook)
https://doi.org/10.1007/978-3-032-10440-3

This Springer imprint is published by the registered company Springer Nature Switzerland AG
The registered company address is: Gewerbestrasse 11, 6330 Cham, Switzerland

Dedicated to my parents
Abdullah & Songül Küfeoğlu

Foreword

The world is facing a deep, wide-ranging, and multidimensional environmental challenge that requires a systematic transition and an extraordinary shift to sustainable and green development and economic growth. This transition is crucial to the future of the planet and survival of human civilisation.

The scientific and technological solutions required for a rapid shift are ready and continuously evolving, with many of them already exhibiting competitive costs. However, financing the green transition remains an insurmountable challenge, particularly for developing and emerging economies.

For instance, the International Energy Agency has indicated that to finance a just power sector transition aligned with a net-zero emissions pathway by 2050, lower- and middle-income countries—excluding China—must increase annual investment from an average of US$240 billion (2016–2020) to US$ 1 trillion in 2030.

Understanding how to address the spectrum and magnitude of risk, as well as how to structure financing schemes that optimise the use of scarce public resources to mobilise large volumes of private and concessional capital is imperative.

In this book, Sinan Kufeoglu delivers a comprehensive review of the principles of sustainable finance, the methodologies for understanding and measuring sustainable finance impact, and the instruments and mechanisms that have emerged over the last few years to effectively deliver a green planetary transition.

This book is an authoritative reference and practical guide for the international and academic communities on how to finance the green transition at a time in which accelerating and mobilising financing for a major global environmental shift is crucial.

Gabriela Elizondo

Practice Manager Energy Sector,
Latin America and the Caribbean
The World Bank,
Washington, DC, USA

Preface

Achieving global net-zero emissions by 2050 will require an estimated US$5–10 trillion in annual investment worldwide. A considerable portion of this funding is expected to come from sustainable finance mechanisms, which play a vital role in supporting the green transition. However, despite its increasing importance, sustainable finance remains a complex and rapidly evolving field. This book, *Financing the Green Transition: Principles and Instruments*, is written to bridge the gap between academia and industry, linking theory with practice. This is a descriptive and practical book, albeit not purely theoretical. With its professional-facing content and structure, it aims to target fresh graduates, finance professionals, and career switchers, but not primarily academic researchers.

The green transition, aimed at reducing carbon emissions and achieving net-zero goals, requires significant financing and expertise. Understanding the financial mechanisms and strategies that support this transition is crucial for those working in finance, policy, and business. This book provides a road map for navigating the financial complexities of sustainability, empowering readers to support and drive meaningful change towards environmental sustainability and social impact.

The green transition has gained traction within the finance and banking sectors, with a marked increase in job opportunities related to sustainability reporting, impact assessment, impact investing, and green financing. Employers are seeking candidates equipped with specific technical knowledge and skills that align with financing this vast transformation. By closely examining job market trends and skill requirements, this book aims to equip readers with the competencies and expertise sought by industry leaders, thereby positioning them to contribute effectively to sustainability goals.

While a range of academic literature exists on environmental economics and finance, a notable gap remains in resources that combine both a foundational understanding and practical application within a single volume. This book seeks to fill this gap by covering essential topics, ranging from the principles of sustainable finance to the latest sustainable finance instruments and impact assessment methodologies. Through structured chapters and real-world case studies, it offers readers insights into both the theoretical and operational aspects of sustainable finance.

Vaasa, Finland Sinan Küfeoğlu

Acknowledgement

I would like to express my heartfelt gratitude to the following people for their invaluable support and encouragement. Your help and unwavering belief in this book project have been a constant source of support.

Aybala Dalmış
Eyüp Ahmet Başaran
Hikmet Başdaş
İdil Aydın
Kübra Yavuz Günay
Oğuz Öztekin
Taha Yıldırım

I should also stress that the content of this book does not necessarily represent or reflect the official or unofficial stances or views of Ofgem or the UK Civil Service.

Competing Interests The author has no competing interests to declare that are relevant to the content of this manuscript.

Contents

Principles of Sustainable Finance

1

Abstract

This chapter explores the basic principles of sustainable finance. It discusses how it has changed over time and how it is different from green finance. This chapter also explains why sustainable finance is becoming more important worldwide. It addresses urgent environmental issues, especially climate change, and stresses the need to reach net-zero emissions by assessing risks and integrating ESG factors. Furthermore, it examines the economic effects of climate change, the expected costs of transitioning to net zero, and the need for new financial solutions to tackle these issues. By identifying important stakeholders and sharing real-world examples, it shows how sustainable finance can promote significant changes in critical infrastructure sectors, contributing to a more resilient and fair future.

1.1 Exploring the Principles and Values of Sustainable Finance

Modern economic models were developed under the presumption that natural resources and goods were abundantly available in the world. In the early stages of the Industrial Revolution, nature and its services were regarded as abundant in economic production, but labour and capital were limited resources that needed to be used efficiently. However, the global ecology, society, and economy were all significantly impacted by the Industrial Revolution. The loss of trees as a fuel source and the need for more energy-intensive fuels to feed rapidly increasing energy demand have made human society more reliant on fossil fuels and other non-renewable resources. As this dependence on fossil fuels grew over time, so did its detrimental effects on the environment. As a result, vivid discussions began on divesting from a carbon-based economy and investing in more sustainable sources. This discussion has evolved into a more inclusive investment philosophy covering social, ethical, and governance aspects, along with caring about environmental concerns.

1.1.1 Defining Sustainable Finance and Green Finance

All aspects of human existence, including economic activity and the social welfare it generates, will gradually be impacted by anthropogenic climate change. Using fossil fuels contributes to global warming, raising the amount of damage produced by the climate about total output before a detrimental occurrence. This situation causes the net output of the economy to decline. Moreover, the growth rate of an economy's capital stock is slowed by rising average temperatures. Eventually, it increases the stochastic risk

S. Küfeoğlu, *Financing the Green Transition*, Sustainable Development Goals Series,
https://doi.org/10.1007/978-3-032-10440-3_1

of natural disasters brought on by climate change (Steuer and Tröger 2022). Therefore, massive financial resources and investments are needed to eradicate poverty, fight climate change, reduce economic inequality, and, more recently, mitigate pandemic hazards. The United Nations Development Programme (UNDP) states that the public and commercial sectors are expected to cost between USD 5 and 7 trillion annually to achieve the Sustainable Development Goals (UNDP 2014). Raising additional funding and allocating existing resources to more sustainable initiatives becomes imperative and indispensable. To overcome problems associated with sustainability, businesses, investors, financial institutions, governments, and other stakeholders need to promote sustainable finance and investment projects (De Souza Cunha et al. 2021).

1.1.1.1 What Is Sustainable Finance?

The financial sector is crucial to the shift to a sustainable economy because of its importance as a growth engine in the economy. Sustainable finance is characterised by financial services and products that are designed to encourage responsible investments that have a positive influence on the environment, society, and governance while taking environmental, social, and governance (ESG) considerations into account during every step of investment, risk management, and decision-making mechanism (Buol et al. 2014). The allocation of investments to sustainable businesses and projects is facilitated by sustainable financing, which helps hasten the shift to a low-carbon, circular economy. This idea supports the concept of the triple bottom line profit, planet, and people, which broadens the concept of shareholder value to stakeholder value. Additionally, from the standpoint of sustainable finance, financial institutions only fund sustainable businesses and initiatives. According to this strategy, finance serves as a tool to promote a sustainable economy, with financial decisions beginning with a thoughtful selection of worthwhile initiatives and businesses by considering shared characteristics (Felicetti and Rizzello 2020).

1.1.1.2 The Main Players in Sustainable Finance

Based on their respective functions, four major participant groups in sustainable finance are identified: supporters, beneficiaries, recipients, and providers. The institutional environment necessary for sustainable finance implementation is created and developed by supporters. Conversely, the environment and society are the primary beneficiaries of sustainable finance. Social engagement is essential to legitimising sustainable finance activities and results since traditional financial participants continue to perceive the environment as an unlimited natural resource (Ebrahim and Rangan 2014; Lehner and Nicholls 2014). The demand fragment of sustainable finance is represented by recipients, companies that operate on sustainable practices in their investment and financial management. These companies assist services and products that create promising environmental and social effects through sustainable business mechanisms. The supply side of sustainable finance is formed by providers, including channel and intermediate financial capital and investments to facilitate maintainable solutions, innovations, and projects. Providers consist of individual investors, institutional investors such as mutual, pension, hedge, seed, and venture capital funds, value-based organisations like charities, and financial institutions like banks, insurance companies, and microfinance institutions. Compared to institutional investors, financial institutions have a far wider reach throughout the financial system; therefore, financial institutions should adopt more comprehensive strategies that integrate sustainability considerations into their procedures, policies, goods, and services to participate in sustainable finance. Many players previously described can participate in more than one of the four categories. For instance, governments can play supporters' roles by issuing sovereign bonds and providers by funding sustainable finance projects through public investment banks. In conclusion, creating and constructing sustainable business models can expedite the limpid implementation of sustainable finance (De Souza Cunha et al. 2021).

1.1.1.3 What Is Green Finance?

The German Development Agency (GIZ) defines Green Finance as a strategic approach that integrates the financial sector in the transition to resource-efficient, low-carbon economies as well as in the framework of climate change adaptation (GIZ 2011). Green finance aims to increase economic growth while decreasing greenhouse gas emissions and pollution, eliminating waste, and maximising the efficiency with which natural resources are utilised. Green finance should be considered an inherent element of sustainable finance since it is the accumulation and distribution of financial resources and assets across investment, banking, and insurance industries. Besides, sustainable finance corresponds explicitly with the achievement of the SDGs and a wide range of environmental, social, and economic objectives (Migliorelli and Dessertine 2019).

Green finance is a noteworthy method to enhance the commitment of the financial sectors to environmental conservation by offering capital to companies willing to adopt a green economy (Kumar et al. 2022). Furthermore, it has been acknowledged that green financing is a crucial component of SDGs 16 and 17 of the United Nations (UN) (Kumar et al. 2023). Implementing green finance strategies while considering the green economy is one of the most remarkable ways to increase the volume of financial flows (banking, microcredit, insurance, and investment) from the public, private, and non-profit sectors to accomplish the SDGs. For financial institutions to provide those funds vital for individuals navigating an unpredictable climate, they should consider promoting green goods in savings, credit, insurance, money transfers, and contemporary electronic delivery channels. Therefore, the overall investment strategy should be modified to be more environmentally friendly to support the achievement of the UN SDGs, all of which are intended to promote sustainable development (Kumar et al. 2022).

1.1.2 Historical Context: Evolution of Sustainable Finance

1.1.2.1 Early Roots and Development and Mainstream

The sustainable finance framework is ever-changing. The social-environmental component of the integrated value calculation is being increased because of new government regulations or taxes on social and environmental externalities (Schoenmaker and Schramade 2019). Figures 1.1, 1.2, 1.3, 1.4 and 1.5 show the history of sustainable finance between 1994 and 2022 which was published by the United Nations Environment Programme (UNEP). These historical events demonstrate the development and mainstream of sustainable finance.

The trends towards sustainable finance reflect the growing importance of ESG considerations in the global financial system. One key trend is ESG integration in mainstream investment. Traditional asset managers now recognise the importance of ESG factors and incorporate them into their decision-making. This more comprehensive adoption of ESG principles is expected to influence how capital is allocated and drive more sustainable business practices overall. As traditional financial instruments fall short of capturing the full range of sustainability risks and opportunities, there is a growing need for innovative solutions. This has resulted in the creation of new financial instruments, such as green bonds, social bonds, and sustainability-linked loans. These tools offer viable means of financing sustainable initiatives. Driven by environmental concerns and the recognition of ESG factors, investors increasingly seek financial returns alongside positive social and environmental impact. This trend transforms the financial landscape and promotes more sustainable business practices globally (Kandpal et al. 2024).

1.1.2.2 Transition Finance

Sustainable finance refers to funding current environmentally friendly (green) and long-term projects that gradually move towards environmentally friendly performance levels (a.k.a transition finance). To attain a sustainable and

Fig. 1.1 Development of sustainable finance from 1994 to 2004. (UNEP 2024)

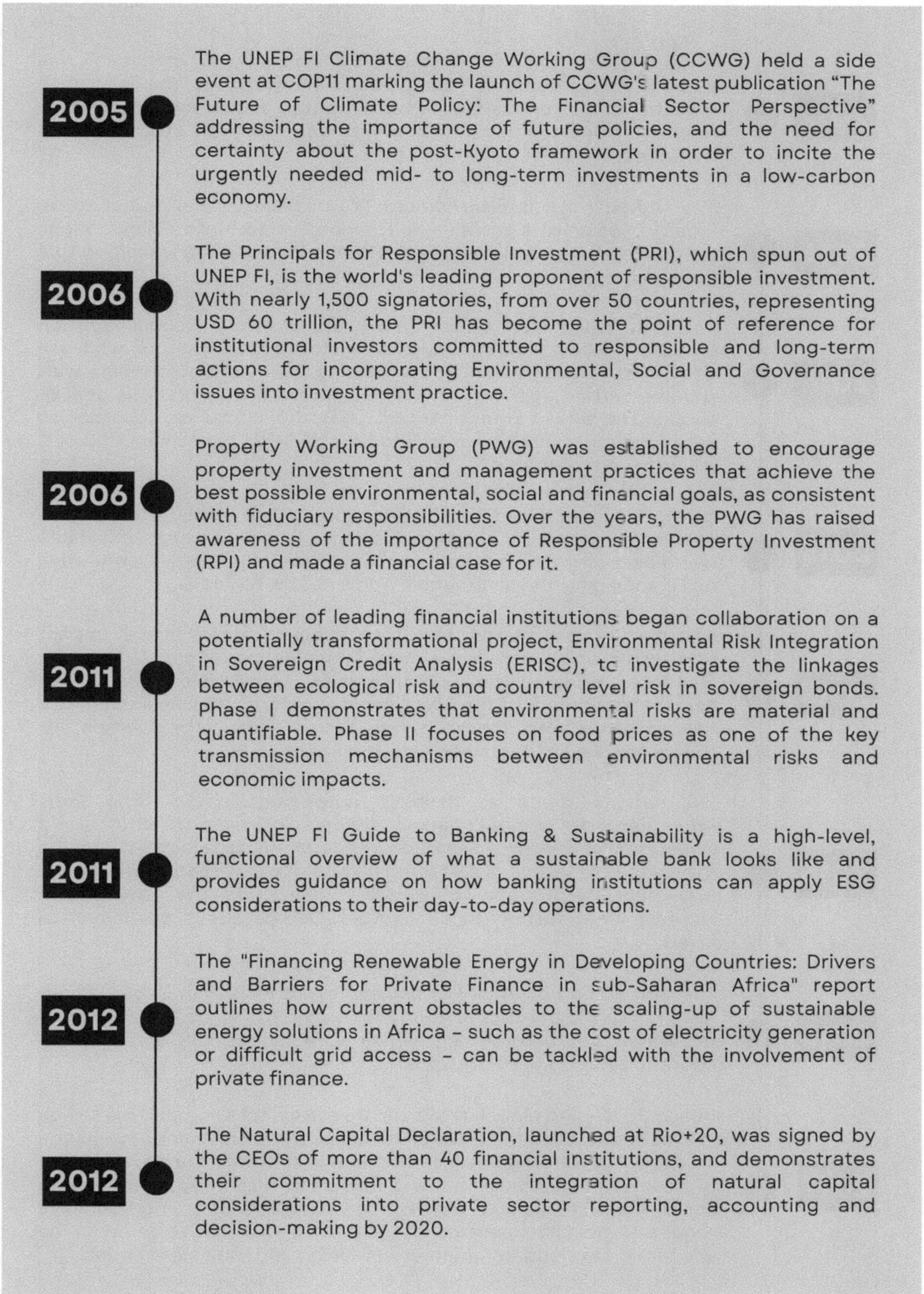

Fig. 1.2 Development of sustainable finance from 2005 to 2012. (UNEP 2024)

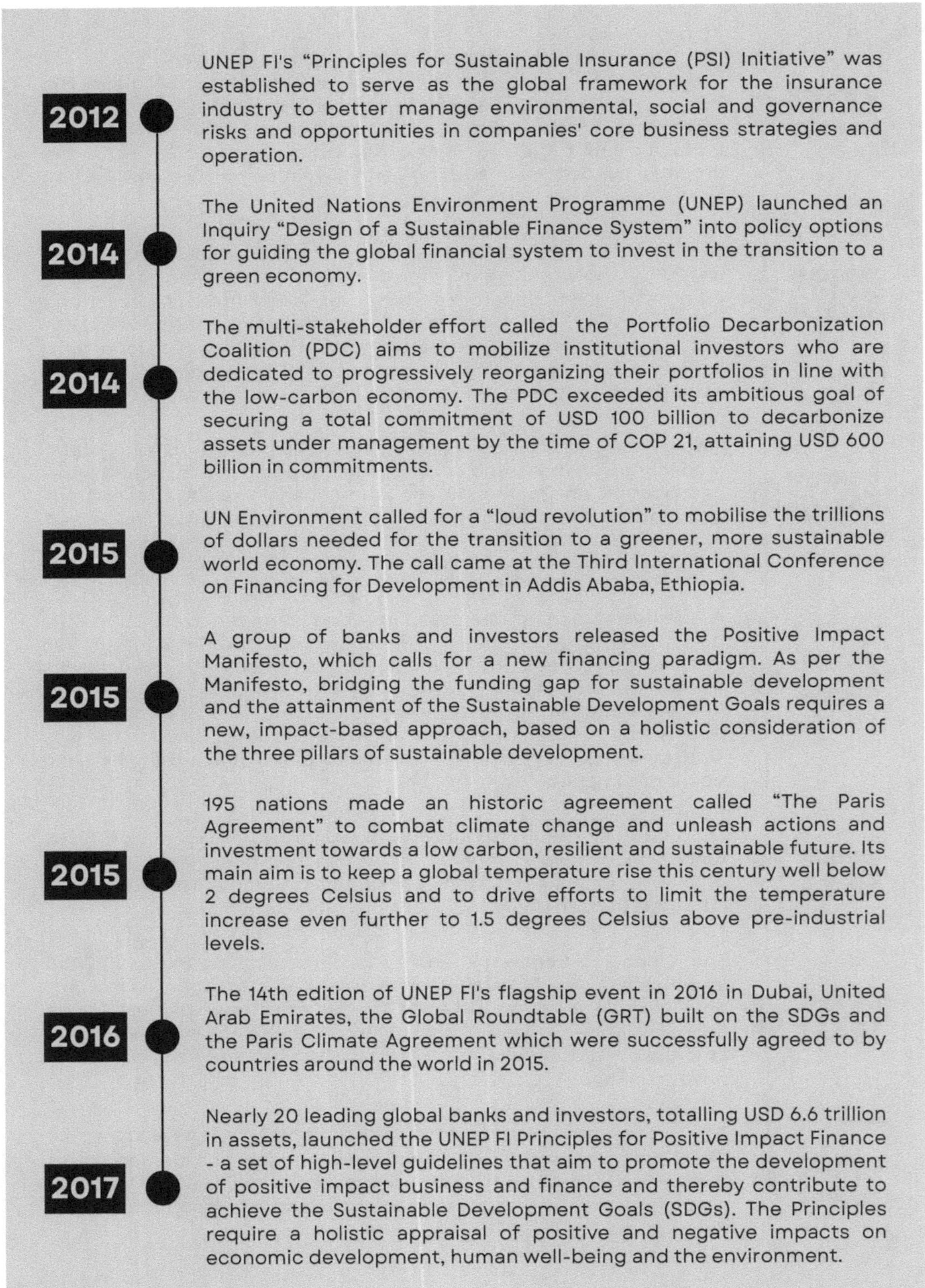

Fig. 1.3 Development of sustainable finance from 2012 to 2017. (UNEP 2024)

2017
UNEP FI's first series of 5 regional roundtables set up to mark our 25th Anniversary were agenda-setting events which showcased best practices in integrating environmental, social and governance factors into financial products and services, and provided insights into how a strategic approach to sustainability can position organisations for the 2030 Agenda. Financial institutions were able to: explore actions to mitigate risks and scale up green business lines; enable peer-to-peer learning and knowledge sharing across our regional hub, and interact with professionals and policymakers from their own region.

2018
At TCFD Pilot Project nine investors from six countries representing close to USD 3 trillion formed a leadership group with UNEP FI to promote climate transparency by the investor community.

2018
Developed by the Natural Capital Finance Alliance (NCFA), the web-based tool, Exploring Natural Capital Opportunities, Risks and Exposure (ENCORE), is the world's first comprehensive tool linking environmental change with its consequences for the economy. It helps global banks, investors and insurance firms assess the risks that environmental degradation, such as the pollution of oceans or destruction of forests, causes for financial institutions.

2019
UNEP FI hosted 5 Regional Roundtables. UNEP FI welcomed 2,000+ delegates and 500+ speakers across the series of events to help advance the sustainable finance agenda across the banking, insurance and investment industries.The Regional Roundtable for Africa and Middle East was held in Cairo, Egypt in April 2019. The Regional Roundtable for Asia Pacific was held in Shanghai, China in June 2019. The Regional Roundtable for North America was held in New York, USA in September 2019. The Regional Roundtable for Latin America and the Caribbean was held in Sao Paulo, Brazil in October 2019. The Regional Roundtable for Europe was held in Europe in Luxembourg in November 2019.

2019
The Principles for Responsible Banking were launched by 130 banks from 49 countries, representing more than USD 47 trillion in assets, on 22 and 23 September 2019 in New York City, during the annual United Nations General Assembly. More than 45 CEOs together with the UN Secretary-General attended the launch ceremony. The Principles provide the banking industry with a single framework that embeds sustainability at the strategic, portfolio and transactional levels and across all business areas. The Principles align banks with society's goals as expressed in the Sustainable Development Goals and the Paris Climate Agreement.

Fig. 1.4 Development of sustainable finance from 2017 to 2019. (UNEP 2024)

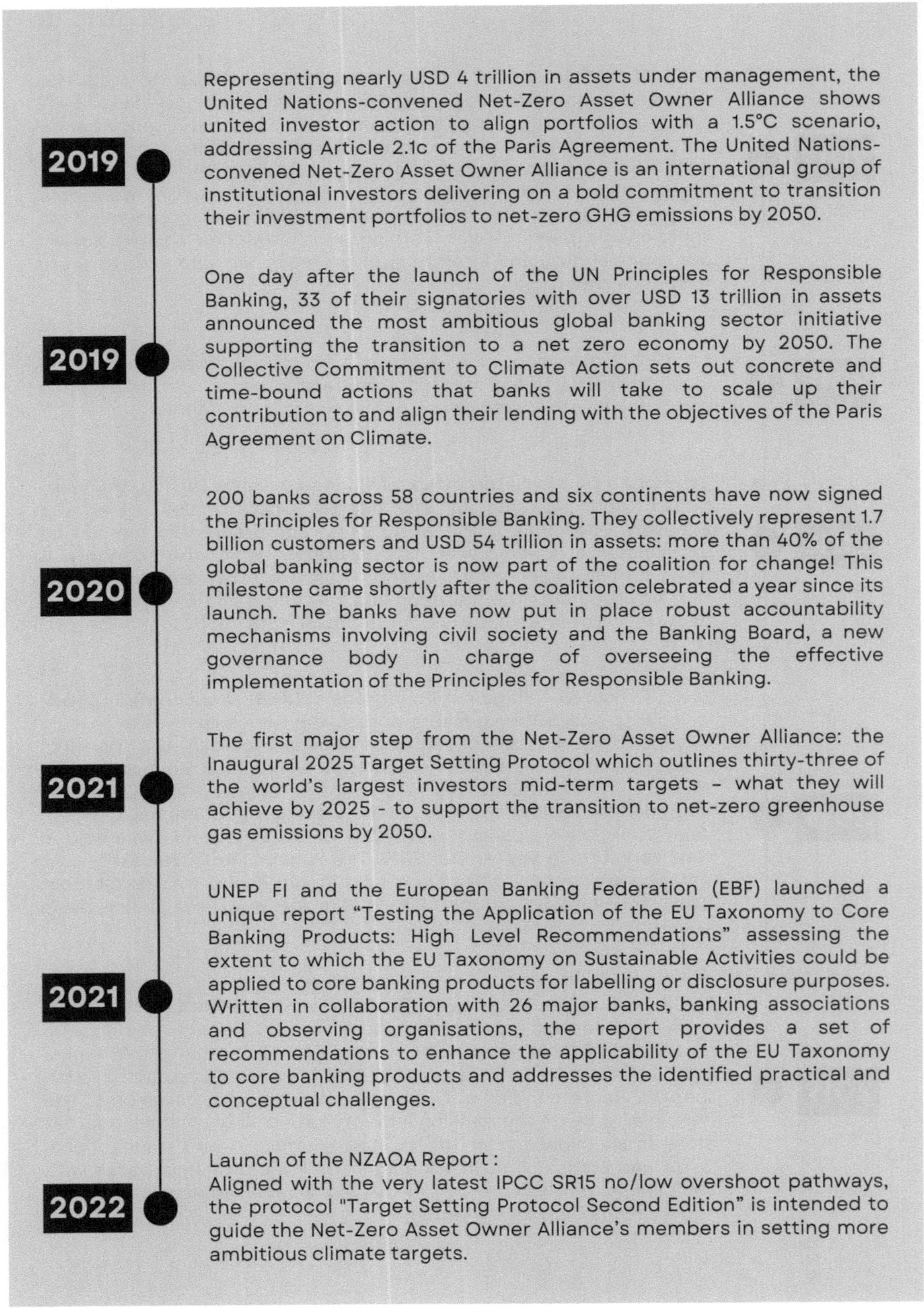

Fig. 1.5 Development of sustainable finance from 2019 to 2022. (UNEP 2024)

climate-neutral economy, transition finance aims to finance private initiatives to lower the high levels of greenhouse gas emissions and other environmental problems. If no green technologies are available, these could take the form of investing in environmentally friendly production techniques or making every effort to minimise the environmental impact (European Commission 2024a). Figure 1.6 shows the development of the shifting European Union (EU) sustainable finance debate.

Figure 1.6 underlines the two shifts in sustainable finance between 2009 and 2019. The first shift is from a voluntary to a legislative approach, and the second provides sustainability to climate finance. Also, three conflicting phases are indicated, and significant events on this timeline are given.

In October 2011, the European Commission proclaimed that it would take a closer look at disclosing responsible investing criteria mandatory for all investors. This served as the foundation for creating an EU policy on sustainable finance, which was supported by the SRI community, which saw sustainable finance as a viable substitute for the flawed and narrowly focused financial logic. Socially Responsible Investment (SRI) gradually entered the EU policy discourse to strengthen financial sustainability. An essential piece of legislation that was part of the historic 2011 Single Market Act was the establishment of a European framework that aided in the growth of social investment funds (Ahlström and Monciardini 2021).

The commission released non-binding guidelines on June 13, 2023, regarding non-financial and financial enterprises' voluntary use of EU sustainable finance tools to obtain or supply transition finance. The EU's sustainable finance toolbox supports companies with distinct sustainability aims and varying starting positions, as are those with the best sustainability records. Additionally, it enables smaller businesses to raise funds proportionately for their transformation. The particular domestic environment of each nation will determine the premises of the net-zero transition. The shape and ambition of the transition will depend on several factors, including resource endowment, economic structure, socio-economic priorities, fiscal capacity, present emission levels, the potential and costs

Fig. 1.6 Shifts and phases of sustainable finance for ten years. (Ahlström and Monciardini 2021)

of mitigation, and the socio-political acceptability of possible climate and environmental policies. In practical terms, this means that sectors and actions deemed to be "supporting the transition" may alter over time and between jurisdictions, as well as based on the capabilities of various nations (European Commission 2024a).

The enormous financial power of the European asset management sector (EUR 7000 billion in 2009) ought to be channelled towards the advancement of companies that have decided to prioritise social, ethical, or environmental development over the justifiable pursuit of profit (European Commission 2011). A Financial Transaction Tax plan was suggested for taxing all financial market transactions in 2013. The idea's proponents claimed this would generate significant public funds for crucial social and environmental initiatives (Ahlström and Monciardini 2021). Within this framework, defining "transition investments" has become a more pressing concern for a number of national and local government agencies, business associations, financiers, and civil society organisations. According to the Organisation for Economic Co-operation and Development's (OECD) 2021 assessment of methods connected to transition finance, there has been a growing emphasis from several national and local government agencies, business groupings, financiers, and civic society. Typically, transition finance is meant to help reduce the carbon footprint of businesses or economic activities that produce high emissions levels. These activities may not currently have a cost-effective or viable low- or zero-emission alternative available in all relevant situations, and they are crucial for future socio-economic development, as stated in the OECD's 2021 assessment of transition finance-related strategies. However, as of right now, neither a widely accepted concept of transition finance nor a set of technical requirements nor qualifying industries or technologies exist (OECD 2022).

1.1.2.3 Challenges and Barriers to Adoption

The conventional view of finance aligns with the assertion that "the business of business is business" and that a company's main social obligation is to use its resources and participate in profitable ventures as long as it plays by the rules. It is believed that the government should be responsible for social and environmental objectives and establish sustainability principles. Nonetheless, it is to be contended that governments, businesses, and citizens all have a shared duty for sustainable development in accordance with the SDGs of the UN (Schoenmaker 2017a).

More than half of the money from European investment goods comes from sustainable financial products. This is in line with the global trend. However, despite many efforts, the financial markets still lack a precise definition that systematically and clearly classifies an investment product as sustainable. As such, while assessing the sustainability level of a financial product, investors are presented with a convoluted and confusing array of facts. Making wise investment selections is hampered by this lack of transparency, necessitating a financial understanding of sustainable investments beyond basic financial literacy (Filippini et al. 2021).

Increasing regulation involves several contentious technological challenges. Beyond regulation, a major difficulty is getting individual companies and service providers to follow voluntary norms and principles for sustainable finance. Many financial firms sign sustainability pledges to protect their reputation but rarely follow through. For sustainability standards and principles to be more credible and effective, there is a need for institutionalised, independent monitoring and evaluation. The supervision of national authorities may take on some of these procedures. Beyond supervision, though, there should be reliable reporting and tracking of the application of agreed-upon sustainability standards (Gerster 2012). The spread of sustainability standards may result in more obstacles to funding and investment in developing nations, raising the cost of compliance. The diversity of sustainability standards, which has resulted in more than 200 coalitions of actors or sustainability initiatives, not only confuses markets but also makes it difficult for nations to draw in sustainable investors from around the world. Systems that consider local circumstances should be interoperable without standardised definitions and standards. To ensure that international standards are suitable for all, developing nations should contribute to their development without sacrificing their aspirations (Cattaneo 2022).

1.1.3 Why Sustainable Finance Is Important: Motivations

First and foremost, the structure of our world is primarily influenced by the financial sector. Fostering entrepreneurship and innovation is the backbone of trade and commerce and links nearly every nation through sectors (HM Government 2023). Finance, predominantly through socially conscious investments, is emphasised by Scholtens (2006) as a driver of sustainability. According to the "Report of the World Commission on Environment and Development: Our Common Future" also known as the Brundtland Report (UN 1987), sustainability ensures that current development satisfies needs without endangering the capacity of future generations to satiate their own. Using economic activity to safeguard the environment and resources for future generations is the aim of this paradigm shift (Rodrigo-González et al. 2021). In this sense, sustainable finance appears to represent a major change in how financial markets function. It signifies a notable shift away from passing trends and towards long-term sustainability and responsibility (British Embassy of Luxembourg 2023).

Even though it is frequently linked to the shift to carbon-neutral economies and alignment with the goals of international agreements such as the Paris Agreement, sustainable finance's scope extends far beyond these aspects (British Embassy of Luxembourg 2023). Roncalli (2022) categorises the motivations for adopting socially responsible investment into two primary groups. The first group encompasses economic sustainability, moral values, and social pressure, which align with the "do no harm" principle and broadly apply across various situations and decisions. The second group consists of factors such as financial performance, fiduciary duty, and risk management, which are rooted in investment principles. The central premise is that ESG risks must be addressed and cannot be overlooked when constructing investment portfolios. Incorporating ESG considerations into investment decisions, sustainable finance represents tomorrow's economy, and by investing in tomorrow's economy, the path towards a more livable future is paved, leading the way to prosperity, sustainability, and societal well-being, encompassing all essential aspects. A key component of sustainable finance is investors' capacity to create investment plans that have the power to shift the paradigm of how business is conducted in the future. Investors may be potent agents of sustainable finance by directing their investment flows towards economic sectors that can bring about that type of transformation (Amon et al. 2021). Figure 1.7 illustrates some critical areas where sustainable finance has an impact.

The concept of sustainable finance signifies a shift in the role of financial services towards fostering a more equitable, sustainable, and environmentally friendly future, as depicted in Fig. 1.7, considering notions such as climate change, diversity, SDGs, global prosperity, and economic growth, among others. Traditional finance prioritises financial return and views the financial industry as existing independently of both the environment and society in which it is entrenched. In contrast, sustainable finance takes into account the interplay of financial, social, and environmental rewards (Schoenmaker 2017a). This shift from traditional finance to sustainable finance yields significant rewards. Figure 1.8 illustrates the key changes in perspectives from traditional finance to sustainable finance, summarising the main aspects of this transformation.

As shown in Fig. 1.8, traditional finance usually prioritises short-term profits within legal bounds while possibly ignoring its potential contribution to tackling crucial global sustainability concerns for future generations. Conversely, sustainable finance actively participates in international sustainability initiatives and demonstrates a progressive strategy for protecting the future by

Fig. 1.7 Key areas related to and impacted by sustainable finance. (Kumar et al. 2022)

fusing financial, social, and environmental concerns (Schoenmaker 2017b).

The creation and execution of sustainable finance projects and the standards and regulatory frameworks that go along with them are now the focus of several international entities. While some of these institutions finance projects with a global aim, others have a more regional concentration. In addition to the necessary legislative frameworks and standards, numerous multinational entities and organisations are now developing and executing sustainable finance initiatives. Some of these institutions are regionally focused on the projects they support, while others have a global focus (Bhattacharyya 2021). The number of people adopting sustainable financing shows how dedicated public and commercial stakeholders are to the shift towards a greener economy and the widespread establishment of sustainable finance road maps around the globe (British Embassy of Luxembourg 2023).

1.1.3.1 Future Value Creation Through Sustainable Finance

The term "value" is used in various ways in today's commercial and economic environments, and value creation has become a significant topic of discussion across all industries. When value is discussed in today's world, especially in the business sector, people usually think of a product or commodities' monetary value. This is typical in a world where realised profit is used to determine a company's or business's level of success. There has been a paradigm change in the standards of success in global business and society since the United Nations unveiled the SDGs. The social and environmental effects of a company's activities are receiving more and more attention these days, as is the emphasis on evaluating a company's success using a range of criteria and measurements (Küfeoğlu 2022). According to Amon et al. (2021), investors should align their efforts with the principle of "doing well while doing

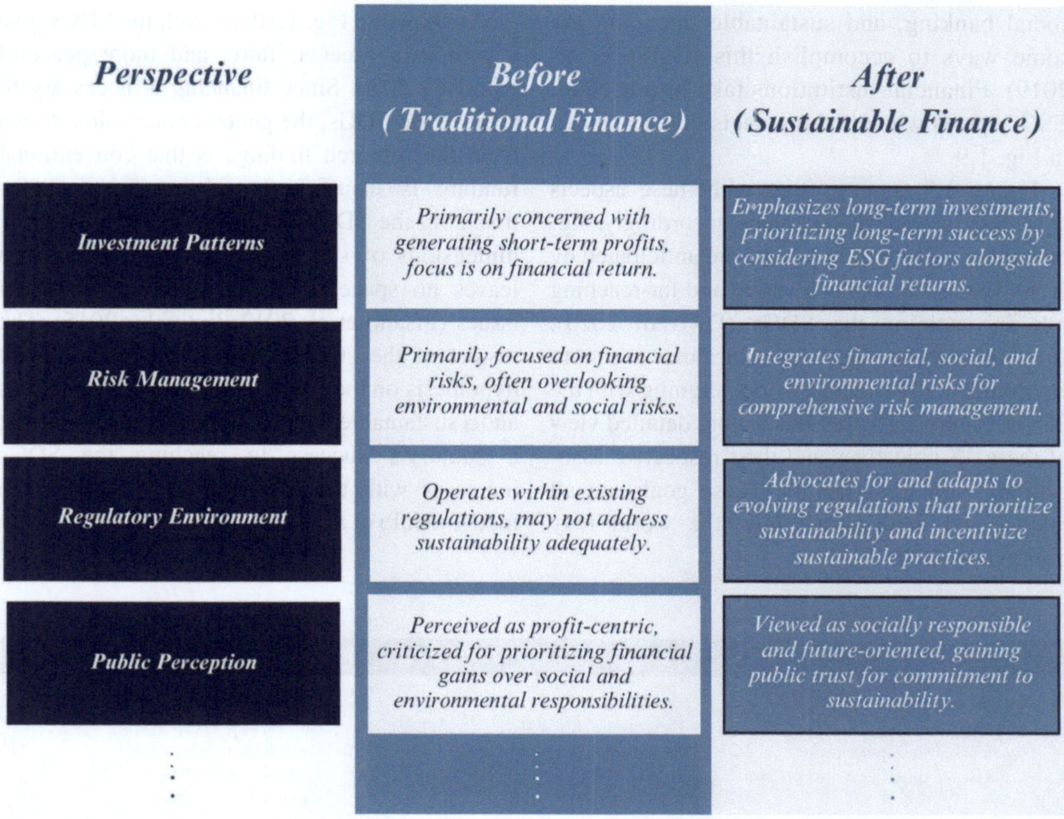

Fig. 1.8 Perspective shift: Traditional finance to sustainable finance. (Schoenmaker 2017b)

good". In this section, sustainable finance, which has various impacts across different fields, will be discussed in terms of its contributions to environmental, social, and governance enhancement, its role in economic, monetary, and business value, as well as its significance in mitigating risks and enhancing resilience.

The Role of Sustainable Finance in Environmental, Social, and Governance Enhancement

The UN (2015) has identified 17 interconnected and integrated SDGs in the 2030 Agenda for Sustainable Development. These SDGs represent a globally accepted, shared development vision towards a secure, equitable, and sustainable environment where all people may live and prosper (Osborn et al. 2015). Financial paths must be created to protect the environment; therefore, sustainable finance aims to ensure environmental protection. The worldwide cost of climate change damage is expected to be between USD 1.7 trillion and USD 3.1 trillion annually by 2050. In contrast, the World Economic Forum (WEF) estimates that the average cost of extreme weather losses from 2000 to 2019 was around USD 143 billion, or almost USD 16.3 million per hour (WEF 2019).

The financial sector and sustainable development have a contentious relationship. On the one hand, the financial sector lends to and invests in sustainable development. However, it has also contributed to the global financial crisis that has wreaked havoc on society. Despite this, the financial industry can be a crucial player in helping to finance the SDGs. Since it is evident that more than state money will be needed to achieve these goals, the banking industry may encourage private financing. Socially conscious investing, impact investing, social banking, and sustainable financing are some ways to accomplish this (Dalby et al. 2019). Financial institutions take into account ESG aspects when making decisions, as shown in Fig. 1.9.

Figure 1.9 suggests that with these aspects considered and decisions made accordingly, significant positive improvements are anticipated by 2030. ESG factors play a varied and far-reaching role in attaining the SDGs (Dwivedi 2023). Combining those objectives contributes to a more environmentally friendly world, aligning with the SDGs. Figure 1.10 provides a more detailed view of these 17 objectives and their respective associations with ESG factors. These goals are all related and demonstrate how the world may improve by 2030.

As shown in Fig. 1.10, overall, the SDGs give a path to a greener, faire, and more peaceful world by 2030. Since financing is necessary to achieve the SDGs, the general conclusion drawn from the research findings is that conventional funding is insufficient and inappropriate for financing the SDGs because it ignores the three dimensions of sustainable development, which leaves no space for social and environmental issues (Pisano et al. 2012; Fullwiler 2015). The results of the studies support the predominant hypothesis on the relationship between the SDGs and a sustainable finance model, which holds that a country's success in reaching the SDGs increases with the sustainability of its finance model (Ziolo et al. 2020).

Environment

- Carbon intensity emissions
- Climate change
- Control of environmental impacts
- Eco-design (financial green products and services)
- Eco efficiency
- Emissions
- Energy consumptions
- Environmental policy management
- Environmental reporting
- Environmental risk management
- Hazardous waste
- Materials recycled and reused
- Packaging
- Pollution management/recourses
- Protection of biodiversity
- Raw material sourcing
- Renewable energy consumption
- Travel and transport impact
- Waste management reduction
- Water use and management
- Industry-specific criteria

Social

- Business behavior
- Community relations
- Corporate citizenship/philanthropy
- Customer relationship management
- Customer and product responsibility
- Diversity
- Human capital development and training
- Human rights criteria
- Labor management
- Local suppliers
- Market ethics
- Non-discrimination, promotion equality
- Privacy and data security
- Protection of children
- Exclusion of children labor
- Quality of working conditions
- Respect of trade unions
- Responsible investing
- Rights of indigenous people
- Social reporting
- Stakeholder engagement
- Supply chain management
- Talent attraction/retention
- Work-life balance
- Industry-specific criteria

Governence

- Antitakeover policy
- Audit and control system
- Board diversity
- Board structure
- Brand management
- Business ethics and fraud
- Codes of conduct/compliance
- Corporate government functions and commitments
- Prevention of corruption and bribery
- Remuneration of members of the executive team
- Respect of shareholders rights
- Risk and crisis management
- Transparency
- Vision and strategy
- Antitrust policy
- Industry specific criteria

Fig. 1.9 Key ESG factors in financial decision-making. (Ziolo et al. 2019)

#	*Name*	*Description*	E	S	G
1	No poverty	End poverty in all its forms everywhere		S	
2	Zero hunger	End hunger, achieve food security and improved nutrition and promote sustainable agriculture		S	
3	Good health and well-being	Ensure healthy lives and promote well-being for all at all ages		S	
4	Quality education	Ensure inclusive and equitable quality education and promote lifelong learning opportunities for all		S	
5	Gender equality	Achieve gender equality and empower all women and girls		S	G
6	Clean water and sanitation	Ensure availability and sustainable management of water and sanitation for all	E	S	
7	Affordable and clean energy	Ensure access to affordable, reliable, sustainable and modern energy for all	E		
8	Decent work and economic growth	Promote sustained, inclusive and sustainable economic growth, full and productive employment and decent work for all		S	G
9	Industry, innovation and infrastructure	Build resilient infrastructure, promote inclusive and sustainable industrialization and foster innovation	E	S	G
10	Reduced inequalities	Reduce inequality within and among countries		S	
11	Sustainable cities and communities	Make cities and human settlements inclusive, safe, resilient, and sustainable	E		G
12	Responsible consumption and production	Ensure sustainable consumption and production patterns	E	S	G
13	Climate action	Take urgent action to combat climate change and its impacts	E		G
14	Life below water	Conserve and sustainably use the oceans, seas and marine resources for sustainable development	E		
15	Life on land	Protect, restore, and promote sustainable use of terrestrial ecosystems, sustainablymanage forests, combat desertification, and halt and reverse land degredation, and halt biodiversity loss	E		
16	Peace, justice and strong institutions	Promote peaceful and inclusive societies for sustainable development, provide access to justice for all and build effective, accountable and inclusive institutions at all levels		S	G
17	Partnerships for the goals	Strengthen the means of implementation and revitalize the Global Partnership for Sustainable Development			G

Fig. 1.10 Sustainable Development Goals with their descriptions. (Roncalli 2022)

The Role of Sustainable Finance in Economic, Monetary, and Business Value

Looking towards the future, sustainable finance emerges as a cornerstone in reshaping the global economy. The simultaneous and explosive growth of sustainable finance instruments and investment strategies is a result of two distinct trends: first, financial institutions are becoming more aware of the impact that sustainability risks, like climate change, can have on financial returns, financial stability, and a company's creditworthiness; second, there is a growing demand for "impact", with stakeholders ranging from individual investors to major institutional asset owners hoping to positively impact sustainability goals through their investment choices (Caldecott et al. 2022). The phrase "sustainable finance" refers to a broad category of investing strategies that go beyond analysing risk and return. Impact investing is one of these strategies. According to the well-known network in the industry, the impact investing market is valued at USD 502 billion (Dithrich and Mudaliar 2020). In contrast, according to a comprehensive definition, the total value of assets managed sustainably at the beginning of 2018 was USD 30.7 trillion (Global Sustainable Investment Alliance 2019).

The Role of Sustainable Finance in Mitigating Risks and Enhancing Resilience

A crucial aspect of sustainable finance involves managing risks through effective corporate and financial governance and transparency regarding ESG-related risks that could affect the financial system (European Commission 2024a). According to Alshubiri and Alraja (2022), sustainable finance aims to mitigate market risk, adjust to long-term environmental changes, and lessen financial strain by saving money for emergencies.

The Global Risks Report 2024, published by the World Economic Forum, presents insights from the Global Risks Perception Survey, gathering perspectives from 1490 experts across academia, business, government, the international community, and civil society (WEF 2024). The report examines global risks across three periods to assist decision-makers in striking a balance between immediate emergencies and long-term goals. According to the report, environmental risks are prominent across all time periods. Two-thirds of respondents identify extreme weather as the primary risk likely to trigger a significant global crisis in 2024. The current risk landscape anticipates a substantial global crisis in 2024, led by extreme weather, followed by AI-generated misinformation, social and political polarisation, rising living costs, and cyberattacks. In addition, extreme weather remains among the top 10 risks for both the short-term (2-year) and long-term (10-year) periods, consistent with the findings for 2024 (WEF 2024). It emerges as the primary concern over the next decade, ranking second only to misinformation and disinformation in the immediate future. Following extreme weather, the 10-year prediction emphasises issues, including depletion of natural resources, loss of biodiversity, collapse of ecosystems, and major modifications to Earth systems (WEF 2024). Incorporating sustainability into the mainstream financial system is essential for medium- to long-term economic stability, efficient risk management, and social and environmental responsibility. According to Jensen (2002), the traditional shareholder model frequently prioritises shareholder profit over social and environmental considerations, leaving these matters up to government regulation. Yet, sustainability programmes may reduce financial risks and preserve company and brand reputations by building good connections with stakeholders, gaining support from the community, and encouraging employee loyalty. Furthermore, they lessen the possibility of unanticipated shareholder resolutions and delayed product launches (Stankeviciene and Nikonorova 2014).

Adopting sustainable finance seems to be the best approach with clear financial benefits, despite the ethical requirements that companies commit to sustainable finance and take financial stability factors into account being disregarded (Boston Consulting Group 2023). In addition, 2019–2021 publications of the International Monetary Fund (IMF) indicate that sustainability is critical in managing financial risks from different angles. Financial instability may happen through social perils, failures in governance, or exposure to environmental hazards, among other channels. Hence, it is imperative to adopt sustainable financing approaches for effective control over such risk reduction (IMF 2021).

1.1.3.2 Funding Net Zero and Sustainable Finance

Net zero, to put it simply, is the balance between greenhouse gas emissions and atmospheric greenhouse gas removal. According to Levin et al. (2023), net-zero emissions are attained when all emissions caused by human activity are balanced by removing carbon from the atmosphere through carbon removal techniques. Reaching net zero will result in a net-zero atmospheric concentration of greenhouse gases resulting from human activity. Air quality is declining, and global temperatures are rising as a result of increased emissions. Nowadays, the main goal of net zero is human greenhouse gas emissions.

The rise of sustainable finance has been largely attributed to shifts in investor preferences, especially among millennials. Over the next 20 years, this generation will profit from the greatest transfer of inherited wealth in human history, totalling USD 68 trillion (Kelly 2021). Forty-five percent of millennials emphasise the significance of social responsibility in investing decisions and say they would like to invest their money for the benefit of society. They contend that long-term risk models for investment portfolios ought to consider governance, social, and environmental concerns (Nicholls 2021).

1.2 Global Environmental Challenges and Net Zero

1.2.1 Climate Change

This section provides a thorough overview of climate change and its financial impacts. First, a brief introduction to socio-economic and climate-forcing scenarios is provided, detailing how different routes affect sustainability and resource utilisation. The financial indications of climate change are then examined, considering the price of inaction, alterations to the world supply chain, and infrastructure damage. The section examines climate change scenarios such as Representative Concentration Pathways (RCPs) and Shared Socioeconomic Pathways (SSPs) and their practical consequences in the finance industry. Lastly, it covers risk assessment frameworks, how to include risk assessments in investment decisions, and sustainable finance's essential role in addressing climate challenges.

1.2.1.1 Understanding Climate Challenges: Climate Change Overview

Socioeconomic Scenarios

Climate change poses a growing threat to the world. Extreme weather events, rising temperatures, and shifting rainfall patterns will become more common, posing a unique threat to developing countries with fragile social, economic, and political institutions. Climate change can potentially impoverish millions in developing economies due to inadequate food, water, health, and infrastructure. It may make the United Nations SDGs unattainable and reverse previous development gains in these economies (Adom 2024).

Emissions, Concentration, and Climate-Forcing Scenarios

The rising concentrations of CO_2 and other greenhouse gases (GHG) in the atmosphere have resulted in rapid climate change. CO_2 emissions and other greenhouse gases contribute to global warming (Palocz-Andresen 2012). The best course of action should consider the potential to address both direct and indirect emissions when considering particular technologies or initiatives to minimise GHG emissions. A widely recognised standard known as the GHG Protocol outlines the methods for calculating emissions linked to direct energy consumption (scope 1), emissions related to the production of heat, power, and cooling (scope 2), and emissions from the value chain that are not connected to the (direct) purchase of energy (scope 3) (Hertwich and Wood 2018). The illustration and details of Scope 1, 2, and 3 emissions are demonstrated in Fig. 1.11.

Figure 1.11 visualises the Scope 1, 2, and 3 emissions, which are defined by International Financial Institutions (2021) as:

1. Direct greenhouse gas emissions from sources under the investor's ownership or control impacted by the investment project are called Scope 1 emissions.
2. Indirect greenhouse gas emissions from energy sources that the investment project directly uses but does not own or control are referred to as Scope 2 emissions.
3. Indirect greenhouse gas emissions from sources other than the investor's ownership or control, such as electricity, heating, or cooling used in their operations, are called Scope 3 emissions.

1.2.1.2 Climate Change Scenarios

Understanding the various projections of climate change is essential for anticipating and mitigating its wide-ranging impacts on ecosystems and human societies. Sustainable finance is a field that aims to address the difficulties and prospects brought forward by climate change by applying the knowledge. creativity, and power of the financial sector. Many countries now understand how revolutionary sustainable finance can be in helping people, families, and companies manage the new risks associated with climate change (Government of Canada 2019). This section will present how climate scenarios are used in the finance industry, how they can be incorporated into financial decision-making procedures, how

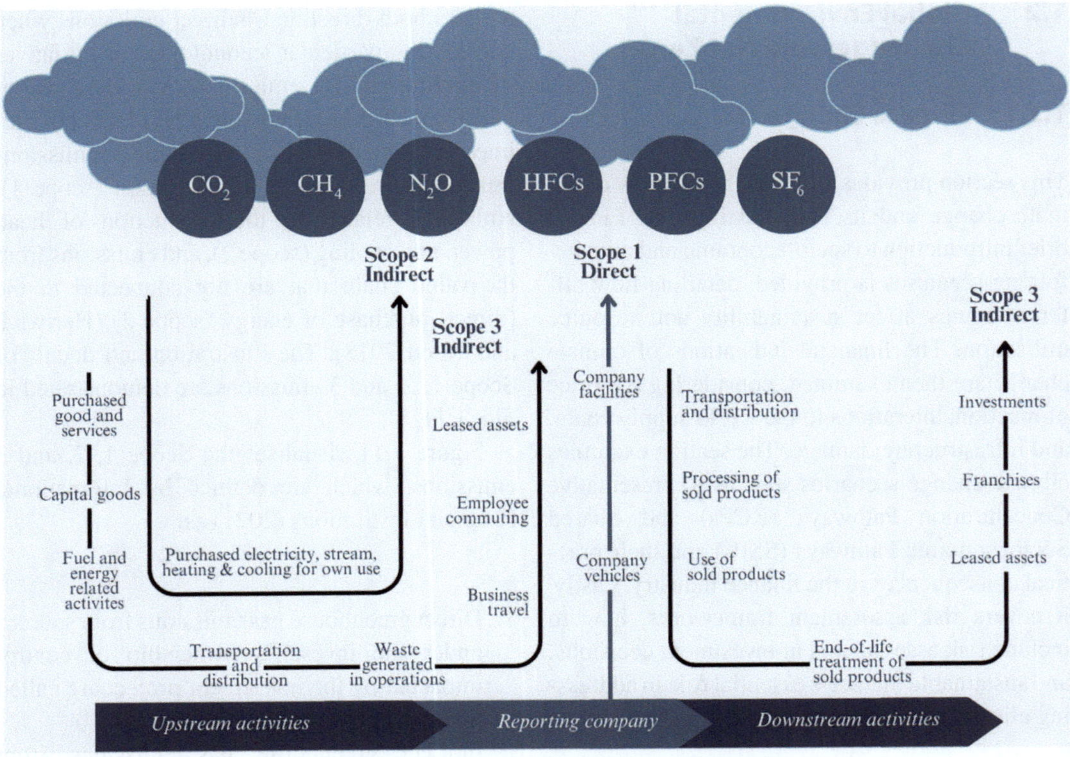

Fig. 1.11 Scope 1, 2, and 3 emissions. (EPA 2024a)

to evaluate the opportunities and risks that come with climate change, and the significance of sustainable finance in addressing climate change issues.

Representative Concentration Pathways (RCPs)

The RCPs are the outcome of an innovative collaboration involving specialists in emission inventories and modelling types associated with climate, terrestrial ecosystem, and integrated assessment. RCPs are valuable tools for estimating the emissions associated with different socioeconomic and policy modifications (Rott and Thomas 2023). As a consequence, a complete data collection with high sectoral and geographical resolutions for the period up to 2100 is produced (Van Vuuren et al. 2011). Scientists use the RCPs to forecast climate change and create impact scenarios (Coastadapt 2017). The figure below demonstrates representative concentration pathways in a short context.

The RCPs are summed up in Fig. 1.12. By the end of the century, 1.5 °C of global warming is the target set by RCP 1.9. The goal of RCP 2.6 is to limit temperature rise to less than 2 °C. The more permissive RCP 4.5 and the more rigorous RCP 2.6 are connected by the RCP 3.4. The moderate and high stabilisation paths are delineated by RCP 4.5 and RCP 6.0, respectively. RCP 8.5 depicts the maximum emission scenario with a continuous rise in emissions, whereas RCP 7.0 depicts a baseline scenario without further climate policy. These scenarios show many possible outcomes depending on differing greenhouse gas emission levels and attempts to slow down climate change (CDP 2023).

Shared Socioeconomic Pathways (SSPs)

The SSPs are a component of a framework for scenarios that delineate five potential paths for future socioeconomic growth and are established by the climate change research community. These paths, which span various difficulties for

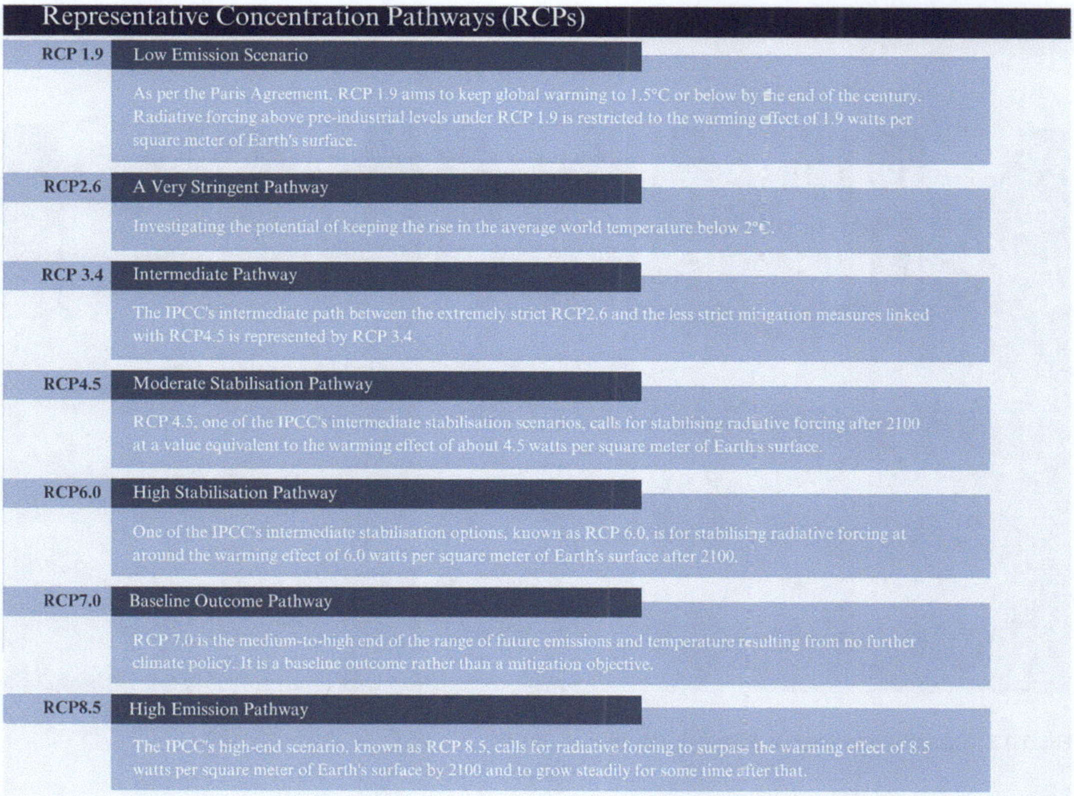

Fig. 1.12 Representative concentration pathways. (CDP 2023)

climate change adaptation and mitigation, are intended to promote policy analysis and research on climate change. They provide a simple comparison by describing the future regarding broad demographic, economic, and policy trends (Riahi et al. 2017). The figure below demonstrates shared socioeconomic pathways in a short context.

Figure 1.13 points out five shared socioeconomic pathways. SSP1, "Taking the Green Road", shows a society where inequality and resource intensity decline as it moves towards sustainability. SSP2, or "Middle of the Road", envisions a future in which attempts at sustainable development are met with uneven progress and enduring environmental issues. SSP3, "Regional Rivalry", emphasises how rising nationalism and regional strife impede economic advancement and cause environmental deterioration. In SSP4, "Inequality", a split planet with major social discontent and power dynamics and investments that are not equal, is depicted. SSP5, or "Fossil-fuelled Development", aims to balance population expansion and environmental concerns by emphasising economic growth through innovation and reliance on fossil fuels. These scenarios show many possible outcomes depending on various amounts of environmental, social, and economic variables (Riahi et al. 2017). RCPs assist in converting SSPs into anticipated changes in the global temperature. They provide potential trajectories for greenhouse gas (GHG) concentrations given varying amounts of GHG emissions over the twenty-first century (Rott and Thomas 2023). According to Frame et al. (2018), these couplings (SSPs and RCPs) can be used to characterise alternate configurations of human society and the natural world on a large scale.

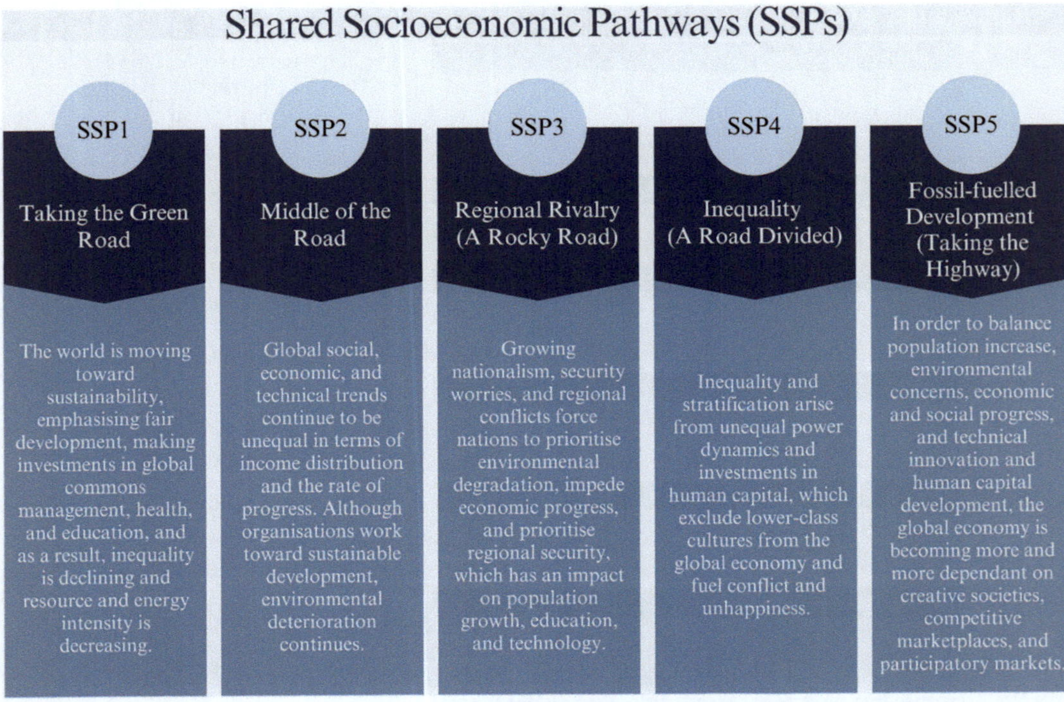

Fig. 1.13 Shared socioeconomic pathways. (Riahi et al. 2017)

Practical Usage of Climate Scenarios in the Finance Sector

The financial sector needs to know how the climate transition will impact economic operations, financial risk, and investment returns. Climate transition scenarios should be included in risk management and strategic planning endeavours. Scenario analysis is a helpful technique for financial companies to investigate the possible effects of climate action and climate change (Köberle et al. 2021). Analysis of climate scenarios is becoming essential in finance. Regulators seek to determine how prepared large financial firms are for systemic risk. External stakeholders closely examine it to assess the institutions' dedication to achieving climate objectives and shifting to a decarbonised economy. In the guide published by the United Nations Environment Programme (UNEP) Finance Initiative and Centre for International Climate Research (CICERO), it is reported that financial institutions understand the strategic importance of these studies in an ever-changing environment, as well as the dual pressure to do them (UNEP Finance Initiative and CICERO 2021). Scenarios should aid in a more profound knowledge of the factors that influence market, credit, operational, and other risks associated with climate change for the goal of risk management (Köberle et al. 2021). Financial institutions may discover a range of risks and possibilities connected with alternative climate futures by analysing various climate scenarios. This allows them to make well-informed decisions and implement the necessary steps (Flynn 2024).

1.2.2 Assessing Environmental Risks

More people increasingly recognise that environmental challenges are one of the greatest risk factors for the world economy. The term "environmental risk" describes the danger brought on by both climatic and non-climatic environmental factors. An industry's environmental hazards can directly influence the financial environment it operates in or the entire financial market. Environmental hazards have the potential to destroy a business as a whole, have an impact on the industry's ability to grow sustainably, or have an influence on human production and well-being. Financial companies may face strategic, credit, legal, and interest rate risks if they fail to manage environmental risks in the industry. It is stated that enhancing the identification of environmental hazards inside businesses, classifying environmental risks across industries, and proactively avoiding progressiveness risks are the noteworthy components financial institutions can perform to improve environmental risk management (Tang and Li 2024). According to a report of the World Economic Forum's 2017 Global Risks, water problems, significant catastrophes caused by nature, severe weather conditions, and the inability to effectively manage adaptation and mitigation of climate change are the main four hazards in terms of their potential effect on the planet. Effective and robust financial markets necessitate accurate management, pricing, and risk identification. Future developments are anticipated to make environmental disasters and externalities that result in financial risk even more crucial for financial regulators and organisations. Inadequate evaluation of environmental hazards can bring about mispricing of assets, dependence on stranded assets, and inefficient capital allocation, all of which might funnel money into polluting industries and abandon green industries underfunded (UNEP 2017b). The more comprehensive national and international transparency agenda strongly connects with attempts to enhance environmental risk assessment in procedures for making financial decisions. It is indicated that banking institutions should strive to perpetually boost the quality and capacity of their services for green finance (Tang and Li 2024).

1.2.2.1 Introduction to Environmental Risk Assessment

The initial action towards guaranteeing the security and profitability of financial assets is to execute an extensive approach to environmental problems (UNEP 2017b). Green finance is essential for raising capital and protecting against environmental risk to create a financially sustainable system. Green financial instruments offer investors and regulators options for managing portfolios and reducing risk (Naifar and Elsayed 2023). There is a growing awareness of how climate change impacts economic and financial variables, as seen by the focus on the sustainability profiles related to financial investments. Transition risk is the effect of shifting to a low-carbon economy on the financial system and the economy. Transition risks stem from human efforts to overcome environmental and climate concerns, such as modifications to laws and regulations, advancements in technology, changes in public or investor opinion, and the creation of novel business models. Risks related to the effects of climate change are classified as physical. According to the Network for Greening the Financial System (NGFS), physical risks are the harm that climate-related events cause to infrastructure and the properties of businesses and households (NGFS 2020). Despite their close connection to climate change risks, environmental risks are rooted in broader phenomena. These include the financial and economic consequences of overusing or degrading natural resources, such as pollution, deforestation, resource depletion, decline of ecological services, and loss of biodiversity (Faiella et al. 2023). Within the physical and transition risk categories, Tables 1.1 and 1.2 provide a concise taxonomy of environmental and climatic causes of risk.

Examples of physical and transition hazards affecting businesses and the financial institutions funding their operations are numerous and could have a negative financial impact. Table 1.1 represents the examples of physical risks in detail. To illustrate, "ecosystem pollution" is a physical risk that includes soil degradation, air and water pollution, and environmental accidents. Table 1.2 indicates the category of transition risks and solution examples related to possible risks. For instance, public policy changes comprise laws and regulations associated with energy transition and resource preservation, whereas technological changes include green technologies and clean energy projects Figure 1.14 points out the relationship between environment-related and

Table 1.1 Examples of physical risks (NGFS 2020)

Physical risks	Examples
Severe weather conditions	Hurricanes and tropical cyclones, heatwaves, droughts, wildfires, and hailstorms
Pollution of ecosystem	Soil degradation, soil/air/water/marine pollution, environmental accidents
Rise in sea level	Sea surges and persistent sea-level rise
Water scarcity	A shortage of water or a drought
Deforestation/ desertification	Population displacement, altered climate conditions, and the extinction of species were all brought on by deforestation

Table 1.2 Examples of transition risks (NGFS 2020)

Transition risks	Examples
Public policy changes	Policies for the energy transition, laws governing pollution, and rules for resource conservation
Technological changes	Green technologies include energy-saving, clean energy, and clean transportation
Shifting sentiment	Shifts in investor attitude towards specific asset classes and shifts in consumer preferences for particular products
Disruptive business models	Innovative business models that can overtake established ones quickly in terms of market share

climate-related risks, economic transmission channels, and financial risks.

Financial risks for lenders and investors are created by transition risks, which impact household wealth and corporate operations, as shown in Fig. 1.14. Via channels related to productivity, investment, and relative prices, they will also impact the larger macroeconomy. This is especially true if the transition results in stranded assets. Physical hazards have two effects on the economy. Extreme weather occurrences can have acute effects that cause property damage and economic interruption. These effects were previously thought to be transitory, but this will no longer be the case as global warming increases. These occurrences may raise the underwriting risks for insurers and lower the value of their assets. Chronic effects may impact labour, capital, and agricultural output. These effects are most likely from rising sea levels, temperature increases, and precipitation. Businesses, individuals, and governments will all need to make large investments in and adjustments to these developments (NGFS 2020).

1.2.2.2 The Role of Environmental Risks in Sustainable Finance

Finance may lead in directing investment into environmentally friendly businesses and initiatives, hastening the shift to a low-carbon and more circular economy. The study of sustainable finance examines the relationship between lending and investment and social, economic, and environmental challenges. Furthermore, finance may assist in addressing the inherent ambiguity surrounding environmental concerns, such as the effect of carbon emissions on climate change, because it excels at pricing risk for valuation purposes. Both sustainability and finance focus on the future (Schoenmaker and Schramade 2019). Adopting the sustainability concept in the banking industry has influenced banks' transformation towards increased corporate eco-efficiency and the creation of banking services and products that support sustainable development. While banks can find new opportunities in sustainable financing, they are also becoming increasingly concerned about environmental harm. Understanding how environmental hazards affect bank operations, central banks and supervisors are implementing various measures to lessen these risks' detrimental effects on bank operations and financial stability (Jemović and Radojičić 2021).

1.2.2.3 Frameworks for Assessing Environmental Risks

Environmental risk management is a crucial component of sustainable market activity principles, and both banks and non-bank financial organisations should consider this when granting loans. It is important to consider environ-

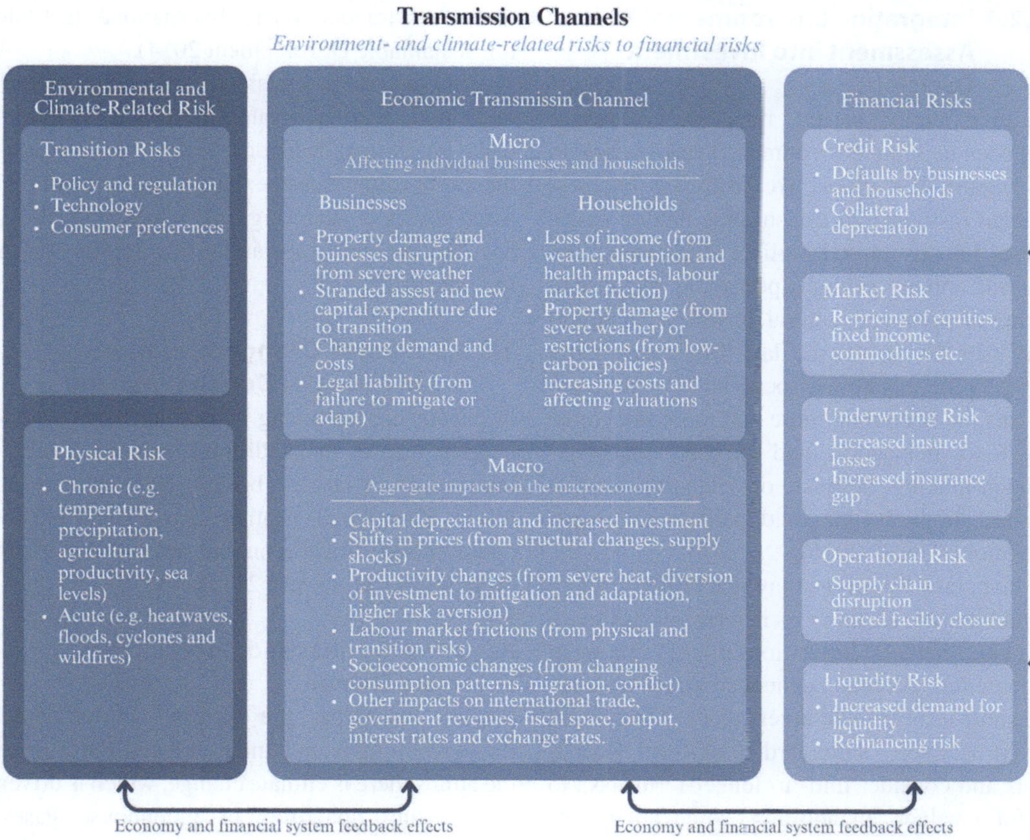

Fig. 1.14 Diagrammatic representation of the flow from financial to environmental hazards. (NGFS 2020)

mental risks to maintain stability in both money and finances. Owing to its objective of maintaining monetary and financial stability, which has a major impact on the inflation of consumer prices, green finance policies are an essential component. The macroprudential framework is considered the inclusion of bank green credit ratings in the assessment of macroprudential risk by central banks, considering climate and environmental risk (Naifar and Elsayed 2023). It is stated that a systematic and progressive strategy for green financing requires a successful assessment of environmental risks. There are four standard steps in the concept of analysis and control of environmental risks:

- Risk Identification: It strategically evaluates the several environmental conditions that could result in financial hazards.
- Risk Exposure: The vulnerability of financial institutions to various hazards is measured in this stage.
- Risk Assessment: In this stage, scenario analysis and stress testing are implemented to evaluate the probability and extent of monetary losses resulting from environmental risks.
- Risk Mitigation: This phase implements internal policies and procedures that forbid exposure to environmentally hazardous assets to lower hazards. Financial institutions can mitigate the risks associated with holding stranded assets by reducing their exposure to carbon-intensive infrastructure assets. They can also promote non-financial companies' green transition and environmental risk management by encouraging better information disclosure, offering risk management services, and engaging with shareholders more actively (NGFS 2020).

1.2.2.4 Integrating Environmental Risk Assessment into Investment Decisions

For an extended period, financial institutions have been tackling environmental risk factors. In this regard, the global (re)insurance business included resilience to catastrophic natural catastrophes into its capital regime, strengthened by scientific and modelling expertise, in response to a string of significant natural disasters that threatened its solvency in the late 1980s and early 1990s. In the meantime, social and environmental risks are becoming more and more integrated into fund managers' and project financiers' decision-making because of changing market standards in the banking and institutional investing sectors (UNEP 2016).

Financial institutions may react differently to the findings of risk analysis in terms of risk management. Financial institutions might implement fundamental risk management strategies to mitigate short-term risks to asset values. They might also proactively safeguard long-term investments and consider mid- to longer-term risks to portfolio values. In addition, some businesses are starting to respond to the long-term macroenvironment more strategically and are changing their investment choices to facilitate a smooth market transition to a low-carbon economy, which may lower risk and yield higher profits (UNEP 2017a, b).

The commitment of investors to incorporate environmental considerations is growing. Institutional investors are increasingly dedicated to integrating ESG factors into their investment plans to fulfil their ownership obligations and provide risk-adjusted returns. Organisations managing assets valued between USD 20 trillion and USD 87 trillion have committed to different degrees of integration; these commitments range from calling for more corporate disclosure to adjusting asset allocation to using ESG factors in valuations. The move to a low-carbon economy will probably disrupt market values. Market values have historically not considered long-term carbon targets, which has led to an overinvestment in fossil fuels. With changes in legislation, market, technical, and social factors, these could end up as stranded assets (International Institute for Sustainable Development 2014).

ESG factors are considered when both value-motivated and non-value-motivated investors make investment decisions. Since its inception as the moral values-based exclusionary screening of listed stocks, considering ESG issues in investing has undergone substantial evolution (CFA Institute 2015).

1.2.2.5 Climate Change Scenarios in Financial Decision-Making

Decisions are becoming increasingly more challenging due to the additional profound uncertainty climate change brings (Hallegatte et al. 2012). Even small weather changes can impact operations in critical economic sectors (American Meteorological Society 2013).

Assessing the Risks and Opportunities of Climate Change

Experts state that the biggest environmental threat resulting from human-induced warming of the atmosphere is climate change, which is driven by growing quantities of greenhouse gases, mainly CO_2 (Gołasa et al. 2021). Climate change brings both risks and opportunities. It provides financial markets with several obstacles but presents numerous possibilities for improving financial performance and guiding the economy towards sustainable growth. The two categories of climate risks are transition risks and physical hazards. Transitional risks are associated with the shift to a reduced carbon economy. In addition to risks, climate change presents possibilities. Innovation in products and services, market expansion, adoption of low-emission energy sources, and bolstering supply chain resilience are all opportunities brought about by climate change. These are just a few examples of how climate change may be mitigated and adapted to (EPA 2024b). The Task Force on Climate-related Financial Disclosures (TCFD) has established the relationship between climate-related risks, opportunities, and financial effects. Figure 1.15 presents this relationship.

Tables 1.3, 1.4, and 1.5 explain the climate-related transition, physical hazards, and opportunities together with their possible financial implications, as shown below, based on the classification provided in Fig. 1.15. This allows for the observation of a wide variety of different financial effects for each risk and opportunity. However, it is essential to note that many more risks and opportunities exist beyond those listed.

Tables 1.3, 1.4, and 1.5 highlight the financial implications of climate-related risks and opportunities as well as their financial consequences. The ways that regulatory, legal, technological, market, and reputational transition risks can lead to higher operational costs, asset depreciation, unstable financial circumstances, and a drop in consumer confidence are shown in Table 1.3. The physical risks that might disrupt operations, raise costs, and lower economic production are listed in Table 1.4, distinguishing between acute risks (like extreme weather) and chronic risks (like long-term climate changes). Table 1.5, which focuses on climate-related opportunities, demonstrates how resilience planning, resource efficiency, adoption of renewable energy, innovative products and services, creation of new markets, and enhanced competitive advantage can lower costs, improve financial performance, and give businesses a path to prosper in a low-carbon economy.

The Crucial Function of Sustainable Finance in Resolving Climate Change Issues

Corporations, financial institutions, and governments have a daily decision: focus on developing green solutions that promote a resilient, stable, and egalitarian economy or invest in tangible assets that damage the environment and release greenhouse gases (WEF 2023). The consequences of climate change on the financial and economic systems are becoming increasingly widespread, detrimental, and dangerous. A growing number of people are becoming concerned about these threats, which has led to a greater interest in environmental issues, including pollution, biodiversity loss, and resource exploitation. In addition, these variables may impact the financial system's stability, the economy's capacity for sustainable growth, and the social and governance components of businesses and organisations (Bank of Italy 2024).

The need for a sustainable future is becoming increasingly evident as civilisations grapple with the harmful effects of climate change (WEF 2023). Finance and investment flows from ecologically damaging economic activities must be redirected towards green initiatives and more sustainable alternatives. Eriksson et al. (2022)

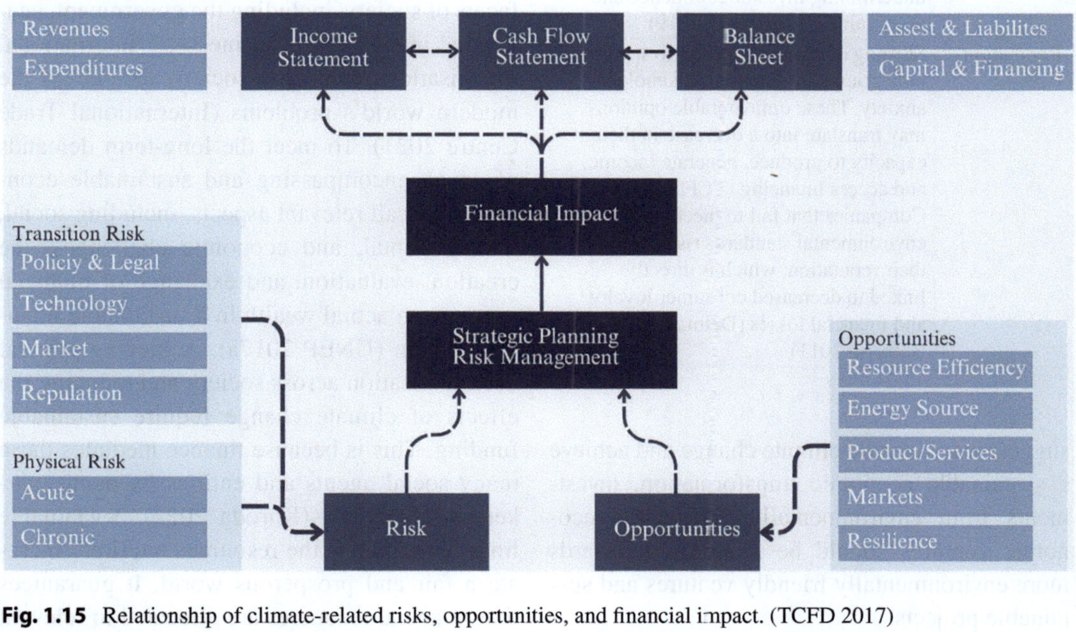

Fig. 1.15 Relationship of climate-related risks, opportunities, and financial impact. (TCFD 2017)

Table 1.3 The potential financial consequences of climate change-related transition risks

Climate-related risks	Potential financial impacts
Policy and legal risks	Higher operational expenses, impairments, and a decline in the demand from financial institutions for specific non-compliant goods and services (TCFD, 2017). Spiralling operating costs and asset devaluation concerns (Busch and Lewandowski, 2018)
Technology risks	Switching to technology with fewer emissions and decreased demand for old products can result in high expenses (TCFD, 2017), high upfront research and development expenses, uncertain returns (Luderer et al. 2019), and significant financial losses due to delayed technological innovation (Caldecott et al. 2016)
Market risks	Changing customer preferences, a lack of clarity in market signals, and rising raw material costs may reduce the market for products and services, drive up production costs, and cause energy prices to fluctuate quickly (TCFD, 2017). Changes in asset prices and severe financial instability in the financial sector might be brought on by fluctuating market circumstances (Dietz et al. 2016). Combining these climate change issues and other ex-ante market risks might amplify volatility and unpredictability in financial markets, undermining investor confidence and asset values (Hong et al. 2019)
Reputation risks	Shifting consumer aversion to specific industries and increased stakeholder anxiety. These unfavourable opinions may translate into a decrease in their capacity to produce, generate income, and access financing (TCFD, 2017). Companies that fail to meet environmental standards risk damaging their reputation, which is directly linked to decreased consumer loyalty and financial losses (Delmas and Burbano 2011)

suggest that to tackle climate change and achieve a sustainable economic transformation, investments from environmentally detrimental economic ventures should be reallocated towards more environmentally friendly ventures and sustainable projects.

Table 1.4 The potential financial consequences of the physical risks linked to climate change

Acute physical risks	Floods and cyclones are extreme weather occurrences that can cause supply chain disruptions, significant property damage, insurance costs, and operating expenses (TCFD, 2017). These incidents cause disruptions to company operations and increase insurance claims, which influences the bottom line of both businesses and insurers (Kousky 2019). Furthermore, there is a greater chance of financial risk, damage exposure, and insolvency due to the notable physical hazards connected to weather catastrophes (Hudson et al. 2019)
Chronic physical risks	Long-term climate risks have the potential to negatively impact economies by decreasing production capacity and raising capital and operational expenditures (TCFD, 2017). Furthermore, Hsiang and Kopp (2018) have shown that abrupt increases in average temperature can harm infrastructure, lower economic production, and incur enormous financial consequences. Burke et al. (2015) state that enduring physical climate-related hazards impede growth prospects and cause long-term economic disturbance

Addressing climate change is a complex challenge that necessitates collaboration across various sectors. Finance alone cannot solve all these issues by itself. It should work with all facets of society, including the government, academic institutions, businesses, international organisations, and civil society, to address the modern world's problems (International Trade Centre 2021). To meet the long-term demands of an all-encompassing and sustainable economy along all relevant aspects, including social, environmental, and economic challenges, the creation, evaluation, and exchange of financial assets form actual wealth in a sustainable financial system (UNEP 2017a). Achieving gradual decarbonisation across society and reducing the effects of climate change require sustainable funding. This is because finance mediates these many social agents and entities by using markets and services (Kuroda 2022). Sustainable finance facilitates the resources required to create a fair and prosperous world. It guarantees that financial resources are allocated to relevant

Table 1.5 Opportunities relating to climate change and its possible financial effects

Climate-related opportunities	Potential financial impacts
Resource efficiency	Improved production procedures, recycling, and energy-efficient construction exemplify resource efficiency. Implementing these practices (TCFD, 2017) can enhance production capacity, reduce operating expenses, and increase fixed asset value. Businesses can achieve significant cost savings and gain competitive advantages through more efficient resource utilisation, as noted by Porter and Van der Linde (Porter and van Linde 1995)
Energy source	New technologies make it possible to embrace lower-emission energy sources, resulting in cheaper operational costs, less susceptibility to changes in the price of fossil fuels, and improved company reputation (TCFD, 2017). Investments in renewable energy and carbon markets may have financial benefits. Switching to renewable energy sources can result in considerable cost savings and less reliance on unstable fossil fuel prices (Jacobson et al. 2017)
Products and services	Products with low emissions and solutions for climate adaptation satisfy changing consumer needs and provide new business opportunities. These innovations let companies diversify their product and service offerings and compete successfully in the low-carbon economy (TCFD, 2017). Chesbrough (2010) noted that novel company concepts offer chances to break into untapped areas and produce extra revenue. Companies specialising in products that reduce global warming can obtain a competitive edge and draw in eco-aware clients (Berkhout 2012)
Markets	Creating insurable assets, tapping into new markets, and taking advantage of public sector incentives are all ways to increase revenue streams. Green bonds and sustainable infrastructure investments provide fresh opportunities for financial gain (TCFD, 2017). According to Flammer (2021), corporate green bonds enhance financing options for growth and draw in new investment flows
Resilience	Engaging in resilience planning, diversity of resources, and renewable energy projects may improve market value and operational dependability. In addition to fostering opportunity in new markets, robust supply chains and infrastructure support company continuity (TCFD, 2017). Engaging in corporate sustainability initiatives may increase a business's value and resilience (Linnenluecke and Griffiths 2010).

people and sectors, directing important shifts and aligning financial frameworks with national, regional, and sectoral development and climate goals (ClimateWorks 2024). Due to the inability of our existing financial and economic institutions to fairly distribute economic gain, past injustices have been maintained and even made worse, further marginalising already marginalised communities (WEF 2023). Numerous initiatives worldwide focus on sustainable finance, which essentially involves actions taken by various institutions and organisations to address climate change and broader environmental issues. Figure 1.16 represents an example framework for addressing climate change through responsible investment

The world can achieve the SDGs by efficiently implementing a fair shift towards more environmentally and economically sustainable behaviours. Not only is this kind of advancement ethically correct, but it will also win over the public, establish political legitimacy, and create the social stability needed to complete the transformation. To ensure a habitable and flourishing world, it is crucial to amplify funding for environmental and climate change initiatives, phase out high-emission assets, and guarantee that the benefits of a sustainable economy reach the most disadvantaged and marginalised populations (WEF 2023).

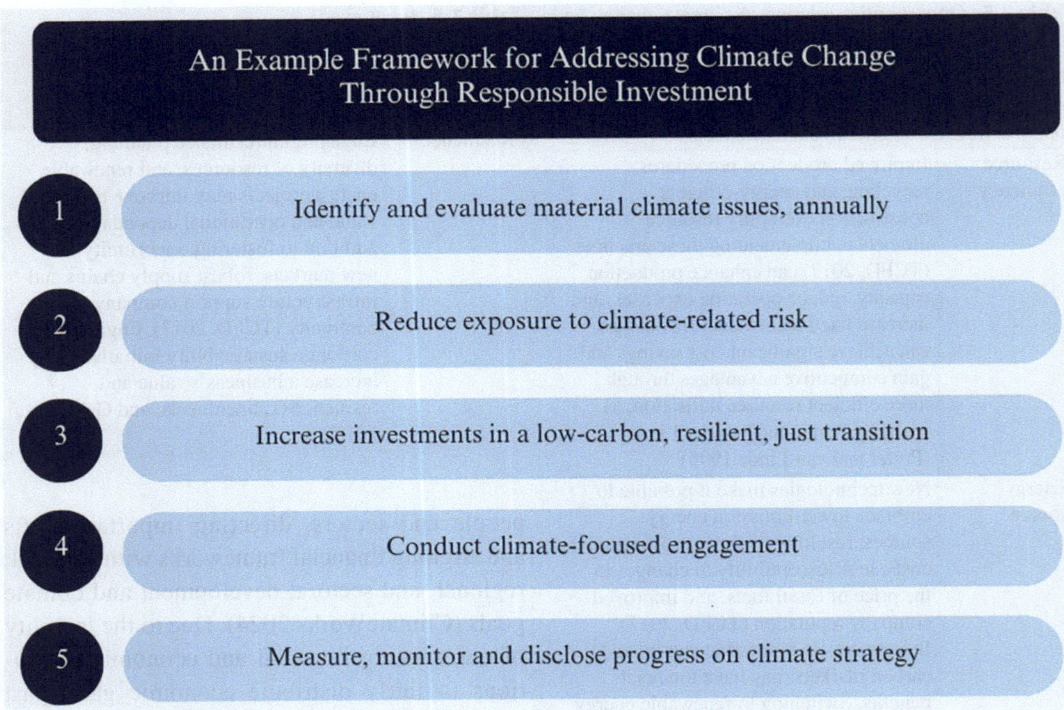

Fig. 1.16 World Resources Institute's (WRI) climate change investment strategy framework. (WRI 2022)

1.3 Net Zero and ESG

1.3.1 Net Zero

Net zero can be defined as the balance between the removal of greenhouse gases from the atmosphere and emissions of them to the atmosphere. Achieving net-zero emissions necessitates a two-pronged strategy: cutting emissions generated by humans as near to zero as feasible and offsetting any leftover emissions with an equivalent amount of carbon removal, either via technological or natural means (Küfeoğlu 2024; Levin et al. 2023). The net quantity of greenhouse gases released by human activity into the atmosphere will equal zero as a result of reaching net zero (Küfeoğlu 2024). Depicting the transitions that arise from reaching net zero, Fig. 1.17 defines net zero and presents before and after situations.

The planet's temperature is rising due to an escalation in emissions, and the quality of the air is declining. For the time being, net zero is primarily concerned with human GHG emissions. Achieving net zero involves minimising human-caused greenhouse gas emissions and eliminating any leftover gases to bring the total amount of human-caused GHG emissions in the atmosphere to zero. It is critical that all parties involved comprehend precisely what "net zero" entails. It would be difficult to get to a point where human activity emits no greenhouse gases. For this reason, the phrase "net zero" is used instead of "absolute zero". "Net zero" implies that to balance the emissions that humans make, parties must remove adequate emissions from the atmosphere (Küfeoğlu 2024). Net Zero Tracker (2023) reports that over the past three years, there has been a growing emphasis on net-zero target setting across several categories.

Net zero is crucial because GHGs play a vital role in the greenhouse effect, which causes global warming by trapping heat in the Earth's atmosphere. Climate change has been substantial because of human activities, including the combustion of fossil fuels, deforestation, and industrial processes that continuously increase GHG

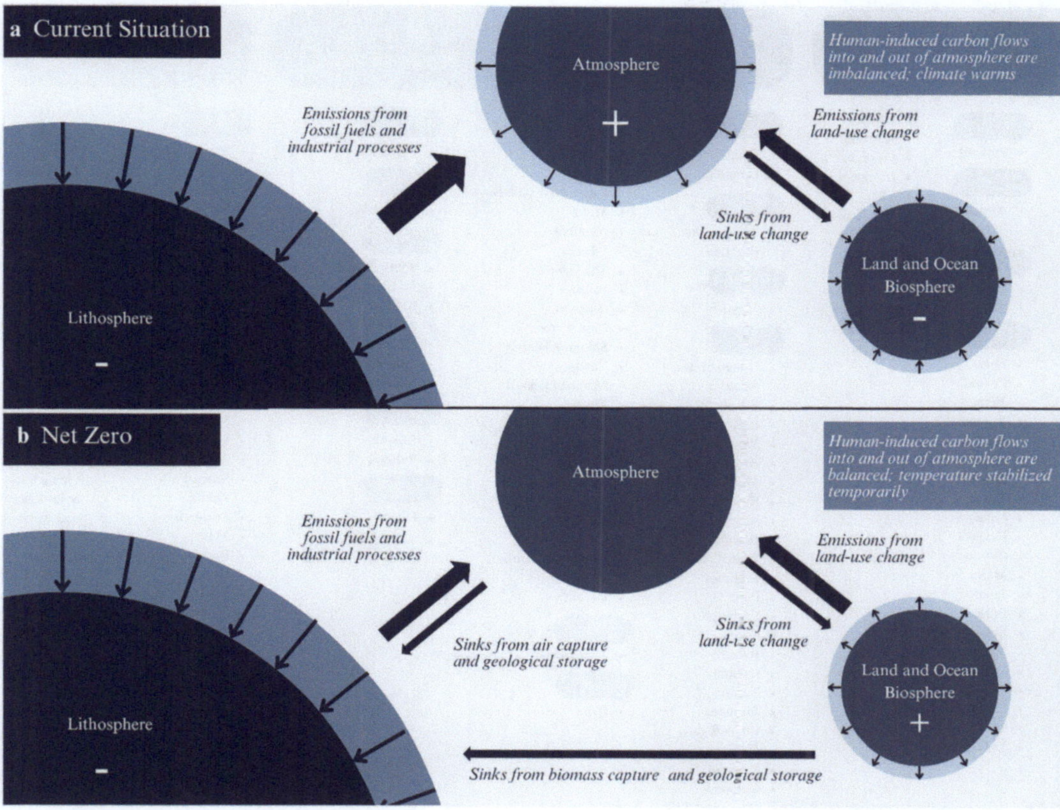

Fig. 1.17 Current situation of emissions and net-zero emissions. (Allen et al. 2022)

emissions. Threats to biodiversity, human health, and global ecosystems are posed by these changes, which show themselves as melting glaciers, droughts, extreme weather events, and rising sea levels. Reaching net zero is necessary to lessen these negative consequences, reduce global warming, and stabilise the climate. For the sake of our planet's sustainability and the welfare of future generations, this objective is essential (Bouckaert et al. 2021; Küfeoğlu 2024; Tracker 2021).

A significant number of both corporate and retail investors today want to make a difference. Solutions that deal with environmental issues must be funded. Hence, the establishment of a financial environment that supports sustainable growth is highly desired by governments and policymakers. The Paris Agreement states that the financial sector has been considered to play a fundamental role in driving the zero-carbon energy transition (Whitley et al. 2018). The anticipated amount of money needed globally to combat climate change is a significant capital investment. The worldwide net-zero transition is heavily dependent on the financial sector since most of these investments will be made possible by the financial system. Given its pivotal role, an expanding alliance of nations, localities, companies, and other establishments is dedicating itself to achieving net-zero emissions. More than 140 nations have established a net-zero aim (UN 2024a). Figure 1.18 represents countries with net-zero goals and their current level in four stages.

Net-Zero In Law, Net-Zero In Policy Document, Net-Zero Declaration/Pledge, and Net-Zero Proposed/In Discussion are the four stages into which nations are divided, as shown in Fig. 1.18. The figure displays the countries' current states and the dates they have set for achieving net zero. Below are more particular explanations of these four stages' meanings and

Fig. 1.18 The list of countries with their net-zero stage. (Net Zero Scorecard 2024)

the extent of each stage. These four steps consist of:

- Net-Zero In Law: Countries that have enacted laws or policies mandating them to achieve net-zero emissions are included in this phase. It is officially accepted that there is a pressing need to combat climate change and that sustainable practices must be adopted with this legal vow (Höhne et al. 2021).
- Net-Zero Policy: This phase includes nations that, while they may not yet be legally binding, have stated clear plans or objectives to achieve net-zero emissions (Jeffery et al. 2018).
- Net-Zero Declaration/Pledge: This phase includes all countries that have publicly committed to achieving net-zero emissions, usually through press releases or announcements. Although these agreements lack legal enforceability, this phase demonstrates a growing recognition of the urgency of taking immediate action in response to the climate calamity (Coppenolle et al. 2022).
- Net-Zero Proposed/In discussion: Countries that have not yet formally committed to net-zero but are thinking about or discussing it are included in this category. In this phase, debates over the viability and consequences of establishing net-zero objectives continue both inside governments and among interested parties. Members of this group of nations may take part in alliances or forums that promote climate action, demonstrating an increasing recognition of the need for coordinated actions to tackle climate change (Soterroni 2023).

1.3.2 ESG

ESG is a multidimensional metric that refers to environmental, social, and governance (ESG) performance assessment. It is essential to assess an organisation's or investment's overall impact in these three dimensions. The growing focus on ESG is associated with a more significant commitment from firms to make investments that are ESG-focused. The main goals of these investments are to reduce carbon and greenhouse gas emissions, deal with climatic and environmental issues, and take on anomalies in governance in addition to urgent social issues, such as human rights violations (Cagli et al. 2023).

The growing significance of sustainability requires the inclusion of ESG factors. This tendency has several causes. First and foremost, the world's problems have made people more aware of how important sustainability is. A "push" effect has been created by people and organisations' increased sensitivity, which is forcing businesses to more firmly incorporate sustainability into their operations.

Concurrently, this need has been further accelerated by the growth of globalisation. The debate on sustainability has become more global due to concerns about resource distribution, global collaborations, and standards. Furthermore, there has been a noticeable increase in managers', stakeholders', and customers' involvement with sustainability-related concerns. Without sustainability, ESG would be like a weak compass that cannot show the actual route towards creating long-term value. Sustainability is the cornerstone of company resilience, protecting against the unpredictability of environmental and social threats (Küfeoğlu 2024). As sustainability becomes an increasingly crucial global goal, many investors are asking for ESG strategies that demonstrate how to achieve net-zero carbon emission objectives. ESG policies and initiatives pave the way for net zero. They place sustainability criteria on institutions and set essential goals (CFA Institute 2020).

The main goal of the environmental part of ESG is maintaining natural ecosystems and the ecological balance. This aspect includes the strategic management of the environmental footprints left by business activities. Environmental benchmarks are evaluated using a range of metrics, such as waste management procedures, energy procurement and use efficiency, programs to reduce carbon and other harmful emissions, laws protecting biodiversity, and early action in responding to climate change phenomena. Monitoring air quality indices, calculating water use and effluent discharge, assessing effects on biodiversity and soil health, and closely examining subsurface resource exploitation in conjunction with energy conservation efficacy are all part of the quantitative assessment of this component (Küfeoğlu 2024; Rajesh 2020). As professional investors include more and more ESG considerations in their analysis and investment processes, a set of standard best practices is starting to develop (CFA Institute 2020).

1.3.2.1 Criticism on ESG

There have been criticisms and challenges against environmental, social, and governance (ESG) practices, which have called for further research. These concerns point to the challenge of implementing ESG principles and the requirement for continuous improvement to help improve credibility and realise actual advances in sustainability. The criticisms against ESG can be classified under several major themes as we can summarise as follows:

Conceptual and Definitional Criticisms

One of the biggest issues in ESG terminology is that no clear consensus definition of sustainable finance and ESG investing (Küfeoğlu 2024; OECD 2025; International Finance Corporation 2023; United Nations 2023). That absence of clarity renders measurement, comparison, and regulation difficult (OECD 2025; United Nations 2023). Variations in organisations' terminology are just an additional complexity (OECD 2025). Even the definition of ESG investing is unclear consensus (OECD 2025).

The informal interchangeability of the terms corporate social responsibility (CSR) and impact investing creates uncertainty regarding distinctions between the two approaches to ESG investing (OECD 2025). A stronger analysis of how integrated sustainability differs from CSR is necessary, particularly in terms of varying implications for performance (Eccles et al. 2017). The ambiguity generated by inaccurate terminology is a barrier to accurate communication and analysis of metrics relevant to sustainable finance.

Criticisms Related to Fiscal Results and Fiduciary Obligations

One central debate concerns the interconnection between integrating environmental, social, and governance (ESG) considerations and better financial performance. Some academics argue that investors and companies should focus on shareholder value maximisation, viewing ESG issues as a distraction (Jacobs and Levy 2022). Empirical results investigating the link between ESG scores and stock price performance have yielded ambiguous results (Vanguard 2023; MSCI 2025). There has been some evidence that the outperformance associated with ESG scores can be driven by other financial attributes of well-rated firms (Patatoukas 2022). There have been worries regarding possible conflict between adopting ESG policies and the fiduciary duties of corporate boards and fund managers (McGowan 2022). Traditionally, fiduciary duty dictates that financial return maximisation is important (McGowan 2022). There is debate regarding whether considering ESG factors, which may involve financial trade-offs, conflicts with this duty (McGowan 2022). It has been stated that U.S. retirement plan fiduciaries may not prefer social and environmental outcomes in a way that would compromise long-term financial performance (Kentucky Attorney General 2023). In other states, legislative and regulatory initiatives seek to restrict ESG investing on the part of state governments due to fiduciary duty concerns (Florida Governor's Office 2023; Texas Legislature 2021; West Virginia State Treasurer 2023).

Methodological Challenges in ESG Analysis and Ratings

ESG rating and scoring processes are subject to strict criticism. One of the root weaknesses is the inconsistency and divergence of the ESG data and scores employed by different agencies (Berg et al. 2022; Christensen et al. 2022; Jacobs and Levy 2022). The agencies have no consensus regarding selecting indicators, weights, and scoring rules (MSCI 2025). This results in huge differences in ESG scores assigned to the same firm, thus low correlations among the scores (Berg et al. 2022). The heterogeneity lowers the effectiveness of ESG scores in making knowledgeable investment decisions. Most ESG rating providers do not reveal their methods and metrics for scoring (Thomson Reuters 2021). Transparency hinders the comparability and transparency of company performance (Thomson Reuters 2021). The difficulty in comparing ESG data across different geographic regions and industries due to different reporting processes constitutes another limitation (Thomson Reuters 2021). The variation in the application of standards and inconsistency in report granularity also comprise this limitation (Thomson Reuters 2021).

Implementation Challenges and Practical Limitations

Implementation challenges undermine effective ESG integration in green finance. Inadequate quality, consistent, and comparable ESG data availability is a significant constraint (Thomson Reuters 2021; Eccles et al. 2017). Some reasons include incomplete data coverage, evolving metrics, and lack of standardised disclosure processes (Thomson Reuters 2021; Fraser Institute 2023). The deficiency of strong specialised institutions that can ensure data reliability is responsible for the issues (Fraser Institute 2023). The deficiency in strong data hinders data-driven regulation by regulators, data-driven business decision-making, and data-driven investment by investors. Regulatory uncertainty and harmonisation shortage across jurisdictions are other issues (Thomson Reuters 2021). Multinational firms are challenged by the lack of international ESG regulations (Thomson Reuters 2021). Differing disclosure requirements cause compliance challenges (Thomson Reuters 2021). Central banks lag in developing climate risk modelling frameworks in most EMDEs (International Finance Corporation 2023). Application of ESG frameworks designed for developed economies in EMDEs is limited (International Finance Corporation 2023). EMDEs are marked by data and capacity shortages, limiting sustainable finance uses (International Finance Corporation 2023). Green finance in EMDEs can be driven by external forces rather than local regulation (International Finance Corporation 2023). EMDEs' financial institutions are more strained by greater climate risk exposure and less developed policy structures (International Finance Corporation 2023).

1.3.2.2 ESG Rating

ESG ratings offer a viewpoint on the sustainability profile or attributes, exposure to sustainability risks, and effects on the environment or society of a business or financial institution. With the purpose of evaluating a company's or financial instrument's creditworthiness, it attempts to guarantee that credit rating agencies take pertinent ESG concerns into account. ESG rating providers utilise a variety of terminology, including valuations, scores, opinions, and ratings. Investor confidence in sustainable investments is bolstered by the growing significance of ESG ratings. It is used by investors more and more to account for the consequences and risks associated with ESG problems as part of their sustainable investing plans. For businesses, it is a tool for identifying and considering operational risks and investment possibilities and evaluating their performance in relation to peers concerning ESG concerns (European Commission 2024b).

To gain a deeper understanding of ESG without going too far, one can look at Fig. 1.19's ESG framework, which summarises the significance and meaning of elements E, S, and G from diverse angles.

FRAMEWORK FOR ESG

Dimension	*Factors*	*Definition*
Environmental	• *GHG Emissions* • *Energy use and effectiveness* • *Air contaminants* • *Utilizing water and recycling* • *Waste generation and management (hazardous, solid, and liquid)* • *Impact of biodiversity and its reliance* • *Influence and reliance on ecosystems* • *Environmentally friendly goods and services are getting better*	*The financial performance or solvency of an entity, sovereign, or individual may be impacted by environmental factors in either a favorable or negative way.*
Social	• *Freedom of association for employees* • *Employing children* • *Forced or required work* • *Safety and health at work* • *Customer safety and health* • *Equal Opportunity, Diversity, and Discrimination* • *Impact of poverty on a community* • *Supply chain administration* • *Education and training* • *Privacy of clients* • *Local effects*	*The financial performance or solvency of an entity, sovereign, or individual may be impacted by social issues in either a favorable or bad way.*
Governance	• *Business ethics and moral guidelines* • *Accountability* • *Openness and disclosure* • *Exec compensation* • *Board composition and diversity* • *Corruption and extortion* • *Participant involvement* • *Ownership rights*	*The financial performance or solvency of an organisation, nation, or individual may be impacted by governance issues in either a favorable or negative way.*

Fig. 1.19 Framework for ESG. (Open Risk Manual 2023; Küfeoğlu 2024)

Competitive long-term financial gains and beneficial social effects can be achieved through responsible investing, which involves considering ESG parameters when assessing investments. A growing number of stakeholders, including asset owners, are concerned about how asset managers such as private equity firms evaluate environmental, social, and governance (ESG) concerns with the goal of offering guidance on purchase agreement decisions and then managing those risks to preserve value and seize opportunities for value creation while the asset is being held. Risks related to ESG encompass those associated with mitigating and adapting to the impacts of climate change, methods of environmental management and social responsibility, labour and safety conditions, human rights respect, anti-bribery and corruption measures, and adherence to pertinent regulations and laws. The effects of evolving laws or voluntary recommendations, megatrends, and the demands for transparency from broader stakeholders should all be considered when making responsible investments (Det Norske Veritas 2024).

ESG ratings and indices assess an organisation's sustainability performance based on a range of particular or broad standards. As sustainability becomes increasingly important to investors, businesses, and stakeholders, these tools provide a comprehensive overview of an organisation's sustainability performance, allowing for pertinent comparisons and evaluations. Table 1.6 shows ten essential ESG indices, raters, and rankers.

1.4 Cost and Financing of Net Zero

1.4.1 Examining Cost Scenarios to Reach Net Zero

Considerations for net zero's costs and benefits both domestically and internationally come in many forms. There are several ways to weigh the advantages and disadvantages of net zero, both nationally and globally (Sasse 2021). There are multiple scenarios about net zero; nevertheless, it is crucial to remember that scenarios are instruments meant to assist policymakers in considering their answers to many possible futures, not forecasts. Rather than offering a single "correct answer", scenarios highlight the effects of several potential modifications (The UK Government Office for Science 2023). The scenarios provide a framework for examining the outcomes of various policy decisions, socio-economic variables, and technological developments on the path to net zero. Key design decisions, such as long-term policy, short-term policy, and technology availability, distinguish these scenarios, which take into account harmonised assumptions about important socio-economic variables, including population and economic development (NGFS 2022a).

1.4.2 The Economic Consequences of Climate Change

Despite being complicated and unpredictable, the economic effects of climate change show extensive, long-term costs that differ between countries and industries (Tol 2009). Figure 1.20 represents the overall summary and projections of the economic outcomes of climate change, divided into three main sections: "Legal and Financial Risks for Businesses", "Impact on Economic Growth", and "Effects of Rising Temperatures". Data-based estimations are provided for the projected economic losses.

Figure 1.20 indicates that global warming will significantly impact economic growth due to property and infrastructure damage, decreased productivity, mass migration, and security threats. As temperatures rise, the balance of winners and losers becomes increasingly adverse. Inflation is projected to rise as shortages materialise, notably in agriculture, and energy costs are expected to increase with the transition to renewables (Wade and Jennings 2016). Climate directly impacts energy systems; high temperatures increase demand while straining supply and transmission. However, climate also plays a vital role in facilitating adaptation by permitting activ-

Table 1.6 Ten considerable ESG rankers, raters, and indices

Name	Objective	Founder	Participant coverage	Participant number	ESG scope	Detail of scope	Rating methodology	Update period	Sector-specific or general
Carbon Disclosure Project (CDP)	CDP strives to inspire and direct companies and cities to transparently display their best environmental efforts by granting points to them	CDP global	Companies Cities States Regions *Note: Companies can be invited by their customers/ investors*	19,800+	ESG	-Climate change -deforestation -water security -supply chain *note: These are separate disclosure reports*	Leadership (A/A-) Management (B/B-) Awareness (C/C-) Disclosure (D/D-)	Yearly	Specific
Morningstar Sustainalytics	Sustainalytics enables companies by rating to recognise, measure, and manage ESG-oriented risks and opportunities. At the firm and fund level, it offers insightful ESG information to investors from all kinds of assets	Morningstar	Investors Companies	20,000+	ESG	-material ESG issues -exposure management-unmanaged risk	Five risk level Negligible (0–10) Low (10–20) Medium (20–30) High (30–40) Severe (40+)	Yearly	Specific
Refinitiv	Based on publicly accessible evidence, Refinitiv's ESG ratings are intended to assess a company's relative ESG performance, commitment, and efficiency clearly and impartially over ten key subcategories	London stock exchange group	Companies	40,000+	ESG	-environmental -social -governance -ESG controversy	4th quartile (> 75 to 100) 3th quartile (> 50 to 75) 2nd quartile (> 25 to 50) 1st quartile (0 to 25)	Weekly	Specific

(continued)

Table 1.6 (continued)

Name	Objective	Founder	Participant coverage	Participant number	ESG scope	Detail of scope	Rating methodology	Update period	Sector-specific or general
EcoVadis	To offer sustainability ratings and insights that assist all businesses in lowering risk and development, EcoVadis evaluates how effectively a company has incorporated the sustainability/CSR fundamentals into its operations and management structure	CVC capital partners General Atlantic Beyond net zero Astorg	Companies	100,000+	ESG	-environment -labour and human rights -ethics -sustainable procurement	Platinum: Top 1% (78–100) Gold: Top 5% (70–77) Silver: Top 25% (59–69) Bronze: Top 50% (50–58) *Note: Sustainability recognition levels are based on EcoVadis score percentile and minimum theme score*	Ongoing	Specific
Bloomberg Gender-Equality Index (BGEI)	Bloomberg created an index designed to monitor the effectiveness of firms dedicated to promoting gender equality. This is achieved through the development of policies, ensuring representation, and maintaining transparency. The Bloomberg gender-equality index (BGEI) offers investors detailed insights into specific data points concerning gender equality at publicly traded companies	Bloomberg	Companies	600+	S	-leadership and talent pipeline -equal pay and gender pay parity -inclusive culture -anti-sexual harassment policies -external brand	0–100% (100% is best) Note: According to Bloomberg's GEI criteria, there are yearly *thresholds for index inclusion*	Yearly	General
Dow Jones Sustainability Indexes (DJSI)	The DJSI is a marketplace capitalisation-weighted float-adjusted index that tracks the performance of businesses chosen based on ESG factors. They are divided into three geographical categories: DJSI world, DJSI regions, and DJSI countries	S&P global	Companies	6000+	ESG	-environment -social capital -human capital -business model and innovation -leadership and governance	0–100 (100 is the best) Note: The DJSI indices' components are chosen from target %	Yearly	Specific

Morgan Stanley Capital International (MSCI) ESG Ratings	MSCI offers institutional investors a more capable ESG integration tool created to reduce ESG exposure and generate value over the long haul. The MSCI materiality map framework is used for assessment	MSCI	Investors Companies Countries	8500+	ESG	-environmental -social -governance	Seven grading categories Leader (AAA and AA) Average (A, BBB, and BB) Laggard (B and CCC)	Yearly	Specific
Standard & Poor's (S&P) Global ESG Scores	Corporate disclosures are actively guided by S&P Global ESG scores and the CSA research process that continually underpins them to raise sustainability standards	S&P global	Companies	13,000+	ESG	-environmental dimension -social dimension -governance and economic dimension	0 to 100 Note: The score impact of subcategories depends on the sectors	Yearly	Specific
Institutional Shareholder Services (ISS) ESG	The complete spectrum of sustainable and accountable investing concerns is covered by ISS ESG, including climate change, SDG-related effects, human rights, labour standards, corruption, debatable weaponry, and several more. There are nine ISS ESG rating types, including ISS ESG corporate rating	ISS Inc.	Companies Countries	1100+	ESG	-environmental -social -governance *Note: Different ISS ESG ratings have specific criteria*	ISS ESG corporate rating: 12-point grading between A+/4.00 and D-/1.00 (excellent to poor) *Note: Companies with ratings above the sector-specific prime criteria are given prime designation*	Ongoing	Specific

(continued)

Table 1.6 (continued)

Name	Objective	Founder	Participant coverage	Participant number	ESG scope	Detail of scope	Rating methodology	Update period	Sector-specific or general
Sustainable Fitch	In accordance with established criteria and principles from third parties, sustainable Fitch's ESG ratings are among the first ESG analysis products to offer a qualitative and quantitative assessment of an entity's or transaction's effect on environmental, social, and governance problems. Individual entities, green, social and sustainable (GSS) and sustainability-linked debt instruments (collectively referred to as "labelled instruments"), and non-labelled instruments are all given an ESG rating by Fitch	Fitch ratings	Companies	700+	ESG	-entity information -business activities -environmental profile -social profile -governance profile	Fully ESG aligned 100–87.5 Mostly ESG aligned 87.5–62.5 Neutral 62.5–37.5 Mostly ESG unaligned 37.5–12.5 ESG unaligned 12.5–0.0	Yearly	Specific
FTSE Russell ESG score	FTSE Russell supports risk management for the firm's ESG dimension. It combines ESG data into securities, and portfolio analysis complies with mandatory stewardship standards. ESG-conscious investment techniques could be implemented with assistance from FTSE Russell. The FTSE ESG index series modifies the market capitalisation weight of its members based on the overall rating from FTSE Russell's ESG ratings and data model	London stock exchange group	Companies	8000+	ESG	-biodiversity -Climate change -pollution and resources supply chain -water security -customer responsibility -health and safety -human rights and community -labour standards -supply chain anti-corruption -corporate governance -risk management -tax transparency	According to more than 300 separate ESG indicator assessments that are tailored to the conditions of each organisation form	Yearly	Specific

Bloomberg (2023), Carbon Disclosure Project (2023), EcoVadis (2023), ISS ESG (2023), Morningstar Sustainalytics (2023), MSCI (2023), Refinitiv (2023), S&P Dow Jones Indices (2023), S&P Global (2023), Fitch Group (2023), Küfeoğlu (2024)

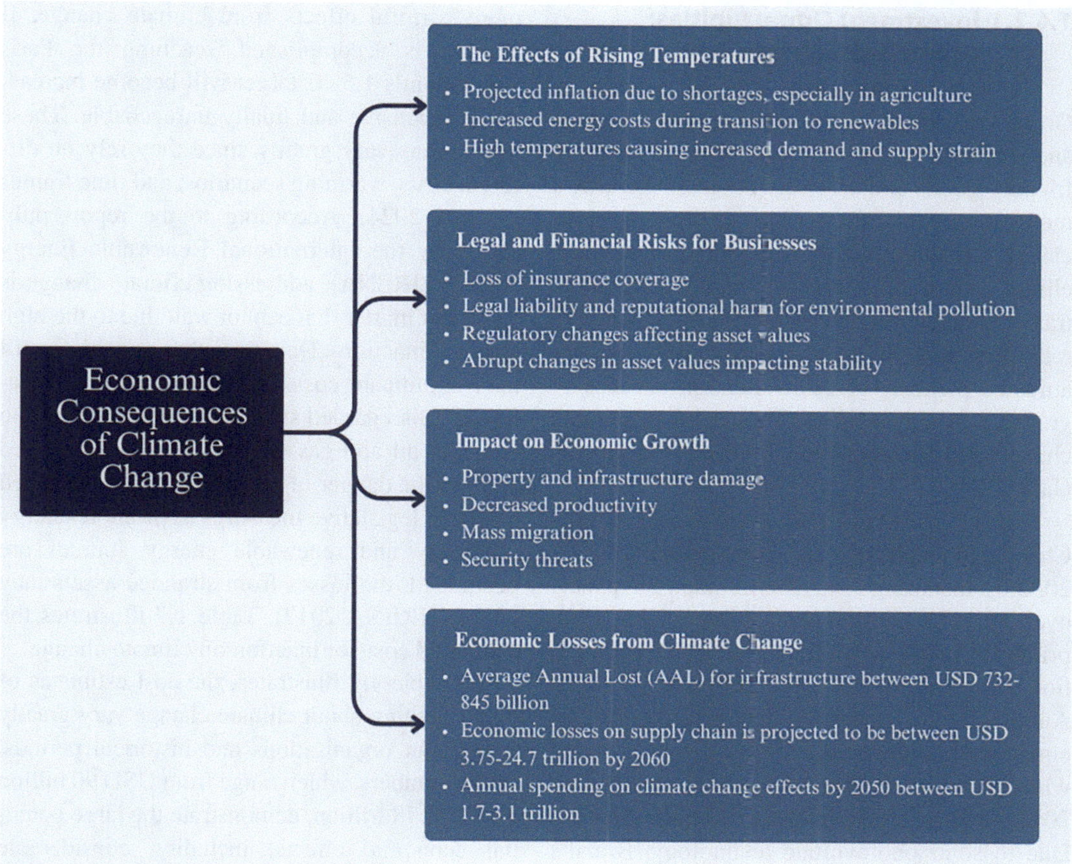

Fig. 1.20 Economic consequences of climate change qualitatively and quantitatively. (Wade and Jennings 2016; Carleton and Hsiang 2016; Debelle 2019; CDRI 2023; Low 2019; Wang 2024; Bennett 2023)

ities like trade, irrigation, heating, and cooling. The primary factor contributing to anthropogenic climate change is concurrent energy usage (Carleton and Hsiang 2016). Businesses and households may occasionally lose their insurance coverage. And the insurance premiums might increase. Businesses that cause environmental pollution may be held legally liable or experience reputational harm due to their operations, and regulatory changes may render once-valuable assets unattractive. These outcomes could lead to abrupt changes in asset values, which would impact the financial system's stability (Debelle 2019).

According to Coalition for Disaster Resilient Infrastructure (CDRI), considering the consequences of climate change, the worldwide Average Annual Loss (AAL) for infrastructure, including buildings, health, and education sectors, ranges from USD 732 to USD 845 billion, or almost 14% of the growth in the global GDP from 2021 to 2022 (CDRI 2023). In addition to this, depending on how much carbon dioxide is released, unmeasured supply chain disruptions might increase already-projected economic losses due to climate change, resulting in a projected net economic loss of between USD 3.75 trillion and USD 24.7 trillion by 2060 (Wang 2024). By 2050, the World Economic Forum anticipates that between USD 1.7 trillion and USD 3.1 trillion will be spent annually on the effects of climate change. Infrastructural, property, agricultural, and human health damages are included in this cost. As the impact of climate change worsens over time, it will be inevitable that this cost will rise (Bennett 2023).

1.4.2.1 Investment Opportunities: Green Technologies and Innovation

Green technology is the application of science and technology to fabricate environmentally friendly goods and services. Progress towards a more sustainable future includes energy efficiency measures, waste management and recycling, renewable energy, and sustainable transportation (Guelfo 2023). Despite being a relatively new industry, green technology has attracted plentiful investor attention because of growing awareness of the influences of climate change and the depletion of natural resources (Jark 2024).

One of the studies in this context is that China can achieve a net-zero power grid by 2050 by installing 7.46 TW of solar PV panels over 1.8% of the country's geographical area, primarily in Western China, for USD 4.55 trillion USD over the next 30 years (Ji et al. 2022). Another study examined the greenhouse gas emissions and financial requirements associated with substituting solar photovoltaic energy by 2030 for fossil fuels in Brazil's energy matrix. Due to solar photovoltaic technology, Brazil's energy mix may include less fossil fuels, which could cut greenhouse gas emissions by 36.9% by 2030. From 2023 to 2030, USD 376,5 billion will be invested in substituting fossil fuels (Saccardo et al. 2023). A U.S. capital investment of more than USD 270 billion has been announced for manufacturing facilities and sustainable energy projects. The 83 recently announced manufacturing sites are projected to considerably boost America's ability to produce key parts for sustainable energy projects. Manufacturers plan to increase manufacturing capacity for all major parts used in utility solar, offshore wind, land-based wind, and battery storage applications (American Clean Power Association 2023).

1.4.2.2 Costs of Inaction

Regarding the issue of climate change, it is widely acknowledged that failing to take action now will lead to higher costs in the future; this is known as the "cost of inaction". If climate action is delayed, the globe will undoubtedly suffer more harmful effects from climate change. If nothing is accomplished, reaching the Paris Agreement's 1.5 °C target will become increasingly expensive and finally unreachable. These estimations vary greatly since they rely on different costs, warming scenarios, and time frames (Alberti 2024). According to the report published by the International Renewable Energy Agency (IRENA), addressing climate change is an urgent matter that cannot wait due to the high cost of inaction. Delaying the transition will have significant costs even if the costs of inaction are disregarded since funding the status quo (such as oil and gas spending) will be retained and run the danger of more losses from stranded assets. If legislative measures to promote energy efficiency and renewable energy sources are postponed, the losses from stranded assets may double (IRENA 2017). Table 1.7 illustrates the estimated costs of inaction on climate change.

As Table 1.7 illustrates, the cost estimates of doing nothing about climate change vary greatly throughout organisations and historical periods. These numbers, which range from USD 90 billion to USD 178 trillion, demonstrate the large potential economic effects, including considerable GDP declines. This variance emphasises how critical it is to implement strong climate measures to reduce these significant financial risks.

1.4.3 Expected Costs of Net Zero

Maintaining the current course to keep global temperatures at 1.5 °C is expected to cost tremendous amounts. In addition to replacing fossil fuel-based infrastructure with renewable energy sources like solar and wind, investment is needed to find solutions for challenging industries like steel, cement, and aviation. The estimated cost of the green transition varies widely; estimates range from USD 100 trillion to 300 trillion between now and 2050. In perspective, today's yearly global GDP is over USD 100 trillion. Based on a conservative estimate, the International Energy Agency (IEA) calculated that the annual worldwide investment in energy infrastructure alone will need to reach the 1.5 °C target. This investment will increase from USD 2 trillion

Table 1.7 Cost of inaction concerning different institutions and time frames

Institution	Time frame	Total cost estimation
Deloitte	2021–2070	USD 178 trillion
IRENA	2017–2050	USD 20 trillion
NGFS	2022–2100	Up to 20% GDP reduction
OECD	2015–2060	1.0–3.3% GDP reduction
OECD	2060–2100	2–10% GDP reduction
Oxford Economics	2022–2050	10–20% GDP reduction
Swiss RE Institute	2021–2050	11–14% GDP reduction
The Economist	2015–2100	USD 4.2–43 trillion
The World Bank	2021–2030	USD 90–225 billion

OECD (2015), IRENA (2017), The Economist (2015), NGFS (2022b), Oxford Economics (2022), Johnson et al. (2021), Swiss RE Institute (2021), Philip et al. (2022)

today (about 2.5% of global GDP) to over USD 5 trillion (4.5%) by 2030, before declining back to 2.5% by 2050. A radical estimate of the average annual energy investment needs between now and 2050 has been determined by consulting firm McKinsey: USD 9.2 trillion (7.5% of world GDP) (Keller and O'Neal 2023). Various international institutions and organisations such as United Nations Framework Convention on Climate Change (UNFCCC) Climate Champions and the IEA have provided estimates for the total cost of achieving net zero across all sectors, as seen in Table 1.8.

As seen from Table 1.8, the costs of achieving net zero differ significantly across reports and institutions because they base their estimates on different assumptions, methodologies, and perspectives. For instance, the McKinsey report uses a 1.5 °C warming scenario, whereas Deloitte uses a 3 °C warming scenario. Additionally, while some organisations concentrate solely on CO_2 emissions, others also take into account other greenhouse gases (GHGs) like methane, nitrous oxide, and fluorinated gases for net-zero emission scenarios. Moreover, if net-zero emissions are to be achieved just for CO_2 or also for other greenhouse gases like methane, nitrous oxide, and fluorinated gases, the timeline for doing so may differ (Levin et al. 2023). The cost projections are also affected by the rate and scope of technology

Table 1.8 Total cost estimations for achieving net zero from various international institutions and organisations

Organisation	Estimated total cost (USD trillion)	Time frame	Estimated cost per year (USD trillion)
IEA (Jaynes 2023)	36	2023–2031	4.5
Barclays (Keller and O'Neal 2023)	100–300	2023–2050	3.7–11.1
Stanford Institute for Economic Policy Research (Crawford 2023)	100–120	2020–2050	3.3–4
Glasgow Financial Alliance for Net Zero (UNFCCC 2021)	125	2023–2050	4.6
Deloitte (2022)	178	2022–2070	3.7
Bloomberg (Gongloff and Denning 2024)	215	2024–2050	8.3
McKinsey & Company (Krishnan et al. 2022)	275	2021–2050	9.5

deployment. Time frames, infrastructure differences, and economic modelling approaches can be counted as other reasons why the costs of achieving net-zero emissions vary greatly.

Compared to the rest of the world, developed markets currently have far greater per capita emissions. Emissions will probably increase if growth cannot be separated from carbon-intensive activities as developing nations strive to catch up with developed nations concerning living standards. As emissions are expected to continue to ascend, developing and emerging market nations are the primary recipients of the majority of green transition expenditures. Therefore, significant investments will be required to prevent a scenario in which decarbonisation efforts in developed economies fail to offset increases in global emissions, given that energy accounts for nearly 75% of global emissions and developing economies will account for nearly all additional energy demand in the future (Keller and O'Neal 2023).

1.4.4 Why Should We Develop Financial Mechanisms for Net Zero?

First, considering the significant influence that financial systems have on global greenhouse gas emissions, reaching net-zero emissions heavily depends on the health and reform of these systems. To achieve the objectives of the Paris Agreement, there is an immediate need for a shift towards sustainable investments, as the finance industry now provides funding for more than 60% of global emissions (Global Ethical Finance Initiative 2022).

The current manner of lending money to nations for climate and sustainable development is inadequate. The effects of injustices in the global financial system are tremendous. Many nations lack the financial resources to invest in climate action that benefits the entire world despite their desire. Fifty-two developing economies are now dealing with serious debt issues. Forty percent of the world's population lives in extreme poverty there, and fifty percent of those countries are among the most vulnerable to climate change (UN 2024b). The main reason why investors are crucial to achieving net zero is because more funding is required than what the public sector can offer. The majority of assets that emit carbon are privately held. Although replacing carbon-emitting assets may be beneficial to society as a whole, there is disagreement over how much public funds should be used to renovate and improve privately held assets, particularly since there are alternative ways for policymakers to encourage decarbonisation. Similar to previous socio-economic shifts, funding for mitigating climate change is expected to come from a mix of public and private sources (CFA Institute 2024).

To turn pledges into real carbon reductions, the creation of net-zero financial systems is essential due to the significant role that financial institutions play in reallocating capital to assist with decarbonisation initiatives. As Taskforce on Scaling Voluntary Carbon Markets (TSVCM) suggested (2021), they should measure and pledge to cut financed emissions, evaluate and publish climate risks, and modify financing requirements to take longer-payback decarbonisation projects into account. Institutions should assist counterparties in their transition plans by providing new financial solutions, offering guidance on ways to reduce emissions, and growing the range of climate-finance products available, such as funding for carbon-neutral projects and voluntary carbon markets, as they align their portfolios with global warming targets.

1.4.5 Key Players and Stakeholders in Sustainable Finance

Sustainable finance sources include government budgets, public sector initiatives, bilateral and multilateral development finance institutions, foreign direct investment, grants from charitable organisations, project developers and corporations, commercial banks, and residential, commercial, and institutional investments (Sinha et al. 2020). Figure 1.21 depicts the fundamental players in sustainable finance and the many funding sources that assist in funding sustainable projects.

A wide variety of organisations are involved in sustainable finance and have a noteworthy impact on the direction and growth of the industry. The sources section represents the numerous contributors to sustainable finance, illustrating the wide range of organisations participating. Taken as a whole, these sources support the development and use of sustainable finance by offering financial resources, legal frameworks, and revolutionary financial products that promote social and environmental impact.

Private financing may account for 70% of primary investments, with corporate balance sheets directly financing nearly half of them. In addition to playing a significant role in direct investment, financial actors such as institutional investors, infrastructure funds, commercial and

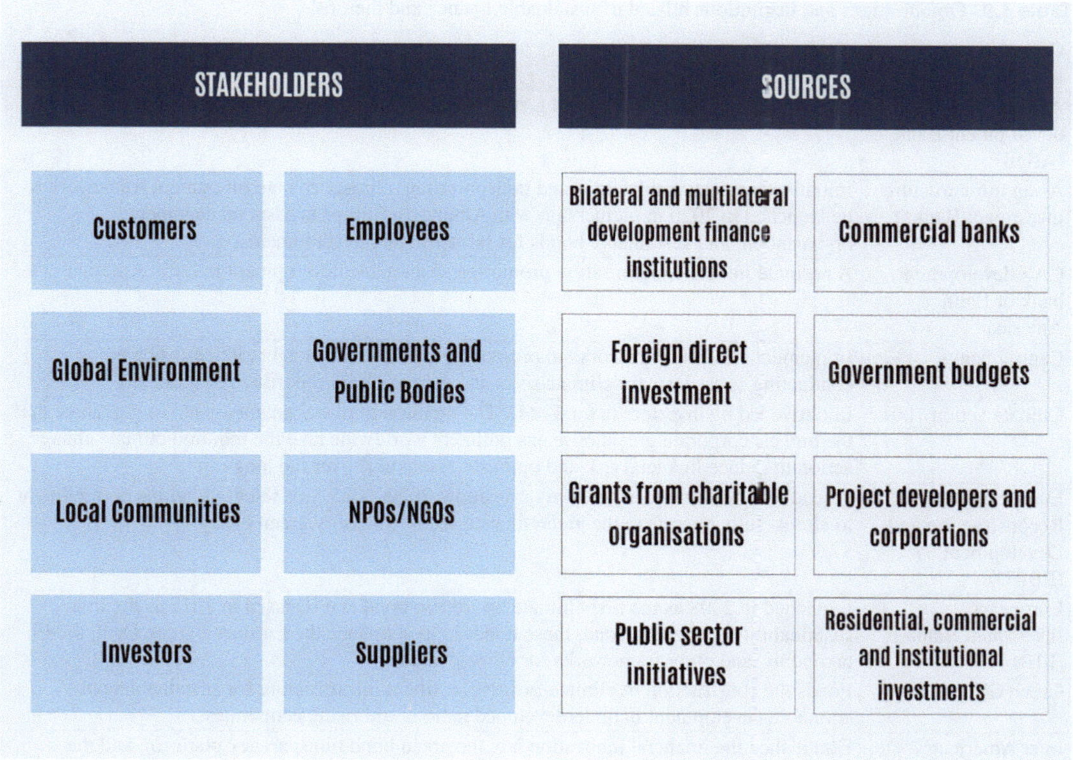

Fig. 1.21 Stakeholders and finance sources in sustainable finance

financial institutions, private equity, and venture capital will also be vital in facilitating and supporting corporate and household investment. However, a variety of financial and non-financial obstacles, such as regulatory limitations, political risk, and limited returns, prevent investors from making significant investments. Decarbonisation will need to be facilitated by public actors and investors through incentives and policy changes, as well as through direct public funding and blended finance that attracts private investment (UNFCCC Climate Champions 2023). Table 1.9 provides an overview of the major institutions and organisations that are part of sustainable finance, summarising their role.

The various financial institutions that contribute to sustainable finance are included in this table. Each institution has a distinct regional or functional focus that aims to address financial difficulties associated with climate change.

Stakeholders with a strong plan for sustainable finance will be more able to draw in capital and obtain green loan options to help them reach net zero. Since banks and investors see the value of encouraging positive changes in ESG parameters, sustainable finance can offer dedicated money to support ESG goals (Deloitte 2021). In this respect, the most reasonable way to finance the cost of achieving net zero is by using sustainable finance mechanisms.

1.4.6 Case Studies in Critical Infrastructure Sectors

This section will provide a selection of global case studies to show how the policies and outcomes of many nations highlight how sustainable finance impacts our future. The global case studies are chosen from nine critical infrastructure sectors: Emergency services, energy, finance,

Table 1.9 Organisations and institutions related to sustainable finance and their role

Organisation/ institution	Role
African development Bank (AfDB)	Programs for climate finance and green bonds, with a particular focus on Africa
Asian infrastructure investment Bank	Initiatives on sustainable cities and transportation; climate change investment framework to be launched in 2020 in partnership with Amundi, a European asset management organisation; and sustainable bonds for infrastructure development
CAF development bank of Latin America	A regional financial organisation promoting sustainable development in Latin America
Central banks	Implement various frameworks to promote responsible financial practices, such as conducting stress tests for climate risks and defining ESG reporting requirements
Climate action 100+	Initiative led by investors (a total of USD 41 trillion in management assets) to guarantee that the biggest corporate greenhouse gas polluters worldwide take the required climate change action to reduce financial risk and optimise asset value over the long run
European Bank for Reconstruction and Development (EBRD)	A road map for the green economy transition, 2020–2025, was released, with a commitment to almost fully operate in the green finance sector (majority green bank)
European Investment Bank (EIB)	Launched in 2008 as the first Climate awareness bond and released in 2017 as the first sustainability awareness bond, these bonds helped finance the European green Deal, the EU taxonomy, and other frameworks for climate action
Green Climate fund	Funds the construction of climate-positive, resilient infrastructure for sustainable and equitable development in underdeveloped nations and rising economies
Inter American development Bank (IDB)	Established the financial innovation lab, the green bond transparency platform, and the green Finance LAC platform to coordinate the deployment of green finance initiatives in the Latin American and Caribbean areas
International Finance corporation (IFC)	The largest development institution collaborates with private financial institutions to encourage development in emerging economies that are socially and environmentally responsible as well as climate resilient. It also develops strategies for ESG integration in financial institutions
International Monetary Fund (IMF)	Developing sustainable finance guidelines and climate-related stress testing; investigating the financial implications of climate change
Islamic Development Bank	Implementation of the Sustainable Finance Framework, 2019 by Islamic principles; 2.5 billion USD was raised via Sustainability Sukuk to fund social and environmental development initiatives
Millennium science Initiative	The World Bank Group's initiative to assist innovations driven by science and technology for Uganda and other African countries
Nongovernmental organisations (NGOs)	Play a vital role in advancing sustainable finance through raising awareness, advocating for policy changes, and providing guidance on best practices
Nordic investment Bank	Take the initiative to issue green bonds and invest in social and sustainability bonds within the Nordic-Baltic region
Private sector organisations	Actively engage in understanding the importance of sustainable finance and developing products and services aligned with ESG principles
Regulatory authorities	Implement various frameworks to promote responsible financial practices, such as defining stress tests for climate risks and ESG reporting requirements
Special Climate change fund (SCCF)	Initiative run by GEF under the UNFCCC that finances climate change adaptation, mitigation, and resilience initiatives
Sustainable banking and Finance network (SBFN)	The SBFN has 95 member institutions as of August 2024, accounting for 72 countries and USD 68 trillion, or 92% of all banking assets in developing economies.
The World Bank	Financing projects contributing to sustainability goals and shaping global sustainable finance practices through policymaking

food, government, health, telecommunications, transport, and water. They highlight the continuity of services in these sectors, which are essential for the everyday lives of societies and economies. The following figure exemplifies a case study for each critical infrastructure sector, providing an overview of the diverse applications and implications within each sector's unique operational framework (Fig. 1.22).

Aspects such as the evaluation of investment impacts, approaches to sustainability assessment, various tools and methods of sustainable finance, and new developments and breakthroughs in technology will all be covered in detail in the following

CASES FOR SELECTED SECTORS

Emergency Services

"The World Bank, Pandemic Fund"

Provides long-term grants for pandemic prevention, preparation, and response;

USD 500 million grant to boost pandemic preparedness; aims for USD 10 billion annually starting from 2023.

Enhances pandemic preparedness; aims to mitigate the impacts of future pandemics.

Energy Sector

"European Commission, LIFE: Clean Energy Transition Programme"

The goal is to provide assistance in implementing policies linked to sustainable energy that help achieve the goals of the European Green Deal.

The budget is approximately EUR 1 billion from 2021 to 2027.

Finance Sector

"Morgan Stanley, Green Investment"

Led a EUR 1.5 billion equity capital round for H2 Green Steel in September 2023, aimed at decarbonizing the European steel sector.

Aims to reduce CO2 emissions by about 95% compared to traditional steel production.

Food Sector

"The World Bank, Financial Intermediary Funds (FIFs) Global Agriculture and Food Security Program (GAFSP)"

The program was founded in 2010. The framework for multilateral finance supports regional and national strategies for food security and agriculture in developing economies.

The total contibutions around the world is USD 2.15 billion as of September 2024

Government Sector

"European Fund for Sustainable Development Plus"

EUR 13 billion is funded

Supports projects in challenging markets by assuming risks and avoiding market distortions.

Healthcare Sector

"International Finance Facility for Immunisation (IFFIm), Vaccine Bond"

Established in 2006; Secured over USD 8 billion in pledges; Disbursed over USD 3 billion to Gavi

Reduced child mortality by 50% in 73 countries since 2006; Benefited over 981 million children through 2023.

Telecommunication Sector

"Verizon Communications Inc. Green Bonds"

Issued 6 green bonds totaling USD 6 billion; funds for renewable energy investments

23.3% reduction in Scope 1 & 2 emissions; 15.1% reduction in scope 3 (2019-2022).

Transportation Sector

"European Union, Connecting Europe Facility (CEF) Funds"

Over EUR 6 billion awarded to 107 transport infrastructure projects; aimed at increasing efficiency and sustainability

Enhances efficiency, sustainability, and connectivity of transport networks (TEN-T).

Water Sector

"NWB Bank (Netherlands), Water Bonds"

Issued 13 water bonds totaling EUR 5.7 billion in 2021.

Outstanding loans of NWB is EUR 7.172 billion

Emission reductions; funding for water infrastructure.

Fig. 1.22 Cases for selected sectors. (The World Bank 2024a, b, c; European Commission 2024c; UNEP Finance Initiative 2023; European Commission 2024d; International Finance Facility for Immunisation Company 2022; Verizon 2024 European Commission 2023; NWB Bank 2021a, b)

chapters. These components, along with others, will be thoroughly examined, illustrating how sustainable finance is shaping both current and future practices and outcomes. The critical role of sustainable finance in achieving net-zero ambitions will be emphasised, as it is one of the key drivers in accelerating the transition to a carbon-neutral and more sustainable global future.

Acknowledgement The author would like to acknowledge the help and contributions of Alper Hasan Kızık, Buse Güngören, Akif Can Filiz, Erşan Gürkan Gültekin, and Boran Karaçalık in completing this chapter.

References

Adom PK (2024) The socioeconomic impact of climate change in Developing Countries in the Next Decades. Available at: https://www.cgdev.org/sites/default/files/socioeconomic-impact-climate-change-developing-countries-next-decades.pdf. Accessed: 22 June 2024

Ahlström H, Monciardini D (2021) The regulatory dynamics of sustainable finance: paradoxical success and limitations of EU Reforms. J Bus Ethics. https://doi.org/10.1007/s10551-021-04763-x

Alberti C (2024) The cost of inaction. CPI. Available at: https://www.climatepolicyinitiative.org/the-cost-of-inaction/#:~:text=Broadly%2C%20costs%20of%20inaction%20fall,people%20and%2For%20their%20environments. Accessed: 12 July 2024

Allen M et al (2022) Net zero: science, origins, and implications. Annu Rev Environ Resour. https://doi.org/10.1146/annurev-environ-112320-105050

Alshubiri F, Alraja M (2022) Guest editorial: green Finance for support emergency services strategies and Sustainable Development Goals (SDGs). Int J Emerg Serv 11(3). https://doi.org/10.1108/ijes-10-2022-087

American Clean Power Association (2023) Clean energy investing in America report. Available at: https://cleanpower.org/wp-content/uploads/gateway/2023/08/CleanEnergyInvestingReport_digital.pdf. Accessed: 28 June 2024)

American Meteorological Society (2013) Climate information needs for financial decision making. American Meteorological Society Policy Program

Amon J, Rammerstorfer M, Weinmayer K (2021) Passive ESG portfolio management—the benchmark strategy for socially responsible investors. Sustainability 13(16):9388. https://doi.org/10.3390/su13169388

Bank of Italy (2024) Sustainable finance. Available at: https://www.bancaditalia.it/focus/finanza-sostenibile. Accessed: 28 June 2024

Bennett P (2023) Climate change is costing the world $16 million per hour: study. World Economic Forum. Available at: https://www.weforum.org/agenda/2023/10/climate-loss-and-damage-cost-16-million-per-hour/?deliveryName=DM203247. Accessed: 11 July 2024

Berg F, Kölbel JF, Rigobon R (2022) Aggregate confusion: the divergence of ESG ratings. Rev Finance 26(6):1315–1346. 11

Berkhout F (2012) Adaptation to climate change by organisations'. WIREs Climate Change 3(1):91–106. https://doi.org/10.1002/wcc.154

Bhattacharyya R (2021) Green Finance for energy transition, climate action and sustainable development: overview of concepts, applications, implementation and challenges. Green Finance 4(1):1–35. https://doi.org/10.3934/gf.2022001

Bloomberg (2023) Bloomberg ESG Scores: Methodology and Data Guide. Bloomberg Finance L.P.

Boston Consulting Group (2023) From bold ideas to exponential impact. Available at: https://media-publications.bcg.com/2023-Annual-Sustainability-Report-April-2024.pdf. Accessed: 15 May 2024

Bouckaert S et al (2021) Net zero by 2050: a roadmap for the global energy sector

British Embassy of Luxembourg (2023) The state of play of sustainable finance in Luxembourg and the United Kingdom. Available at https://assets.publishing.service.gov.uk/media/65b272a926a40f000d7ac194/2024_01_SustFinanceDev-LU-UK_FINAL-2.pdf. Accessed: 26 Apr 2024

Buol J, Buomberger B, Meier K, Retter M, Schmidli M (2014) Overview of the "Green" Swiss Financial Market Commissioned by the Federal Office for the Environment (FOEN). Available at: https://www.bafu.admin.ch/dam/bafu/en/dokumente/wirtschaft-konsum/externe-studien-berichte/overview_of_the_greenswissfinancialmarket.pdf.download.pdf/overview_of_the_greenswissfinancialmarket.pdf. Accessed: 18 Apr 2024

Burke M, Hsiang SM, Miguel E (2015) Global non-linear effect of temperature on economic production. Nature 527(7577):235–239. https://doi.org/10.1038/nature15725

Busch T, & Lewandowski S (2018) Corporate carbon and financial performance: A meta-analysis. Journal of Industrial Ecology, 22(4), 745–759. https://doi.org/10.1111/jiec.12591

Cagli ECC, Mandaci PE, Taşkın D (2023) Environmental, social, and governance (ESG) investing and commodities: dynamic connectedness and risk management strategies. Sustain Account Manag Policy J 14(5):1052–1074

Caldecott B, Kok I, Harnett E, Cojoianu T, Pfeiffer A (2016) Stranded assets: a climate risk challenge [Preprint].:10.18235/0012646

Caldecott BL, Clark A, Harnett E, Koskelo K, Wilson C, Liu F (2022) Sustainable finance and transmission mechanisms to the real economy. University of Oxford Smith School of Enterprise and the Environment Working Paper 22-04

Carleton TA, Hsiang SM (2016) Social and economic impacts of climate. Soc Sci 353(6304). https://doi.org/10.1126/science.aad9837

Carbon Disclosure Project (CDP) (2023) CDP Global Environmental Disclosure Report 2023. CDP Worldwide.

Cattaneo O (2022) Sustainable finance challenges to fix global economy. Available at: https://www.weforum.org/agenda/2022/05/sustainable-finance-challenges-global-inequality/. Accessed: 1 May 2024

CDP (2023) CDP Climate Change 2023 reporting guidance. Version 1.0. Available at: https://guidance.cdp.net/es/guidance/export?cid=46&ctype=theme&export_type=pdf&idtype=ThemeID&incchild=1&microsite=0&otype=Guidance&page=1&tags=TAG-13071%2CTAG-605%2CTAG-599. Accessed: 30 May 2024

CDRI (2023) Global infrastructure resilience—capturing the resilience dividend. CDRI Secretariat doi:https://doi.org/10.59375/biennialreport.ed1

CFA Institute (2015) ESG issues in investing: a guide for investment professionals. [pdf] Available at: https://rpc.cfainstitute.org/-/media/documents/article/position-paper/esg-issues-in-investing-a-guide-for-investment-professionals.pdf. Accessed: 6 July 2024

CFA Institute (2020) ESG integration in Canada. Available at: https://www.cfainstitute.org/-/media/documents/article/position-paper/cfa-esg-integration-canada-web-3pp.ashx. Accessed: 26 July 2024

CFA Institute (2024.) Who is financing net zero? [online] Available at: https://rpc.cfainstitute.org/en/topics/net-zero-investing/who-is-financing-net-zero. Accessed 24 Aug 2024

Chesbrough H (2010) Business model innovation: opportunities and barriers. Long Range Plan 43(2–3):354–363. https://doi.org/10.1016/j.lrp.2009.07.010

Christensen DM, Serafeim G, Sikochi A (2022) ESG ratings: a compass without direction? J Account Econ 73(3):101488. 20

ClimateWorks (2024) Sustainable finance. Available at: https://www.climateworks.org/programs/sustainable-finance/. Accessed: 26 June 2024

Coastadapt (2017.) What are the RCPs? Available at: https://coastadapt.com.au/infographics/what-are-rcps. Accessed: 21 June 2024

Coppenolle H, Blondeel M, Graaf T (2022) Reframing the climate debate: the origins and diffusion of net zero pledges. Global Pol 14(1):48–60. https://doi.org/10.1111/1758-5899.13161

Crawford K (2023) Beyond promises: the $120 trillion path to a 'net zero' world. Stanford Institute for Economic Policy Research (SIEPR). Available at: https://siepr.stanford.edu/news/beyond-promises-120-trillion-path-net-zero-world. Accessed: 24 Aug 2024

Dalby S, Horton S, Mahon R, Thomaz D (eds) (2019) Achieving the sustainable development goals: global governance challenges, 1st edn. Routledge. https://doi.org/10.4324/9780429029622

De Souza Cunha FAF, Meira E, Orsato RJ (2021) Sustainable finance and investment: review and research agenda. Bus Strateg Environ 30(8):3821–3838. https://doi.org/10.1002/bse.2842

Debelle G (2019) Climate change and the economy. Available at: https://www.cpd.org.au/wp-content/uploads/2019/03/sp-dg-2019-03-12.pdf. Accessed: 22 June 2024

Delmas MA, Burbano VC (2011) The drivers of greenwashing. Calif Manag Rev 54(1):64–87. https://doi.org/10.1525/cmr.2011.54.1.64

Deloitte (2021) Financing your net zero transition. Available at: https://www.deloitte.com/uk/en/issues/climate/financing-your-net-zero-transition.html. Accessed: 8 June 2024

Deloitte (2022) Deloitte research reveals inaction on climate change could cost the world's economy USD 178 trillion by 2070: Deloitte Middle East: Press release, Deloitte. Available at: https://www2.deloitte.com/xe/en/pages/about-deloitte/press-releases/deloitte-research-reveals-inaction-on-climate-change-could-cost-the-world-economy-usd-178-trillion-by-2070.html. Accessed: 24 Aug 2024

Det Norske Veritas (DNV) (2024) Environmental, Social and Governance (ESG) risk assessment and due diligence. Available at: https://www.dnv.co.uk/services/environmental-social-and-governance-esg-risk-assessment-and-due-dilligence-76998. Accessed: 24 July 2024

Dietz S et al (2016) "Climate value at risk" of global financial assets. Nat Clim Chang 6(7):676–679. https://doi.org/10.1038/nclimate2972

Dithrich H, Mudaliar A (2020) Sizing the impact investing market. The Global Impact Investing Network (GIIN). Available at: https://thegiin.org/assets/Sizing%20the%20Impact%20Investing%20Market_webfile.pdf. Accessed: 25 May 2024

Dwivedi PP (2023) Role of Environmental, Social and Governance (ESG) in achieving sustainable development goals: a brief review. Public H Open Acc 7(2):000259

Ebrahim A, Rangan VK (2014) What impact? A framework for measuring the scale and scope of social performance. Calif Manage Rev 56(3):118–141. https://doi.org/10.1525/cmr.2014.56.3.118

Eccles RG, Stroehle J, Youmans T (2017) The materiality of ESG factors for corporate bond pricing. J Appl Corp Finance 29(4):37-53. 8

EcoVadis (2023) EcoVadis Sustainability Ratings: Methodology & Scoring Principles. EcoVadis SAS.

EPA (2024a) Scope 1 and Scope 2 inventory guidance. Available at: https://www.epa.gov/climateleadership/scope-1-and-scope-2-inventory-guidance. Accessed: 18 July 2024

EPA (2024b) Climate risks and opportunities defined. Available at: https://www.epa.gov/climateleadership/climate-risks-and-opportunities-defined. Accessed: 21 June 2024

Eriksson K et al. (2022) Policies for sustainable finance to fund the climate transition. A report by the Sustainable Finance Lab (Sweden) Commissioned by the Swedish Climate Policy Council. TRITA-ABE-RPT-226

European Commission (2023) EU invests €6.2 billion in sustainable, safe and efficient transport infrastructure. Available at: https://transport.ec.europa.eu/news-events/news/eu-invests-eu62-billion-sustainable-safe-and-efficient-transport-infrastructure-2023-06-22_en. Accessed: 3 Oct 2024

European Commission (2024a) Overview of sustainable finance. Available at: https://finance.ec.europa.eu/sustainable-finance/overview-sustainable-finance_en#policy-making-timeline. Accessed: 11 Apr 2024

European Commission (2024b) ESG Rating Activities. Available at: https://finance.ec.europa.eu/sustainable-finance/tools-and-standards/esg-rating-activities_en. Accessed: 25 Aug 2024

European Commission (2024c) EU funding possibilities in the energy sector. Available at: https://energy.ec.europa.eu/topics/funding-and-financing/eu-funding-possibilities-energy-sector_en. Accessed: 4 Sept 2024

European Commission (2024d) European fund for sustainable development plus. Available at: https://international-partnerships.ec.europa.eu/funding-and-technical-assistance/funding-instruments/european-fund-sustainable-development-plus_en#:~:text=The%20%E2%82%AC13%20billion%20EFSD%2B,countries%20to%20achieve%20the%20SDGs. Accessed: 3 Oct 2024

Faiella I, Bernardini E, Di Giampaolo J, Letta S, Nasti D, Fruzzetti M (2023) The exposure of investments to climate and environmental risks. In: Scalia A (ed) Financial risk management and climate change risk. Contributions to finance and accounting. Springer, Cham. https://doi.org/10.1007/978-3-031-33882-3_12

Felicetti R, Rizzello A (2020) Financing sustainable goals: economic and legal implications. Palgrave studies in impact finance. In: La Torre M, Chiappini H (eds) Contemporary issues in sustainable finance, pp 5–36. https://doi.org/10.1007/978-3-030-40248-8_2

Filippini M, Leippold M, Wekhof T (2021) Sustainable finance literacy and the determinants of sustainable investing. SSRN Elect J [Preprint]. https://doi.org/10.2139/ssrn.3997285

Fitch Group (2023) ESG ratings methodology: sustainable Fitch. Available at: https://your.fitch.group/rs/732-CKH-767/images/2023_12_20_ESG-Ratings-Methodology-Sustainable-Fitch.pdf. Accessed: 4 Oct 2024

Flammer C (2021) Corporate green bonds. J Financ Econ 142(2):499–516. https://doi.org/10.1016/j.jfineco.2021.01.010

Florida Governor's Office (2023) Governor DeSantis announces actions to protect Floridians from ESG Mandates. Retrieved from https://www.flgov.com/2023/07/25/governor-desantis-announces-actions-to-protect-floridians-from-esg-mandates/14

Flynn A (2024) The essentials of climate scenario analysis for financial institutions. Available at: https://www.climateaction.org/news/the-essentials-of-climate-scenario-analysis-for-financial-institutions#:~:text=By%20analyzing%20different%20climate%20scenarios,decisions%20and%20take%20appropriate%20actions. Accessed: 8 June 2024

Frame B et al (2018) Adapting global shared socio-economic pathways for national and local scenarios. Clim Risk Manag. https://doi.org/10.1016/j.crm.2018.05.001

Fraser Institute (2023) ESG myths and realities: the impracticality of standardizing ESG reporting. Retrieved from https://www.fraserinstitute.org/sites/default/files/ESG-myths-realities-impracticality-of-standardizing-ESG-reporting.pdf 10

Fullwiler ST (2015) Sustainable finance: building a more general theory of finance (Working Paper No. 106). Binzagr Institute for Sustainable Prosperity

Gerster R (2012) Sustainable finance: achievements, challenges, outlook. Available at: http://www.gersterconsulting.ch/docs/sustainable_finance_final_11.02.10.pdf. Accessed: 17 Apr 2024

GIZ (2011) Green Finance an innovative approach to fostering sustainable economic development and adaptation to climate change. Available at: https://www.greenpolicyplatform.org/sites/default/files/downloads/resource/Green_finance_GIZ.pdf. Accessed: 10 Apr 2024

Global Ethical Finance Initiative (2022) Achieving net zero: the crucial role of the financial sector | Accuracy Guest Blog. Available at: https://www.globalethicalfinance.org/2023/08/28/achieving-net-zero-the-crucial-role-of-the-financial-sector-accuracy-guest-blog/. Accessed: 10 Aug 2024

Global Sustainable Investment Alliance (2019) Global sustainable investment review 2018. Available at: http://www.gsi-alliance.org/wp-content/uploads/2019/03/GSIR_Review2018.3.28.pdf. Accessed: 25 Apr 2024

Gołasa P et al (2021) Sources of greenhouse gas emissions in agriculture, with particular emphasis on emissions from energy used. Available at: https://www.mdpi.com/1996-1073/14/13/3784. Accessed: 16 July 2024

Gongloff M, Denning L (2024) Climate change: $215 trillion to save the planet is a bargain. Bloomberg.com. Available at: https://www.bloomberg.com/opinion/articles/2024-05-22/climate-change-215-trillion-to-save-the-planet-is-a-bargain?embedded-checkout=true. Accessed: 24 Aug 2024

Government of Canada (2019) Final report of the Expert Panel on Sustainable Finance—Mobilizing Finance for Sustainable Growth. Available at: http://publications.gc.ca/collections/collection_2019/eccc/En4-350-2-2019-eng.pdf. Accessed: 29 May 2024

Guelfo D (2023) What is green tech? A comprehensive guide to green technology in 2023. Digital Acceleration Company. Available at: https://www.netguru.com/blog/what-is-greentech. Accessed: 17 July 2024

Hallegatte S et al (2012) Investment decision making under deep uncertainty—application to climate change. Policy Res Work Paper. https://doi.org/10.1596/1813-9450-6193
Hertwich EG, Wood R (2018) The growing importance of scope 3 greenhouse gas emissions from industry. Environ Res Lett 13(10):104013. https://doi.org/10.1088/1748-9326/aae19a
HM Government (2023) Mobilising green investment. Available at: https://assets.publishing.service.gov.uk/media/643583fb877741001368d815/mobilising-green-investment-2023-green-finance-strategy.pdf. Accessed: 23 May 2024
Höhne N et al (2021) Wave of net zero emission targets opens window to meeting the Paris Agreement. Nat Clim Chang 11(10):820–822. https://doi.org/10.1038/s41558-021-01142-2
Hong H, Li FW, Xu J (2019) Climate risks and market efficiency. J Econ 208(1):265–281. https://doi.org/10.1016/j.jeconom.2018.09.015
Hsiang S, Kopp RE (2018) An economist's guide to climate change science. J Econ Perspect 32(4):3–32. https://doi.org/10.1257/jep.32.4.3
Hudson P, Botzen WJW, Aerts JCJH (2019) Flood insurance arrangements in the European Union for future flood risk under climate and socioeconomic change. Glob Environ Chang 58:101966. https://doi.org/10.1016/j.gloenvcha.2019.101966
IMF (2021) 2021 IMF annual report: build forward better. Available at: https://www.imf.org/external/pubs/ft/ar/2021/eng/downloads/imf-annual-report-2021.pdf. Accessed: 14 May 2024
International Finance Corporation (2023) Challenges of green finance in emerging markets and developing economies: a survey of private financial firms. Retrieved from https://www.ifc.org/content/dam/ifc/doc/2023/challenges-of-green-finance.pdf 1
International Finance Facility For Immunisation Company (2022) Annual Report of the Trustees and Consolidated Financial Statements. Available at: https://iffim.org/sites/default/files/trustees-reports/IFFIm-2022-Trustees-Report-and-Financial-Statements.pdf. Accessed: 3 Oct 2024
International Financial Institutions (2021) International Financial Institutions guideline for a harmonised approach to greenhouse gas accounting. Available at: https://unfccc.int/sites/default/files/resource/AHG-003_Guideline_on_GHG_Accounting_and_reporting_1Jun_.pdf. Accessed: May 3 2024
International Institute for Sustainable Development (2014) Integrating environmental risks into asset management. Available at: www.iisd.org/integrating_environmental_risks_en.pdf. Accessed: 27 June 2024
International Trade Centre (2021). Available at: https://www.intracen.org/sites/default/files/inline-files/Annual%20Report%202021_ENG.pdf. Accessed: 3 Oct 2024
IRENA (2017) Stranded assets and renewables: how the energy transition affects the value of energy reserves, buildings and capital stock, International Renewable Energy Agency (IRENA). www.irena.org/remap
ISS ESG (2023) ISS ESG Corporate Rating: Methodology Update 2023. Institutional Shareholder Services.
Jacobs BI, Levy KN (2022) The challenge of disparities in ESG ratings. J Environ Invest 13(1):1–10. 11
Jacobson et al (2017) 100% clean and renewable wind, water, and sunlight all-sector energy roadmaps for 139 countries of the world. Joule 1(1):108–121. https://doi.org/10.1016/j.joule.2017.07.005
Jark D (2024) What is Green Tech? How it works, types, adoption, and examples, Investopedia. Available at: https://www.investopedia.com/terms/g/green_tech.asp. Accessed: 17 July 2024
Jaynes CH (2023) IEA: Clean Energy Investment must reach $4.5 trillion per year by 2030 to limit warming to 1.5°C, World Economic Forum. Available at: https://www.weforum.org/agenda/2023/09/iea-clean-energy-investment-global-warming/#:~:text=IEA's%20Net%20Zero%20Roadmap%20says,technologies%20and%20more%20nuclear%20power. Accessed: 24 Aug 2024
Jeffery M, Gütschow J, Rocha M, Gieseke R (2018) Measuring success: improving assessments of aggregate greenhouse gas emissions reduction goals. Earth S Future 6(9):1260–1274. https://doi.org/10.1029/2018ef000865
Jemović M, Radojičić J (2021) Sustainable finance and banking: a challenge for regulators and a risk management system. Facta Univ Ser Econ Organ 1
Jensen MC (2002) Value maximization, stakeholder theory, and the corporate objective function. Bus Ethics Q 12(2):235–256. https://doi.org/10.2307/3857812
Ji L et al (2022) Solar photovoltaics can help China fulfil a net-zero electricity system by 2050 even facing climate change risks. Resour Conserv Recycl 186:106596. https://doi.org/10.1016/j.resconrec.2022.106596
Johnson JA et al (2021) The economic case for nature. Rep. World Bank Group. Available at: https://openknowledge.worldbank.org/server/api/core/bitstreams/9f0d9a3a-83ca-5c96-bd59-9b16f4e936d8/content. Accessed: 11 July 2024
Kandpal V, Jaswal A, Santibanez Gonzalez EDR, Agarwal N (2024) Sustainable energy transition: circular economy and sustainable financing for Environmental, Social and Governance (ESG) practices. Circ Econ Sustain. https://doi.org/10.1007/978-3-030-55654-0
Keller C, O'Neal M (2023) Costing the earth: The costs of the green transition, Barclays. Available at: https://www.ib.barclays/our-insights/3-point-perspective/costs-of-the-green-transition.html. Accessed: 24 Aug 2024
Kelly J (2021) Millennials will become richest generation in American history as Baby Boomers transfer over their wealth. Forbes. Available at: https://www.forbes.com/sites/jackkelly/2019/10/26/millennials-will-become-richest-generation-in-american-history-as-baby-boomers-transfer-over-their-wealth/. Accessed: 27 May 2024

Kentucky Attorney General (2023, August 10) Opinion 23-06. Retrieved from https://ag.ky.gov/Resources/Opinions/OAG%202023-06.pdf 14

Köberle A, Ganguly G, Ostrovnaya A (2021) A guide to building climate-financial scenarios for financial institutions. Available at: https://spiral.imperial.ac.uk/bitstream/10044/1/85497/6/A%20guide%20to%20building%20climate-financial%20scenarios%20for%20financial%20institutions.pdf. Accessed: 8 June 2024

Kousky C (2019) The role of natural disaster insurance in recovery and risk reduction. Ann Rev Resour Econ 11(1):399–418. https://doi.org/10.1146/annurev-resource-100518-094028

Krishnan et al (2022) The net-zero transition what it would cost, what it could bring. McKinsey & Company Available at: https://www.mckinsey.com/~/media/mckinsey/business%20functions/sustainability/our%20insights/the%20net%20zero%20transition%20what%20it%20would%20cost%20what%20it%20could%20bring/the-net-zero-transition-what-it-would-cost-and-what-it-could-bring-final.pdf. Accessed: 07 Sept 2024

Küfeoğlu S (2022) Emerging technologies: value creation for sustainable development. Springer, Cham. https://doi.org/10.1007/978-3-031-07127-0

Küfeoğlu S (2024) Net zero: decarbonising the global economies. Springer, Cham

Kumar S, Sharma D, Rao S, Lim WM, Mangla SK (2022) Past, present, and future of sustainable finance: insights from big data analytics through machine learning of scholarly research. Ann Oper Res [Preprint]. https://doi.org/10.1007/s10479-021-04410-8

Kumar B, Kumar A, Kumari R, Tagar U, Sassanelli C (2023) Green finance in circular economy: a literature review. Environ Dev Sustain. https://doi.org/10.1007/s10668-023-03361-3

Kuroda H (2022) Haruhiko Kuroda: sustainable finance—entering a new phase of implementation. The Bank for International Settlements. Available at: https://www.bis.org/review/r221115k.htm. Accessed: 22 June 2024

Lehner OM, Nicholls A (2014) Social finance and crowdfunding for social enterprises: a public–private case study providing legitimacy and leverage. Ventur Cap 16(3):271–286. https://doi.org/10.1080/13691066.2014.925305

Levin K et al (2023) What does 'net-zero emissions' mean? 8 common questions, answered. World Resources Institute. Available at: https://www.wri.org/insights/net-zero-ghg-emissions-questions-answered. Accessed: 20 Sept 2024

Linnenluecke MK, Griffiths A (2010) Corporate sustainability and organisational culture. J World Bus 45(4):357–366. https://doi.org/10.1016/j.jwb.2009.08.006

Low L (2019) How climate change impacts infrastructure: experts explain. The University of Sydney. Available at: https://www.sydney.edu.au/news-opinion/news/2019/11/27/how-climate-change-impacts-infrastructure%2D%2Dexperts-explain.html#:~:text=Increased%20precipitation%20can%20lead%20to,affect%20both%20the%20availability%20and. Accessed: 26 June 2024

Luderer G et al (2019) Environmental co-benefits and adverse side-effects of alternative power sector decarbonisation strategies. Nat Commun 10(1). https://doi.org/10.1038/s41467-019-13067-8

McGowan JA (2022) The trouble with Tibble: environmental, social, and governance (ESG) and fiduciary duty. Univ Chicago Bus Law Rev Online 2022:13

Migliorelli M, Dessertine P (eds) (2019) The rise of Green Finance in Europe. Palgrave Stud Impact Fin [Preprint]. https://doi.org/10.1007/978-3-030-22510-0

MSCI (2023) MSCI ESG Ratings Methodology 2023. MSCI Inc.

MSCI (2025) ESG ratings. Retrieved from https://www.msci.com/esg-ratings 5

Naifar N, Elsayed A (2023) Green Finance instruments, FinTech, and investment strategies. Sustainable portfolio management in the post-COVID era. https://doi.org/10.1007/978-3-031-29031-2

Net Zero Scorecard (2024) Net zero emissions race. Available at: https://eciu.net/netzerotracker. Accessed: 9 Aug 2024

Net Zero Tracker (2023) Net zero stocktake 2023: assessing the status and trends of net zero target setting across countries, sub-national governments and companies, Net Zero Tracker. Net Zero Tracker

NGFS (2020) Overview of environmental risk analysis by financial institutions Retrieved from https://www.ngfs.net/sites/default/files/medias/documents/overview_of_environmental_risk_analysis_by_financial_institutions.pdf

NGFS (2022a) Scenarios for central banks and supervisors. Network for greening the financial system. Available at: https://www.ngfs.net/sites/default/files/medias/documents/ngfs_climate_scenarios_for_central_banks_and_supervisors_.pdf.pdf. Accessed: 11 July 2024

NGFS (2022b). Climate scenarios database: technical documentation V3.1. Available at: https://www.ngfs.net/sites/default/files/media/2022/11/21/technical_documentation_ngfs_scenarios_phase_3.pdf. Accessed: 24 Aug 2024

Nicholls A (2021) Sustainable finance: a primer and recent developments. Available at: https://www.adb.org/sites/default/files/institutional-document/691951/ado2021bp-sustainable-finance.pdf. Accessed: 27 May 2024

NWB Bank (2021a). Annual report 2021. Available at: https://nwbbank.com/application/files/9816/5468/9535/NWB_Bank_Annual_report_2021.pdf. Accessed: 24 June 2024

NWB Bank (2021b). Water bond report using financial instruments for the benefit of people and planet. Available at: https://nwbbank.com/application/files/7916/5545/2094/NWB_Bank_Newsletter_Waterbonds_2021.pdf. Accessed: 24 June 2024

OECD (2015) The economic consequences of climate change. OECD Publishing, Paris. https://doi.org/10.1787/9789264235410-en
OECD (2022) OECD guidance on transition finance: ensuring credibility of corporate climate transition plans, green finance and investment. OECD Publishing, Paris. https://doi.org/10.1787/7c68a1ee-en
OECD. (2025). Defining sustainable finance. Retrieved from https://www.oecd.org/finance/Defining-Sustainable-Finance.pdf 2
Open Risk Manual (2023) ESG factors. Available at: https://www.openriskmanual.org/wiki/ESG_Factors. Accessed: 30 July 2024
Osborn D, Cutter A, Ullah F (2015) Universal sustainable development goals: understanding the transformational challenge for developed countries. Report of a study by Stakeholder Forum, May 2015
Oxford Economics (2022) Temperature volatility comes at a price to worldwide growth. Available at: https://www.oxfordeconomics.com/resource/temperature-volatility-comes-at-a-price-to-worldwide-growth/. Accessed: 11 July 2024
Palocz-Andresen M (2012) Decreasing fuel consumption and exhaust gas emissions in transportation: sensing, control and reduction of emissions, Google books. Springer. Available at: https://books.google.com.tr/books?hl=en&lr=&id=k6jWxLy1upAC&oi=fnd&pg=PR5&ots=7A5hqIGIHj&sig=IhsgLG1Gy6PVcGtb7mcYEpEHk2Y&redir_esc=y#v=onepage&q&f=false. Accessed: 5 May 2024
Patatoukas PN (2022, April 11) ESG reporting is more than greenwashing. It spurs financial success. Berkeley Haas Newsroom Retrieved from https://newsroom.haas.berkeley.edu/research/research-challenges-relevance-of-popular-esg-scoring-models-for-investment-returns/ 12
Philip P, Ibrahim C, Hodges C (2022) The turning point a global summary. Rep. Deloitte. Available at: https://www.deloitte.com/nz/en/about/story/impact/global-turning-point.html. Accessed: 11 July 2024
Pisano U, Martinuzzi RA, Bruckner B (2012) The financial sector and sustainable development: logics, principles and actors. ESDN quarterly report December 2012, Vienna
Porter ME, van Linde C (1995) Toward a new conception of the environment-competitiveness relationship. J Econ Perspect 9(4):97–118. https://doi.org/10.1257/jep.9.4.97
Rajesh R (2020) Exploring the sustainability performances of firms using environmental, social, and governance scores. J Clean Prod 247:119600
Refinitiv (2023) Refinitiv ESG Scores: Methodology Guide 2023. London Stock Exchange Group (LSEG).
Riahi K et al (2017) The shared socioeconomic pathways and their energy, land use, and greenhouse gas emissions implications: an overview. Glob Environ Chang. https://doi.org/10.1016/j.gloenvcha.2016.05.009
Rodrigo-González A, Grau-Grau A, Bel-Oms I (2021) Circular economy and value creation: sustainable finance with a real options approach sustainability. https://doi.org/10.3390/su13147973
Roncalli T (2022) Handbook of sustainable finance. https://doi.org/10.2139/ssrn.4277875. Accessed: 15 Nov 2022)
Rott R, Thomas R (2023) Climate scenarios. JP Morgan Asset Management. Available at: https://am.jpmorgan.com/content/dam/jpm-am-aem/emea/gb/en/insights/market-insights/climate-scenario.pdf. Accessed: 22 June 2024
S&P Dow Jones Indices (2023) S&P Global ESG Scores Methodology. S&P Global.
S&P Global (2023) S&P Global Corporate Sustainability Assessment (CSA): Methodology & Scoring. S&P Global.
Saccardo RR et al (2023) Investment in photovoltaic energy: an attempt to frame Brazil within the 2030 passage target of the Paris agreement. Cleaner Energy Syst 5:100070. https://doi.org/10.1016/j.cles.2023.100070
Sasse T (2021) Paying for net zero. Available at: https://www.instituteforgovernment.org.uk/article/explainer/paying-net-zero. Accessed: 7 Sept 2024
Schoenmaker D (2017a) From risk to opportunity: a framework for sustainable finance. Rotterdam School of Management, Erasmus University, Rotterdam
Schoenmaker D (2017b) Investing for the common good: a sustainable finance framework. Bruegel Essay and Lecture Series. Available at: https://www.bruegel.org/sites/default/files/wp_attachments/From-traditional-to-sustainable-finance_ONLINE.pdf. Accessed: 7 Jan 2024
Schoenmaker D, Schramade W (2019) Principles of sustainable finance. Oxford University Press, Oxford
Scholtens B (2006) Finance as a driver of corporate social responsibility. J Bus Ethics 68(1). https://doi.org/10.1007/s10551-006-9037-1
Sinha J, Jain S, Padmanabhi R (2020) Landscape of green finance in India. Clim Policy Initiat. https://www.climatepolicyinitiative.org/wp-content/uploads/2020/09/Landscape-of-Green-Finance-in-India-1-2.pdf
Soterroni AC (2023) Environmental policy and land-use dynamics in Brazil: Pathways toward sustainability. Annual Review of Environment and Resources, 48, 1–28. https://doi.org/10.1146/annurev-environ-112320-105050
Stankeviciene J, Nikonorova M (2014) Sustainable value creation in commercial banks during financial crisis. Procedia Soc Behav Sci 110. https://doi.org/10.1016/j.sbspro.2013.12.966
Steuer S, Tröger T (2022) The role of disclosure in green finance. J Fin Regulat (Print) 8(1):1–50. https://doi.org/10.1093/jfr/fjac001
Sustainalytics (2023) ESG Risk Ratings: Methodology Overview. Morningstar Sustainalytics.
Swiss RE Institute (2021) The economics of climate change: no action, not an option. Rep. Swiss RE Institute. Available at: https://www.swissre.com/dam/jcr:e73ee7c3-7f83-4c17-a2b8-8ef23a8d3312/swiss-

re-institute-expertise-publication-economics-of-climate-change.pdf. Accessed: 11 July 2024

Tang M, Li X (2024) Environmental risk identification and green Finance development based on multi-scale fusion recognition network. J Knowl Econ. https://doi.org/10.1007/s13132-024-01996-9

Task Force on Climate-related Financial Disclosures (TCFD) (2017) Final Report: Recommendations of the Task Force on Climate-related Financial Disclosures. Financial Stability Board. https://www.fsb.org/2017/06/tcfd-final-report/

Texas Legislature (2021) Senate Bill No. 13. Retrieved from https://capitol.texas.gov/tlodocs/87R/billtext/html/SB00013F.htm 14

The Economist (2015) The cost of inaction: recognising the value at risk from climate change. The economist intelligence unit limited. Available at: https://impact.economist.com/perspectives/sites/default/files/The%20cost%20of%20inaction.pdf. Accessed: 11 July 2024

The UK Government Office for Science (2023) Net zero society: scenarios and pathways. How could societal changes affect the path to net zero? Available at: https://assets.publishing.service.gov.uk/media/642d85fcfbe620000c17dd71/Net_Zero_Society_Report_2023.pdf. Accessed: 10 Aug 2024

The World Bank (2024a) Global Agriculture and Food Security Program. Available at: https://fiftrustee.worldbank.org/en/about/unit/dfi/fiftrustee/fund-detail/gafsp. Accessed: 4 Sept 2024

The World Bank (2024b). Financial Intermediary Funds (FIFs): the pandemic fund. Available at: https://fiftrustee.worldbank.org/en/about/unit/dfi/fiftrustee/fund-detail/pppr. Accessed: 3 Oct 2024

Thomson Reuters (2021) ESG regulation: the state of play in Asia Pacific. Retrieved from https://www.thomsonreuters.com/en-us/posts/wp-content/uploads/sites/20/2021/09/ESG-2021_A4_Final_web.pdf 19

Tol RSJ (2009) The economic effects of climate change. J Econ Perspect 23(2):29–51. https://doi.org/10.1257/jep.23.2.29

Tracker CA (2021) Warming projections global update. Climate Analytics and New Climate Institute, Berlin. [Preprint]

UN (1987) Report of the World Commission on Environment and Development: our common future (Brundtland Report). Available at: https://www.are.admin.ch/dam/are/en/dokumente/nachhaltige_entwicklung/dokumente/bericht/our_common_futurebrundtlandreport1987.pdf.download.pdf/our_common_futurebrundtlandreport1987.pdf. Accessed: 7 Jan 2024

UN (2015) Transforming our world: the 2030 agenda for sustainable development. Available at: https://sustainabledevelopment.un.org/post2015/transformingourworld. Accessed: 14 Apr 2024

UN (2024a) Net zero coalition. Available at: https://www.un.org/en/climatechange/net-zero-coalition#:~:text=Yes%2C%20a%20growing%20coalition%20of,about%2088%25%20of%20global%20emissions. Accessed: 10 Aug 2024

UN (2024b) Climate Finance. [online] Available at: https://www.un.org/en/climatechange/raising-ambition/climate-finance. Accessed 24 Aug 2024

UNDP (2014) Investing in the SDGs: an action plan for promoting private sector contributions world investment report available at: https://unctad.org/system/files/official-document/wir2014ch4_en.pdf. Accessed: 10 Apr 2024

UNEP (2016) Environmental risk analysis by financial institutions. UNEP Inquiry. Viewed 8 June 2024. https://unepinquiry.org/wp-content/uploads/2016/09/2_Environmental_Risk_Analysis_by_Financial_Institutions.pdf

UNEP (2017a) Roadmap to sustainable financial system. Available at: https://sustainabledevelopment.un.org/content/documents/2451121283121120171533333RoadmapforaSustainableFinancialSystem.pdf. Accessed: 28 June 2024

UNEP (2017b) Enhancing environmental risk assessment in financial decision-making. Available at: unepinquiry.org/enhancing_environmental_risk_assessment_in_financial_decision-making.pdf. Accessed: 27 June 2024

UNEP (2024) From 1992 to 2022: the evolution of sustainable finance. Available at: https://www.unepfi.org/news/timeline/. Accessed: 5 Apr 2024

UNEP Finance Initiative (2023) Green finance case studies in energy and industry. Available at: https://www.unepfi.org/industries/banking/energy-and-industry-green-finance-case-studies/. Accessed: 11 May 2024

UNEP Finance Initiative, and CICERO (2021) Pathways to Paris: a practical guide to climate transition scenarios for financial professionals. Available at: https://www.unepfi.org/wordpress/wp-content/uploads/2021/02/UNEP-FI-Pathways-to-Paris.pdf. Accessed: 8 June 2024

UNFCCC (2021) Net zero financing roadmaps. https://assets.bbhub.io/company/sites/63/2021/10/NZFRs-Key-Messages.pdf

UNFCCC Climate Champions (2023.) What's the cost of net zero? [online] Available at: https://climatechampions.unfccc.int/whats-the-cost-of-net-zero-2/. Accessed 4 Sept 2024

United Nations (2023) The Sustainable Development Goals Report 2023. Retrieved from https://unstats.un.org/sdgs/report/2023/The-Sustainable-Development-Goals-Report-2023.pdf 3

Van Vuuren DP et al (2011) The representative concentration pathways: an overview. Clim Chang 109(1–2):5–31. https://doi.org/10.1007/s10584-011-0148-z

Vanguard (2023) ESG and financial performance. Retrieved from https://www.vanguard.co.uk/content/dam/intl/europe/documents/en/vanguards-approach-to-esg-uk-en-pro.pdf 4

Verizon (2024) Verizon ESG Report 2023. Available at: https://www.verizon.com/about/sites/default/files/

Verizon-2023-ESG-Report.pdf. Accessed: 20 June 2024
Wade K, Jennings M (2016) The impact of climate change on the global economy. Available at: https://mybrand.schroders.com/m/01053abe732aa4a1/original/The-impact-of-climate-change.pdf. Accessed: 22 June 2024
Wang D (2024) Climate disruption to global supply chains could lead to $25 trillion net losses by mid-century. King's College London. Available at: https://www.kcl.ac.uk/news/climate-disruption-to-global-supply-chains-could-lead-to-25-trillion-net-losses-by-mid-century#:~:text=Supply%20chain%20losses%20will%20amount,under%207%20degrees%2C%20by%202060. Accessed: 26 June 2024
WEF (2019) These countries are leading the way in green finance. World Economic Forum. Available at: https://www.weforum.org/agenda/2019/09/these-countries-are-leading-the-way-in-green-finance/. Accessed: 17 May 2024
WEF (2023) 6 shifts the finance system can make to help the planet and people thrive. Available at: https://www.weforum.org/agenda/2023/06/climate-decarbonization-finance-equitable/. Accessed: 29 June 2024
WEF (2024) Global risks report 2024. Available at: https://www3.weforum.org/docs/WEF_The_Global_Risks_Report_2024.pdf (Accessed: 20 Apr 2024
West Virginia State Treasurer (2023, July 28) Treasurer Moore announces further restrictions on banking institutions boycotting energy companies. Retrieved from https://wvtreasury.gov/News/Article/142 14
Whitley S, Thwaites J, Wright H, Ott C (2018) Making finance consistent with climate goals. Available at: https://seors.unfccc.int/applications/seors/attachments/get_attachment?code=326IM5AEC9MNUG3CQVM2VSWT5LR4AYX2. Accessed: 3 Oct 2024
World Bank (2024c) Resilient Infrastructure Financing: Global Case Studies and Frameworks. World Bank Group.
WRI (2022) Climate change investment strategy. Available at: https://www.wri.org/sustainability-wri/climate-change-investment-strategy. Accessed: 29 June 2024
Ziolo M, Filipiak BŻ, Bąk I, Cheba K (2019) How to design more sustainable financial systems: the roles of environmental, social, and governance factors in the decision-making process. Sustainability 11(20):5604
Ziolo M, Bak I, Cheba K (2020) The role of sustainable Finance in achieving sustainable development goals: does it work? Technol Econ Dev Econ 27(1):45–70. https://doi.org/10.3846/tede.2020.13863

Impact Assessment and Sustainable Investing

2

Abstract

This chapter examines impact assessment (IA) as an important tool for guiding sustainable investment. It starts by defining IA and discussing its historical development. This chapter highlights its increasing role in assessing the environmental, social, and economic effects of financial choices. It introduces key IA frameworks, such as life cycle assessment, cost-benefit analysis, and risk management, while examining important values and metrics for measuring impact. This chapter also points out challenges in applying IA and considers its future possibilities. It further discusses the processes of monitoring, evaluation, and reporting, emphasising best practices and common mistakes. Through case studies of organisations like the IFC, EBRD, World Bank, and Intel, it shows how IA applies in the real world. Finally, it moves into sustainable investing, differentiating between socially responsible investing, ESG investing, and impact investing as ways to connect finance with long-term social goals.

2.1 Introduction to Impact Assessment (IA)

Green finance seeks to promote constructive social and environmental transformation in addition to profit-seeking (UNEP 2024). Impact assessment (IA), a critical process that helps us understand how financial decisions affect the world around us, is at the core of it all. IA serves as a guide during decision-making. Stakeholders can evaluate investments' practical effects by examining their environmental, social, and economic elements (Küfeoğlu 2022). This process ensures accountability and openness while assisting in lowering the risks associated with unsustainable activity. This chapter examines the complex area of IA in sustainable finance. From methodologies to indicators, this chapter examines the tools and techniques used to assess the sustainability of investments. Understanding the foundations of IA will help us navigate the constantly shifting financial world more easily, pointing the way towards a sustainable future.

2.1.1 Definition of IA

IA is the process of determining how a project, programme, strategy, or policy will affect individuals and their surroundings while there is still time to change to mitigate harmful and enhance good ones (Mackay et al. 2024; IAIA 2023a). It is used in decision-making across all domains, from specific projects to policies (IAIA 2023a). To forecast social, environmental, and economic consequences, IAs use a variety of data sources and techniques, such as comparative case study material, quantitative modelling, trend predictions, and qualitative data that includes

S. Küfeoğlu, *Financing the Green Transition*, Sustainable Development Goals Series,
https://doi.org/10.1007/978-3-032-10440-3_2

Indigenous and local knowledge (Jolly and Thompson-Fawcett 2021).

IA benefits sustainable finance in several ways. First and foremost, it seeks to provide protocols and techniques for the follow-up, including monitoring and mitigating unfavourable effects during policy, planning, and project cycles. Second, to keep stakeholders informed and involved, IA works to advance transparency and public involvement in decision-making processes. Moreover, IA facilitates well-informed and comprehensive decision-making by assessing suggested actions' biophysical, social, economic, and institutional ramifications. IA promotes environmentally sound and sustainable development by funding projects prioritising social responsibility and environmental stewardship (IAIA 2023a).

The ideal IA would comprehensively address all aspects of the natural and human-modified environments, including people, communities, and well-being. Because of this, many types of IA have emerged to address particular issues brought up by the requirement to evaluate effects within particular environmental sectors. These include ecological, social, health, and cultural IAs. While the specific procedures and techniques used in each of these forms of IA may vary, they all adhere to the same fundamental ideas, methodology, and goal. Strategic environmental assessment (SEA) was created due to the requirement to apply IA to strategic levels of decision-making (such as laws, plans, policies, and programmes) (IAIA 2023a; Desmond 2007). SEA is an IA procedure designed to guarantee the sustainability of strategic choices while integrating environmental, social, economic, and health concerns. Emerging laws about SEA are frequently connected to environmental impact assessment (EIA) organisations and regulations. In 2001, the European Union approved a directive on the environmental assessment of plans and programmes. The SEA is becoming increasingly recognised as a tool that may be used early in the decision-making process to help set the criteria for alternative analysis and help inform decisions at the sectoral and regional levels (IAIA 2023a).

IA is evolving, and many (and not so different) kinds of IAs have existed since the National Environmental Policy Act (NEPA) was enacted. According to Morrison-Saunders et al. (2014), IA is becoming less efficient and effective because it is divided into various categories. They suggest that the answer lies in returning to the fundamentals of IA and advocating for more integration around the objective of sustainable development and improved scoping (Morrison-Saunders et al. 2014). Figure 2.1 shows the proportions of impact captured by different steps of assessment.

The figure above shows different levels of captured impact. Level 1 includes capturing the most visible impacts of the firm and their effect on the ecosystem, extracting the most significant drivers from procurement and corporate social responsibility (CSR) data, and auto-populating impact models. Level 2 is data diving, which involves diving into more key performance indicators (KPI) regarding quality, health, safety and environment (QSHE), supply chain, scope 3 GHG data, and other data. Level 3 is deep data analysis, which involves decomposing impacts from consolidated objects and tailoring them for business. The last level, level 4, captures each small impact and extends the supply chain to granular levels (Simfoni 2023).

2.1.1.1 Economic IA

An economic IA calculates the potential changes in business activity, income, and employment that could arise from a project or programme being considered (IAIA 2023b). It must employ standardised procedures and data, be explicit about all assumptions, describe the economic environment with and without the project, and recommend specific impact management strategies (Queensland Department of State Development 2017). In an economic IA, two distinct assessment types are typically employed: a regional impact analysis (RIA) to characterise the extent and nature of impacts on local, regional, and state economies and a cost-benefit analysis (CBA) to determine the project's costs and benefits (Queensland Department of State Development 2017). Fiscal IA is also included

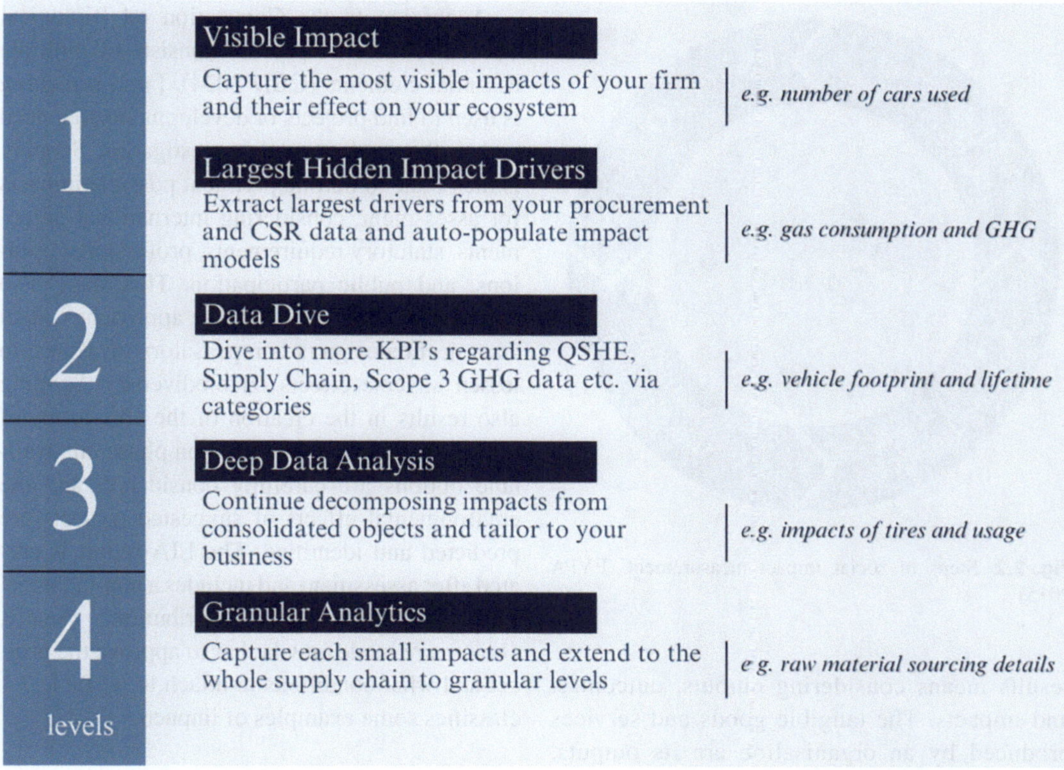

Fig. 2.1 Proportion of impact captured (Simfoni 2023)

within the economic IA in this section. A fiscal impact analysis aims to project the costs and revenues that governments will face due to development, programme, or policy. The financial effects of a new project are dependent upon several variables, including the project's attributes (e.g., the level of investment and the size and schedule of the workforce) and site area attributes (e.g., the state and local tax structure and the capability of the current service delivery systems), in addition to the project's impacts on the population and economy (IAIA 2023b).

2.1.1.2 Social IA

The social impact assessment (SIA) process includes analysing, monitoring, and administering planned interventions (policies, programmes, plans, and initiatives) and any social change processes they may initiate for positive and negative social consequences. Its primary objective is establishing a more equitable and sustainable biophysical and human environment (IAIA 2003). The concerns considered in an SIA should ideally come from the scoping exercise carried out as part of the study. Every location and project will have distinct social effects that are likely to be substantial, and every community and group within a given community will have varied social impacts that are assigned varying weights (Vanclay 2002). Changes to people's way of life, culture, community, political systems, environment, health, personal and property rights, anxieties, and aspirations are convenient ways to conceptualise social consequences (Vanclay 2003). Figure 2.2 shows four steps of social impact measurement.

As can be seen from the figure above, the first step of social impact measurement is setting objectives. This step entails defining the impact measurement's scope and then establishing goals. The next step in analysing stakeholders is to comprehend their expectations, their role in the project, and the activity's possible effects on them. Thirdly, measuring

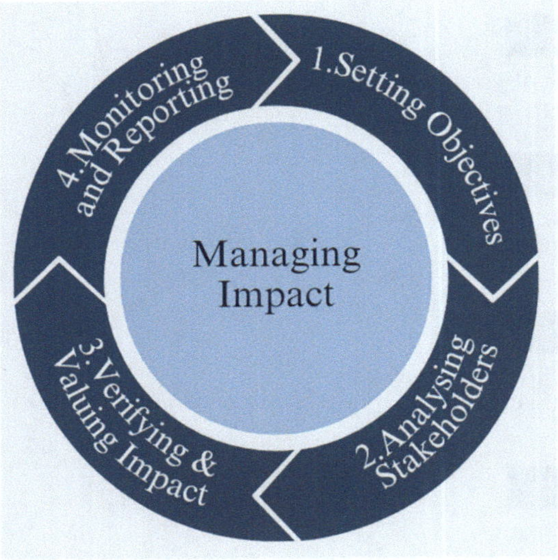

Fig. 2.2 Steps of social impact measurement (EVPA 2015)

results means considering outputs, outcomes, and impacts. The tangible goods and services produced by an organisation are its outputs. Outcomes are the activities' benefits, learnings, changes, or other effects. Verifying and evaluating effects entails determining how vital each impact is about the organisation, its stakeholders, and the people it impacts. The impact measurement process ends with two steps: reporting, which involves presenting data in a way that is understandable to essential stakeholders, and monitoring, which consists of tracking progress towards (or away from) the goals established in the first step and given concrete expression by the indicators set in the third step (EVPA 2015).

2.1.1.3 Environmental IA

An environmental IA (EIA) assesses a project or development's anticipated environmental effects while accounting for both positive and negative interrelated socioeconomic, cultural, and human health effects (UNESCO 2024). EIA evaluates a project's potential direct and indirect impact on various environmental parameters, such as biodiversity, land, soil, water, air quality, climate, landscape, material assets, and cultural heritage (European Commission 2024a).

According to the Convention on Biological Diversity, the EIA process consists of multiple essential elements (CBD 2024). First, screening is used to find projects or developments that need an IA's thorough or partial investigation. Scoping is then done to outline pertinent possible impacts for assessment, considering international agreements, statutory requirements, professional opinions, and public participation. This stage also includes looking into alternate approaches, such as project redesign or compensatory measures, to lessen adverse effects on biodiversity. Scoping also results in the creation of the IA's terms of reference. During the evaluation phase, all available options are carefully considered, and the environmental effects of suggested projects are predicted and identified. The EIA report is created after assessment and includes a non-technical summary for broader distribution. Finally, choices are made on whether to approve the project and what conditions to attach to it. Table 2.1 classifies some examples of impacts.

2.1.1.4 Institutional Impact Assessment

The sustainability of institutions is broadly defined as the frameworks that govern how political decisions are made and carried out. The social capital of societies supplied by the wide range of institutional settings below this point, including orientations and the procedures for making decisions and enforcing rules, is what they refer to instead of general expectations guiding a society's behaviour, values, ideals, and principles. Agenda 21 lists accountability, gender equity, civil society empowerment, and knowledge formation as core institutional objectives (Spangenberg 2004). Agenda 21 is the comprehensive plan of action that should be implemented on a global, national, and local level by governments, major groups, and organisations within the United Nations System in every region where environmental degradation is caused by human activity (UN DESA 2024). Institutional impact assessment (IIA) assesses the potential financial and operational impacts on government institutions arising from legislative proposals (UN 2023; Government of South Australia 2016).

Table 2.1 Examples of different impacts and their classifications (European Commission 2009)

Economic impacts	Social impacts	Environmental impacts
Functioning of the internal market and competition	Employment and labour markets	The climate
Competitiveness, trade, and investment flows	Standards and rights related to job quality	Transport and the use of energy
Operating costs and conduct of business/small- and medium-sized enterprises	Social inclusion and protection of particular groups	Air quality
Administrative burdens on business	Gender equality, equality of treatment and opportunities, non-discrimination	Biodiversity, flora, fauna, and landscapes
Public authorities	Individuals, private and family life, personal data	Water quality and resources
Property rights	Governance, participation, good administration, access to justice, media, and ethics	Soil quality or resources
Innovation and research	Public health and safety	Soil quality or resources
Consumers and households	Crime, terrorism, and security	Renewable or non-renewable resources
Specific regions or sectors	Access to and effects on social protection, health, and educational systems	The environmental consequences of firms and consumers
Third countries and international relations	Culture	Waste production/generation/recycling
Macroeconomic environment	Social impacts in third countries	The likelihood or scale of environmental risks
		Animal welfare
		International environmental impacts

2.1.2 Brief History and Evolution of IA

The search for economic profit has been the primary motivation for investment. Nevertheless, pursuing non-financial goals such as social cohesion, political or military dominance, or even global supremacy is part of a practice that goes back thousands of years (Bose et al. 2019). As far back as the fifth century BC, Confucian principles emphasised the importance of traders striving for responsible and harmonious interaction with nature and society (Wang and Juslin 2009). This consideration of social factors in investment decisions has continued through the ages. In the mid-1700s, Methodism's founder, John Wesley, stressed that the use of money was the second most important topic in New Testament teachings. As Quakers landed in North America, they refused to invest in weaponry or enslaved people (Schueth et al. 1999). Furthermore, establishing the Pioneer Fund by a church group in Boston in 1928 marked the creation of the first investment fund filtered according to ethical criteria (Knoll 2002). Although impact investing has roots in the past, it is essential to emphasise that its significant development occurred mainly in the second half of the twentieth century, beginning with the National Environmental Policy Act (NEPA) in the USA in 1970 (Glasson and Therivel 2012). Figures 2.3–2.6 show the brief history of IA.

2.1.3 Importance of IA

Quantifying and managing impact are crucial to ensure investors achieve their targeted impact outcomes to address the world's most severe social and environmental concerns. The GIIN defines impact measurement and management as core characteristics of impact investing (GIIN 2024a). There are common reasons for the desire

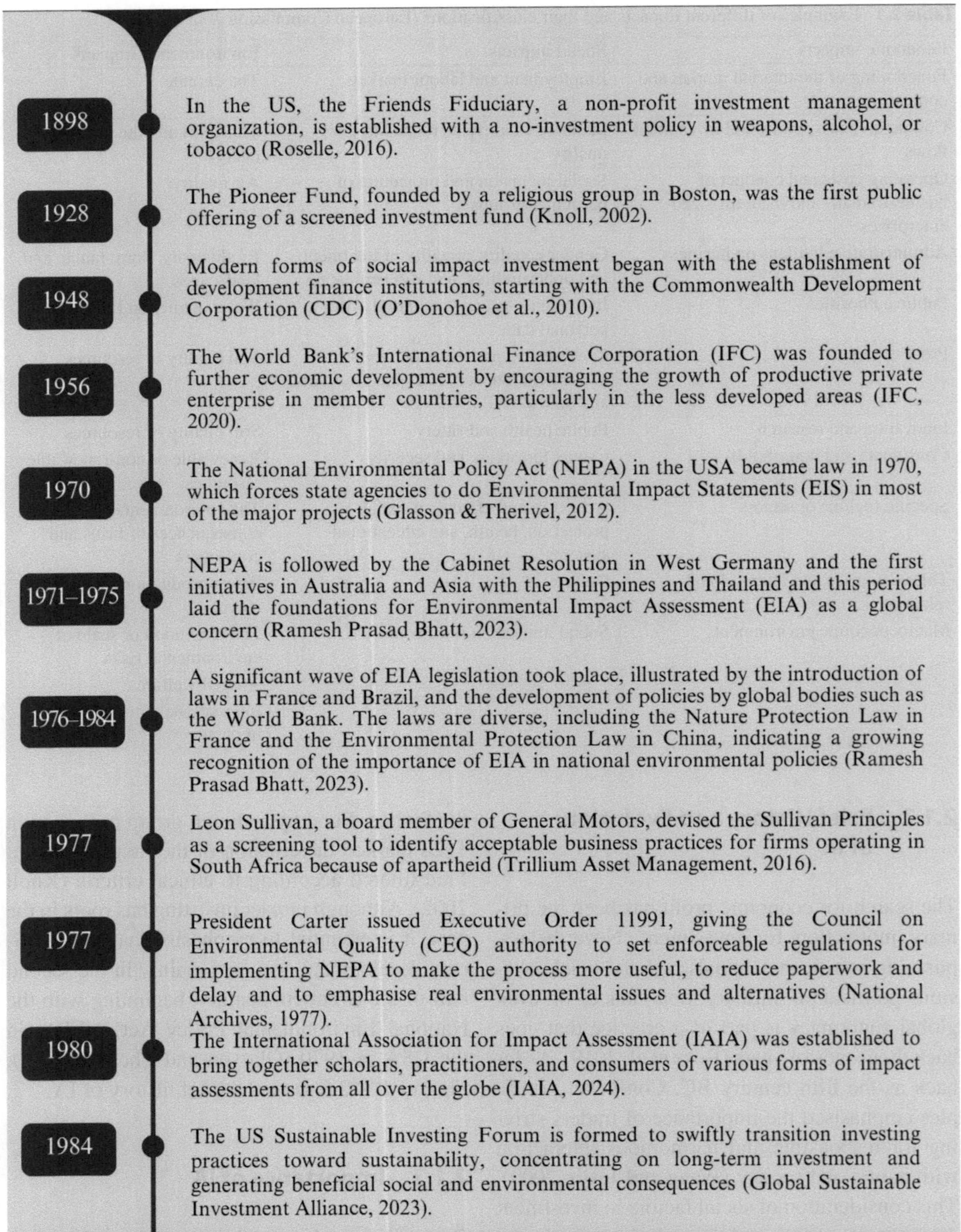

Figs. 2.3–2.6 A brief history of IA

1985 — The ratification of a European Directive on EIA in 1985 triggered the enactment of EIA legislation in numerous European nations in the late 1980s (Glasson & Therivel, 2012).

1985-1987 — EIA gained traction with the implementation of directives by supranational bodies such as the European Union, as well as initiatives in India and the Netherlands. These measures show a shift towards standardizing and institutionalizing EIA inside national and regional regulatory frameworks (Ramesh Prasad Bhatt, 2023).

1986 — The World Bank made a public commitment to include environmental impact assessment in their project appraisal process (Burdge, 1991).

1989 — The World Bank establishes its first operational directive for EIA. This extensive and detailed new policy required an environmental evaluation for all projects with potential substantial environmental implications. Bank employees were expected to assess all prospective loans for any negative environmental issues (The World Bank, 1991).

1990 — The MSCI KLD 400 Social Index was founded as the Domini 400 Social Index, which is one of the first Socially Responsible Investment (SRI) indexes. (Morgan Stanley Capital International, 2024)

1992 — The European Bank for Reconstruction and Development has set up comprehensive EIA guidelines called Environmental Procedures (European Bank for Reconstruction and Development, 1992).

1994 — Clinton's 'Executive Order on Federal Actions to Address Environmental Justice in Minority and Low-Income Populations' is signed. According to this Order, each federal agency must analyze the environmental, health, economic, and social repercussions of federal activities, especially those on minority and low-income communities (White House, 1994).

1996 — UK Overseas Development Administration sets up EIA guidelines (Glasson & Therivel, 2012).

1996 — In the early 1990s, the number of EIA laws and guidelines issued in Africa and South America increased significantly. By 1996, over 100 countries had EIA systems installed (Sadler, 1996).

1997 — CEQ found that federal agencies in the United States estimated that approximately 50,000 Environmental Assessments (EAs) were produced each year (CEQ, 1997).

1999 — Dow Jones Indexes (now S&P Dow Jones Indices) and SAM established the world's first global sustainability benchmark, the Dow Jones Sustainability Group Index (DJSGI) (Knoepfel, 2001).

2002 — The motto 'people, planet, and prosperity' was endorsed during the 2002 World Summit on Sustainable Development in Johannesburg. These three pillars demonstrate the need for sustainable development that stresses social development, and environmental conservation and focuses on the economy (White and Lee, 2009).

Figs. 2.3–2.6 (continued)

2003 The Equator Principles are released. These principles set a risk management framework used by financial institutions to identify, analyze, and manage environmental and social risk in project funding (Wörsdörfer, 2015).

2006 Principles for Responsible Investment (PRI) which is an investor initiative in partnership with the United Nations Environment Programme (UNEP) Finance Initiative and UN Global Compact is launched (PRI, 2021).

2006 The Nobel Peace Prize is handed to the founder of Grameen Bank, Muhammad Yunus, "for their efforts through microcredit to create economic and social development from below" (Clarkin and Cangioni, 2016).

2007 The term 'impact investing' was coined in 2007, when the Rockefeller Foundation held a meeting of chosen investors at its Bellagio Center in Italy to examine ways to better integrate an investment's social and environmental performance with its financial return. (Harji and Jackson, 2012).

2008 The World Bank offers the first designated green bond. The bond laid the groundwork for today's green bond market. It established the requirements for projects qualifying for green bond funding (PRI, 2021; The World Bank, 2019).

2009 The Impact Reporting and Investment Standards (IRIS) and Global Impact Investment Rating System (GIIRS) were created as a reaction to the Monitor Institute's study, Investing for Social and Environmental Impact (Bouri et al., 2018).

2010 J.P. Morgan Social Finance, the GIIN, and the Rockefeller Foundation released Impact Investments: An Emerging Asset Class, providing the first comprehensive study of the subject (Bouri et al., 2018).

2011 The Global Reporting Initiative (GRI) was established. It is a non-profit organization that offers all businesses and organizations with a complete sustainability reporting methodology extensively utilized across the world (PwC, 2013).

2012 Morgan Stanley develops the Investing with Impact Platform, which aims to provide market-rate financial returns simultaneously showing good environmental and/or social impacts (Morgan Stanley, 2018).

2013 The World Economic Forum's From the Margins to the Mainstream: Assessment of the Impact Investing Sector and Opportunities to Engage Mainstream Investors explores the limits that traditional investors face when devoting cash to impact investments (Bouri et al., 2018).

2013 The India Companies Act establishes new Corporate Social Responsibility requirements, requiring firms to pay 2% of their net revenues to address societal challenges (The Economic Times, 2022).

2014 The Organization for Economic Cooperation and Development (OECD) has published New Investment Approaches for Addressing Social and Environmental Challenges, its first paper on impact investing, offering a review of the market for OECD member countries and emphasizes the role policymakers can play in advancing the industry (Bouri et al., 2018).

Figs. 2.3–2.6 (continued)

2015 BlackRock, the world's biggest investment company, and Bain Capital in 2015 that they would launch impact investing funds (Trelstad, 2016).

2015 The GIIN and Cambridge Associates have released Introducing the Impact Investing Benchmark, the first comprehensive financial performance statistics on impact investments, with a focus on private equity funds. Wharton Social Impact Initiative announces Great Expectations: Mission Preservation and Financial Performance in Impact Investing, adding to the body of evidence on the performance of impact investments in private equity (Bouri et al., 2018).

2015 The United Nations establishes the Sustainable Development Goals (SDGs) through the 2030 Agenda. 193 nations have officially embraced a total of 17 SDGs (UN General Assembly, 2015).

2016 The World Economic Forum launched the Accelerating Impact Measurement and Management project, a multistakeholder endeavour to define and collaboratively address critical actions to minimize fragmentation in methodologies and hurdles investors encounter when incorporating effect into decision-making (Bouri et al., 2018).

2017 The GIIN introduces Navigating Impact to support the adoption of standardized core indicators across investors with similar impact expectations (Bouri et al., 2018).

2017 The European Venture Philanthropy Association and Social Value International released the Impact Management Principles, which highlight the important processes for investors to follow in their impact assessment and management approach (Bouri et al., 2018).

2018 The leaders of the G20 came to the conclusion that impact investment promotes long-term growth and is a vital solution for financing and, ultimately, achieving the SDGs (Bass et al., 2020).

2019 The GIIN has released the IRIS+ System, which was created with input from hundreds of well-known impact investing practitioners worldwide. The GIIN released the IRIS+ with a common understanding of how to assess and manage their impact efficiently as well as direction on how to grow that impact over time (Global Impact Investing Network, 2024).

2021 The GIIN estimates that over 3,349 entities are presently managing 1.164 trillion USD in impact investment (Hand et al., 2022).

2021 Green bond issuance has reached USD 578 billion by increasing at a 43% annual pace since 2008 (Hand et al., 2022).

2022 The GIIN has launched the IRIS+ Impact Performance Benchmarks, a groundbreaking instrument for impact investment. The first benchmark to be issued, with more to come, is focused on financial inclusion. This tool allows investors to assess and measure their impact relative to their colleagues and the larger market in which they invest (Global Impact Investing Network, 2024).

2023 GIIN presented statistics proving that impact investors can achieve market-rate returns, with virtually all investors meeting or exceeding their financial and effect performance targets (Hand et al., 2023).

Figs. 2.3–2.6 (continued)

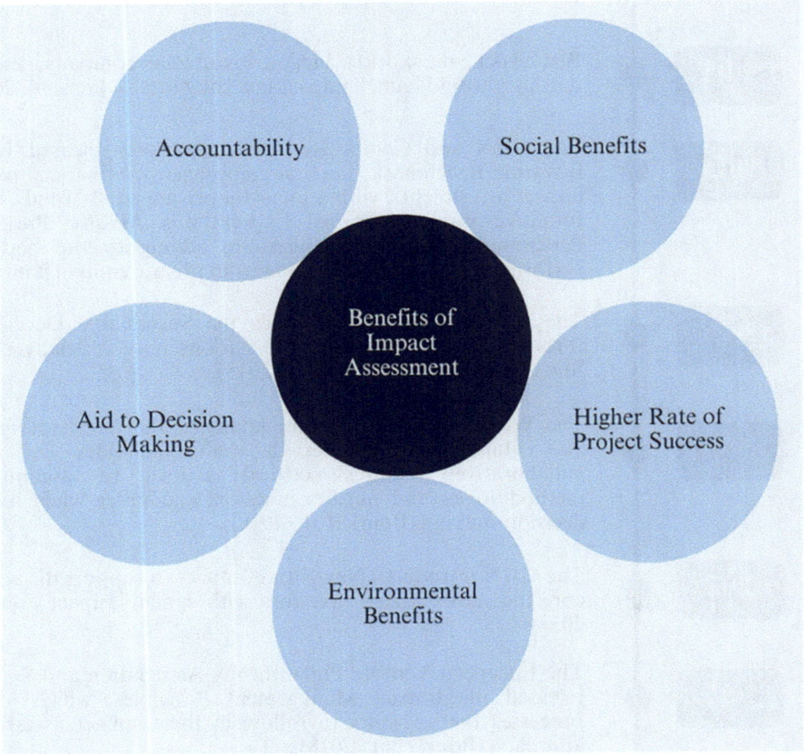

Fig. 2.7 Benefits of IA

to conduct IA. First and most frequently, actors can avoid losing their credibility and being disqualified from receiving any more sponsorship because they work diligently to generate the money. IA is an ideal instrument for proving accomplishment and taking responsibility. Second, improvements are necessary to demonstrate their accomplishments when awarded further funding. Participating actors can also use the results of IA to understand the impact of their work and achieve breakthroughs. Third, performers must recruit supporters for their trip because the public voice is crucial to the commercial process. They may increase this support by showcasing the IA outcomes (Küfeoğlu 2022).

It is essential to evaluate development initiatives for several reasons, many of which have related but separate goals. First, it acts as evidence of achievements, which validates the funds obtained and makes it easier for donors, stakeholders, and members of the public to request more financing. This maintains support for continuing projects in addition to helping to illustrate contributions made towards achieving the Millennium Development Goals (MDGs). To increase the effectiveness of adopted interventions and provide natural, noticeable benefits to people's lives, the evaluation also aims to learn more about how these initiatives affect local communities. Additionally, it emphasises the duty to stakeholders by supporting cooperative methods as an alternative to forcing development on local populations. This means talking to local communities about their desired improvements and working together to evaluate any advancements or setbacks. Last but not least, evaluation outcomes support, through the use of IA results as evidence, changes in behaviour, attitudes, policies, and laws at different levels. Therefore, IAs are essential since they can demonstrate efficacy while pointing out areas that require improvement, emphasising how interrelated development initiatives are (O'Flynn 2010). Figure 2.7 shows the benefits of IA.

2.1.3.1 An Aid to Decision-Making

IA is a tool for decision support. Before making a choice, it gives the decision-maker—a local authority, for instance—a systematic analysis of the potential actions' effects on the environment and, occasionally, alternative options. The decision-maker may consider the Environmental Impact Statement (EIS) and other documents about the intended activity. Comparing IA to other methodologies like cost-benefit analysis (CBA), the former is typically less quantitative and has a broader reach (Glasson and Therivel 2019).

2.1.3.2 Accountability

Accountability generally refers to public evaluation and documentation systems that allow people to hold governments and businesses answerable (Li 2009). IA must look beyond project logic and concentrate on changes in connection to target groups to benefit learning and accountability. Instead of asking, "What has been achieved?", "What has changed about efforts" will help assess the impact of the projects more. The parts that should be checked to achieve this are the actors involved in change, strategies used in change, context of the change, process or pathway of change, and connection of efforts to these changes (O'Flynn 2010).

2.1.3.3 Higher Rate of Project Success

Understanding how a project's physical components interact with social organisations and culture gives supporters the chance to make sure that the project and the people are compatible. Conflicts may make it more challenging to accomplish the intended institutional and economic goals. The disregard for social factors in the past has, as one development expert notes, "...led to resistance to change, total loss of capital investment in big agricultural projects, and the failure of many technical assistance programs to make the initial breakthrough in recipient cultures" (Okediji 1965). In these situations, the use of IA would prove to be a helpful tool, enabling assessment of the possibility that the projects' target groups would be effectively reached, lowering the likelihood of unforeseen adverse effects arising, and generally leading to a higher rate of project success (Burdge 1990). Many plans have failed or proved overly expensive due to insufficient consideration of social concerns. For example, the 3000 residents of Penol refused to be relocated to a hydropower project in Columbia, causing a three-year delay and consequent loss of at least USD 100 million in 1980 since they had not been included in the initial decision-making process over their lives (Scudder 1968). To their satisfaction, those Colombians from the countryside were eventually moved. However, by recognising local concerns and wants beforehand, the application of IA would have made planners aware of this issue and eradicated it. Locals could also offer feedback and be informed about the project and resettlement opportunities (Burdge 1990).

2.1.3.4 Social Benefits of IA

One significant advantage of using IA at the planning stage is that it can uncover consequences before a project is implemented, which allows for changes to the project design or mitigation strategies to minimise or eliminate adverse effects. Using public involvement to learn more about potential social effects can yield important insights into how particular changes affect people and social groups and whether or not they are likely to align with local objectives and values (Burdge 1990). For example, an initiative in Tanzania to generate revenue for women was incredibly successful since it gave the women money to spend on their children's education and sustenance. The project's accomplishment delighted the women in the neighbourhood. Reports from monitoring the initiative confirmed the achievement of its objectives. The project was deemed adequate and well-conceived based on the findings of the mid-term assessment and initial evaluation. Although the women were satisfied, others in the society were not. The project had to end after some time because too many of the women discovered that having their source of money caused conflict and violence in their homes. One of this project's unforeseen and detrimental effects was that some women were being coerced—sometimes violently—to give their husbands their earnings, who then spent the money any way they saw fit. This was overlooked during the evaluation since the impact questions were not examined (O'Flynn 2010).

2.1.3.5 Environmental Benefits of IA

The management of environmentally harmful developments currently underway must be optimised. Even in severe circumstances leading to shutdowns, these developments may still pose lingering environmental issues for years. It is preferable to mitigate adverse effects beforehand, during the planning phase, or, when feasible, to avoid specific developments altogether. The groundbreaking EIA laws in the USA and the EU revolve around this principle. This emphasises the core function of EIA as a tool for promoting sustainable development, which some authors have highlighted as frequently underappreciated when assessing EIA effectiveness (Glasson and Therivel 2019). While the primary focus of EIA inquiries is monitoring and controlling environmental effects, in certain situations, information obtained or lessons learned through follow-up may be applied to future initiatives. Since 1990, the UK's promotion of renewable energy, particularly wind power, has sparked discussion about the potential effects on noise, the environment, and wildlife conservation that were not previously considered in traditional energy development siting schemes. Regulators will be motivated to follow up on EIAs to manage compliance, lower uncertainty, validate forecasts, and eventually enhance decision-making in subsequent EIA procedures. The urge for follow-up from an individual or community will centre on handling problematic circumstances (Marshall 2005).

2.1.4 Overview of IA Frameworks

Three pillars are typically used to define sustainability: the economy, the environment, and society. An integrating framework is required due to the attempt to incorporate social, economic, and environmental factors when evaluating plans, projects, programmes, and policies (Slootweg and Mollinga 2009). The IA framework provides a model for thinking about the impact of decision alternatives and a methodology to apply this model in specific decision-making contexts (Miller 2019). The framework aims to illuminate the interactions between the biophysical environment and human society. This conceptualisation centres on identifying, categorising, and evaluating the roles of the biophysical environment in supporting human activities. This method begins with the premise that society uses goods and services that the biophysical environment provides. In economics, the environment is the supply side, and society is the demand side. In short, sustainability is about maintaining a balance between supply and demand today and in the future. Institutions respond to perceived disruptions in this equilibrium by controlling the natural resources or the social demands (Slootweg et al. 2001). Figure 2.8 shows six important frameworks of IA.

The figure above shows six important IA frameworks, which are explained below. Cost-benefit analysis and life cycle assessment are presented in their sections.

2.1.4.1 Theory of Change

The theory of change (ToC) is a methodological technique that programme planners and evaluators frequently utilise to direct the development, execution, and assessment of change initiatives (Funnell and Rogers 2011). This method entails creating a narrative and visual map that theories how a project or series of activities works to bring about change, including who benefits, under what conditions, and with what presumptions (van Es et al. 2015). Van Es et al. (2015) define the ToC as "a guiding framework for all stages of thinking, action, and sense-making when we intervene intentionally in social change processes". Developing a theory of constraints (ToC) typically requires bringing together the perspectives of researchers, practitioners, and the community to align and model change processes that can guide practices and provide a framework for study and assessment (Gates et al. 2024).

The concept of the ToC method appears to have originated in the U SA during the 1990s to enhance assessment theory and application in the domain of community initiatives. Nonetheless, "two streams of development and social programme practice—informed social practice and evaluation—are drawn upon in the current evolution". ToC is a component of more comprehen-

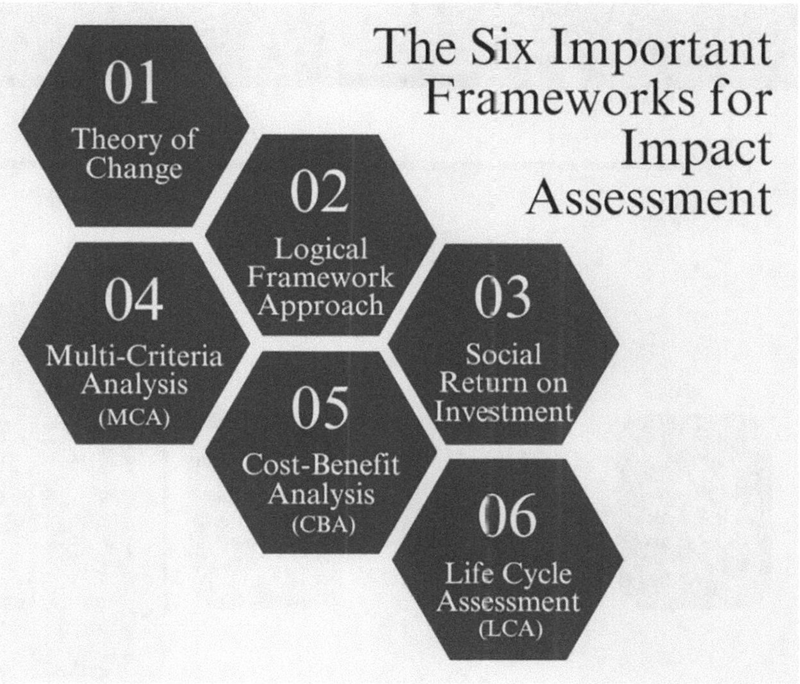

Fig. 2.8 Six important frameworks for IA

sive programme analysis or programme theory from the evaluation standpoint. It also emerged from the development field's history of logic planning models, such as the logical framework method, created in the 1970s. Creating informed social practices is not new; practitioners have long sought (and employed) instruments to actively consider the guiding theories of development practice. Since being employed in the community development sector, ToC approaches have gained widespread acceptance. The expectations of important donors, whose emphasis on TOCs has increased over the past few years, are mostly to blame. ToC appears to indicate a rising desire for organisations to be able to investigate and depict change in a way that reflects a sophisticated and systemic understanding of development, even though some may merely see it as a "buzzword". This ambition is influenced, at least in part, by the "results agenda" since the ToC is viewed as a credible means of demonstrating impact in areas of the world that are unstable or impacted by violence (Stein et al. 2012; Gates et al. 2024).

2.1.4.2 Logical Framework Approach

The logical framework is a fundamental tool for project planning and evaluation within international development. It aims to encompass all aspects of the planning process for an intervention that implies intangible outcomes. In practice, the chronology and causation could be more precise, and the intricate interactions of numerous sources and (hidden) factors frequently produce the impact. Chance is a decisive factor sometimes. The "logic model" is one of the most widely used analytical models of these impact-generating processes, particularly one intended for the performance evaluation of non-profit programmes (Jonkers et al. 2018). Figure 2.9 represents the logic model.

The logic model is graphically represented in the above figure, which places it within the larger framework of mission-oriented programmes motivated by social needs, challenges, and difficulties. This one-directional "linear" model has an implicit timeline. The displayed version clarifies the crucial distinction between effects, results, and outputs. This model identifies shorter-term "outputs" and "results" that show progress towards long-term goals, even though it

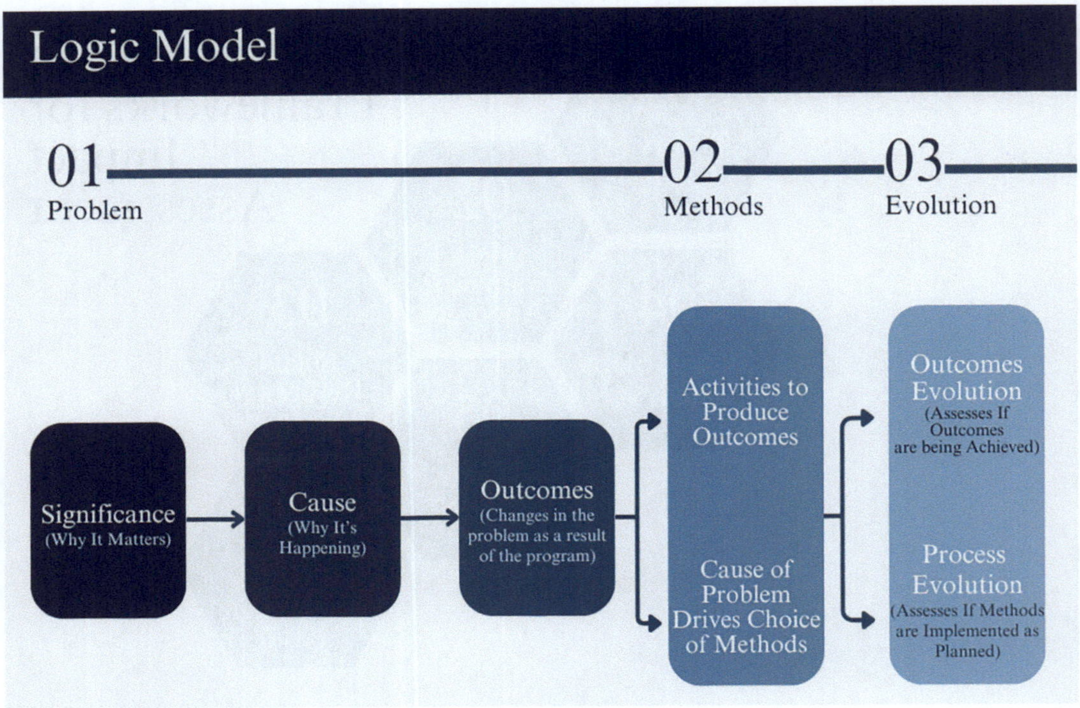

Fig. 2.9 The logic model (Impact Canada 2020)

also suggests that "impacts" may be further in the future (Jonkers et al. 2018). In several domains, the logic model has shown to be an effective tool for programme planning, execution, and performance management. A logic model is a written or graphical depiction of how a program is meant to operate that connects results to procedures and the program's theoretical presumptions. This represents a project or programme that outlines its goals and what it will do. It is a sequence of "if-then" relationships that result in the intended results when carried out according to plan. Put another way, it is a framework for explaining how resources, actions, and outcomes relate to a particular programme or project objective. Making underlying assumptions about the project or programme apparent is another benefit of using a logic model. It offers a standard method for combining preparation, execution, and assessment (Hayes et al. 2011).

Esteemed development agencies, including the United States Agency for International Development (USAID), the German Corporation for International Cooperation GmbH (GIZ), the World Bank, and the United Nations Development Project (UNDP), have utilised this framework. Additionally, local and global NGOs, such as Oxfam and CARE, have adopted variants of the LFA, reflecting its broad acceptance and adaptability in facilitating project support (Martinez and Cooper 2020).

2.1.4.3 Social Return on Investment

A framework for measuring and accounting for the broader concept of value" is how social return on investment (SROI) is described. Considering social, environmental, and economic costs and benefits, it aims to lessen inequality, increase well-being, and stop environmental deterioration. Identifying, quantifying, and summarising the social, economic, and environmental value produced by a programme, organisation, policy, or intervention is called return on investment. SROI has two measurement capabilities: evaluative-type (measures outcomes that have happened) and forecast-type how much value will be generated if the intervention achieves its goals. Through data gathering and analysis, a benefits-to-costs ratio can be computed. A 4:1 ratio, for instance, shows that a GBP 1 investment generates GBP 4 in social benefit. Per the latest guidelines, conducting a study on return on investment necessitates moving through six steps (Banke-Thomas et al. 2015).

2.1.4.4 Multi-Criteria Analysis

Multi-criteria analysis (MCA) approaches to assess or evaluate a plan by explicitly considering the numerous aspects of interest and the interactions between multiple, frequently conflicting objectives and distinct metrics and decision criteria. This contrasts with mono-criteria methods. Therefore, unlike popular belief, MCA is not a single unique procedure. Instead, it should be used as a catch-all phrase for various techniques and instruments, allowing multiple goals and selection criteria (or qualities) to be explicitly included in a problem's examination (Dean 2020).

2.1.5 Life Cycle Assessment

Life cycle assessment (LCA) is an internationally recognised environmental assessment method governed by the ISO 14040 and ISO 14044 standards (ISO 2006a, b). It is used to assess the environmental impact of a product or service throughout its life cycle, from the extraction of raw materials through manufacture, use, and recycling to end-of-life. The LCA is a comprehensive tool that considers all the stages in a product's life cycle and the different environmental impacts, such as climate change, resource depletion, and toxicity to humans and the environment. It thus makes it possible to identify the stages in a product's life cycle that have the most significant impact and to take targeted action to reduce these impacts. LCA is used in many sectors worldwide, including industry, agriculture, energy, and construction. It also assesses the environmental impact of public policies or development projects. Because of the scientific rigour required, an LCA is technically complex and time-consuming. However, if well planned, the benefits can outweigh the costs. It is recommended that ISO 14040 standards be carried out for any LCA and that the results be subjected to peer review before public communication (ISO 2006a, b; PWC and FPAC 2010). Finally, it should be noted that LCA is a standardised, transparent, verifiable, and internationally recognised approach to assessing a product's or service's environmental impacts throughout its life cycle. ISO standards 14,040 and 14,044 provide methods for developing and verifying LCAs, guaranteeing the credibility and comparability of the results (PWC and FPAC 2010). Figure 2.10 shows the LCA methodology diagram.

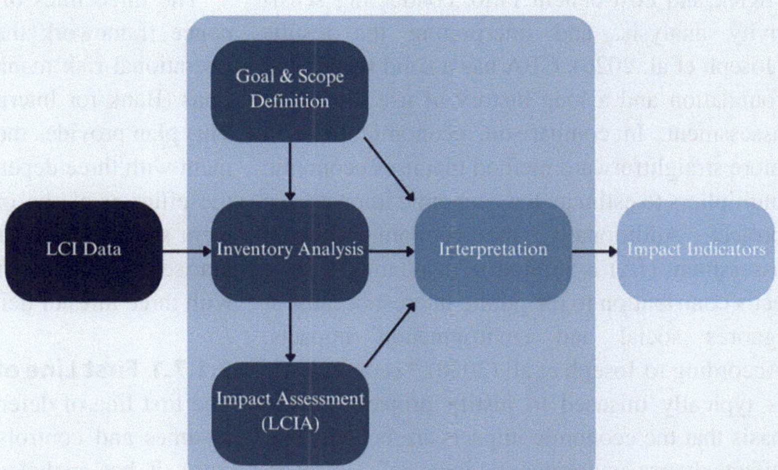

Fig. 2.10 LCA methodology diagram (PWC and FPAC 2010)

The figure above illustrates the different stages in the LCA methodology according to ISO standards 14,040 and 14,044. The first step is to define the objectives and scope of the study, which includes determining the system to be studied, the system boundaries, and the assumptions to be considered. The second step is to carry out the inventory analysis, which involves collecting and calculating life cycle inventory (LCI) data on the inputs and outputs of the system. The third step is to assess the potential impacts on the environment using standardised IA methods. Finally, the fourth stage involves interpreting the study results, identifying the hot spots and weak spots in the system studied, and making recommendations for improving environmental performance.

2.1.6 Cost-Benefit Analysis

Cost-benefit analysis (CBA) is commonly used to evaluate a project's economic, environmental, and social impacts as part of IA. Unlike economic IA, which focuses solely on economic impacts, CBA can determine the net impacts of a project, considering both costs and benefits. CBA follows a series of logical steps, including scoping the analysis, predicting costs and benefits, converting non-monetary impacts into monetary terms, discounting monetised impacts, calculating the net present value (NPV), internal rate of return (IRR), and cost-benefit ratio, conducting sensitivity analysis, and interpreting the results (Joseph et al. 2020). CBA has a solid theoretical foundation and a long history of use in project assessment. In comparison, economic IA is a more straightforward method that uses economic multipliers to estimate the economic impacts of a project. Additionally, the economic impact assessment (IA) is limited in evaluating a project's contribution to the public interest because it ignores social and environmental impacts. According to Joseph et al. (2020), "economic IA is typically misused to justify projects on the basis that the economic impacts are benefits that offset adverse environmental impacts". However, the IA does help forecast the economic impacts of projects to support community planning (Joseph et al. 2020).

The CBA is more comprehensive and rigorous for assessing project impacts within the EA framework. Although it has limitations, such as the difficulty of monetising specific impacts and sensitivity to assumptions, CBA can provide a more accurate assessment of a project's net impacts. Economic IA, however, is a more straightforward and limited method that cannot estimate the net effects of a project. For this reason, it is recommended to use CBA rather than economic IA for IA (Joseph et al. 2020).

2.1.7 Risk Management

Risk management is the process of identifying, assessing, and prioritising risks. It is then followed by the efficient and well-coordinated use of resources to reduce the likelihood or effect of unfavourable occurrences or to optimise the realisation of opportunities (Hubbard 2009). Risk management seeks to minimise the possibility that the project's goals and the interests of its stakeholders would not be met, as well as to spot and seize opportunities. Specifically, risk management helps project managers with resource allocation, priority setting, and implementing procedures and actions that lower the likelihood that the project will not meet its goals (Cooper et al. 2005).

The three lines of defence are a risk governance framework that splits responsibility for operational risk management across three functions (Bank for International Settlements 2011). This plan provides the board and senior management with three dependable sources to guarantee the efficacy of the organisation's risk management system (Bateman 2018). Figure 2.11 summarises risk management in finance institutions with three lines of defences.

2.1.7.1 First Line of Defence

The first line of defence (1LOD) is the line that assumes and controls the risk. Though historically, it has included the business lines, each function is responsible for the risks that it creates

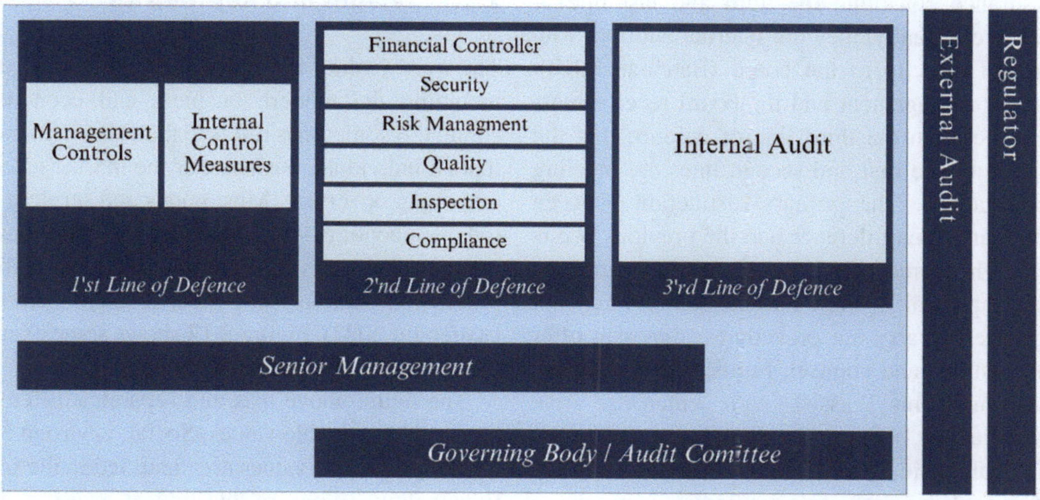

Fig. 2.11 Risk management and line of defences in finance institutions (IIA 2013)

(Dantons 2020). The people who own the risk and control are part of the 1LOD. These individuals work in the company throughout the day and, in addition to their other duties, would be thinking about risk and controls (Bateman 2018).

Operational management is responsible for the daily execution of risk and control processes and maintaining efficient internal controls. This includes determining and evaluating controls and reducing risks. Business and process owners also oversee the creation and application of internal rules and procedures and guarantee that operations align with the university's aims and objectives. Mid-level managers can create and execute intricate protocols that function as controls and oversee how their staff members carry out those protocols (Office of Internal Audit 2015).

2.1.7.2 Second Line of Defence

Risk management and control are the purview of the second line of defence (2LoD). It often involves departments like risk, compliance, HR, legal, etc. (Dantons 2020). The people who supervise or specialise in risk management and compliance constitute the second line of defence (2LOD). These individuals are strongly committed to risk and control and the necessary training to help implement efficient risk management techniques. They are also in charge of reporting and compiling risk across the board for the company (Bateman 2018).

To guarantee that risk and controls are adequately handled, the second line supports management. To aid in developing and monitoring the first line of defence measures, management sets up various risk management and compliance functions. This second line of defence typically consists of the risk management function (and committee) that helps risk owners define the target risk exposure and provide sufficient risk-related information to the organisation, as well as facilitates and oversees the implementation of effective risk management practices by operational management. Additionally, it has a compliance function to monitor certain particular hazards, such as breaking applicable rules and regulations. The distinct function answers directly to high management in this capacity. It finally makes the controllership function operational, keeping an eye on financial reporting problems and dangers (Office of Internal Audit 2015).

2.1.7.3 The Third Line of Defence

The third line of defence (3LoD) provides independent assurance and internal audit (Dantons 2020). The individuals who offer independent assurance constitute the third and last line of defence. Usually, they are internal auditors who answer directly to the board (Bateman 2018). Senior management and the board receive reassurance from the third line of defence that the efforts of the first and second lines are meeting expectations. The primary distinction between this third line of defence and the previous two is its high degree of organisational independence and impartiality. While internal auditors may not oversee or carry out procedures, they can offer suggestions and counsel. Furthermore, an internal audit may assist with enterprise risk management but cannot be implemented or carried out outside its role. Internal auditors achieve their goals by applying a methodical approach to assessing and enhancing the efficacy of risk management, control, and governance procedures (Office of Internal Audit 2015).

Even if the 3LoD model is mostly standardised, there is still room for disagreement and interpretation about several parts of the general structure. For example, does the discovery of control inadequacies in the first line of defence by the third line of defence (internal audit) indicate problems just in that line of defence, or does it also highlight a deficient second line of defence? One way to look at it is that 2LoD might have caused the 1LoD flaws because of a weak overall framework, ambiguous rules and minimal control requirements, inadequate monitoring of the 1LoD framework's execution, and weak 2LoD controls that failed to identify and resolve the issues (Dantons 2020).

2.2 Value, Metrics, and Challenges

As the global financial landscape shifts towards sustainability, assessing the true value of green finance requires comprehensive metrics that track social, environmental, and economic impacts. This section introduces social, environmental, and economic value, examines the key metrics used to assess impacts, and elaborates on challenges. The economic value section also defines monetary and business values.

2.2.1 Values and Key Metrics

The word "value" is employed in many contexts, including the modern business and economic context. It sometimes refers to the value in money that an individual, business, or the market places on a thing or service. Most goods and services in today's economy—including labour, land, businesses, commodities, tangible and intangible possessions, and services—are valued monetarily (Küfeoğlu 2022). Figure 2.12 shows some examples of forms of sustainable value.

The figure above lists and separates different forms of sustainable values. Social, environmental, and economic values are identified in the sections below. To ensure that impact investors are achieving the intended impact results to address the most urgent social and environmental concerns facing the globe, it is crucial to have a comprehensive understanding of IA and management techniques (GIIN 2024a). Key metrics in IA play a pivotal role in this, as they are essential for assessing and measuring how projects affect social, environmental, and economic factors. When supported by research, grounded in best practices, and standardised, these metrics facilitate data comparability and guide decision-making towards more accountable and fair results. They also enable stakeholders to evaluate potential risks and rewards and identify areas needing development, thereby reassuring the reliability of the process.

2.2.1.1 Social Value and Key Metrics

An increasing number of organisations view their operations holistically, considering the broader social, economic, and environmental effects of their operations. Although the precise definition of "social value" is elusive, businesses that strive to ensure their positive results could be seen as contributing to the long-term resilience and well-being of people, organisations, and humankind. A worldwide social and moral charter is essentially what the Sustainable Development Goals (SDGs) of the United Nations (UN) are. Public sector organisations may be able to provide more excellent community service by incorporating social value into policy and spending decisions.

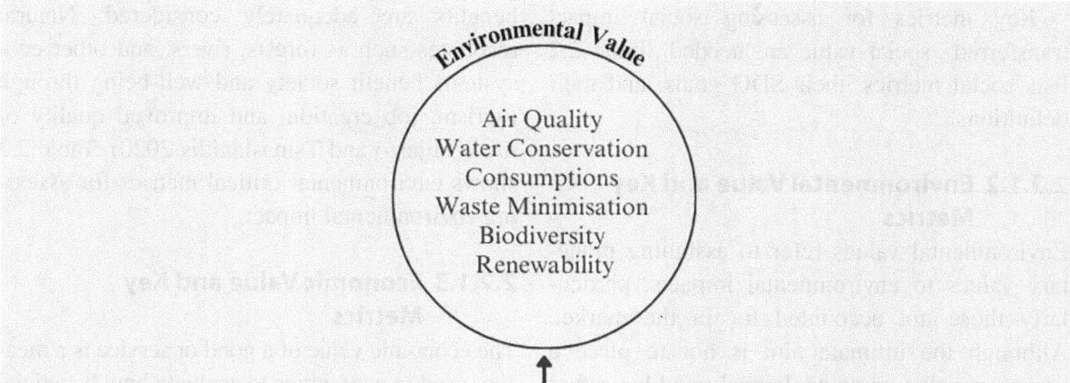

Fig. 2.12 Forms of sustainable values

A company's actions, as well as its methods, may benefit society. Organisations can externalise their social value initiatives by providing precise, measurable results and presenting them understandably to help stakeholders and customers (Küfeoğlu 2022).

Businesses can use their activities to communicate social ideals. One option is to charge a reasonable price for goods and services while prioritising customer support. Offer fair working conditions and seek out employees aligned with the company's or business's values. There are many simple methods for producing social value, which have advantages beyond society. It all comes down to the companies themselves as well. The key to creating this social value is balancing the need to run a successful company, manage it morally, and benefit society (Küfeoğlu 2022). Figure 2.13 summarises the benefits of social value.

Today's social life places an unparalleled emphasis on metrics and measurements. Social media, digital technologies, and connected surveillance have made it possible for people to become quantified; examples of this include the quantified selves that result from these developments, daily public health briefings on pandemic illness tolls and vaccination rates, measurement and reporting of climate emissions and biodiversity impacts, inflation and popping of economic bubbles in the financialised marketplace, and more (Islam and Greenwood 2022).

Key metrics for assessing social impact transferred social value are needed. Table 2.2 lists social metrics, their SDG goals, and their definitions.

2.2.1.2 Environmental Value and Key Metrics

Environmental values refer to assigning monetary values to environmental impacts, particularly those not accounted for in the market. Although the ultimate aim is not to place a monetary value on an ecological good but rather to provide policymakers with the tools to allocate resources efficiently, impose taxes and design compensation programs, the quantification of human and environmental values remains a matter of debate (Guijarro and Tsinaslanidis 2020).

Environmental values are difficult to understand because they are used in many contexts and interpretations. Moreover, some object to the idea of "putting a price on nature" or "reducing ethics to statistics and numbers" (Küfeoğlu 2022). However, integrating environmental values into the operations, plans, and activities of companies and organisations remains crucial. Incorporating environmental values into decision-making ensures that environmental costs and benefits are adequately considered. Natural resources such as forests, rivers, and other ecosystems benefit society and well-being through tourism, job creation, and improved quality of life (Guijarro and Tsinaslanidis 2020). Table 2.3 shows environmental critical metrics for assessing environmental impact.

2.2.1.3 Economic Value and Key Metrics

The economic value of a good or service is a measure used in economics to evaluate how beneficial it is to a particular economic agent. People's willingness to pay for the product, typically stated in monetary terms, is frequently used to calculate it. Preferences impact the economic worth of a good or service and the trade-offs a person is willing to accept to obtain it (Küfeoğlu 2022). New investments impact the economy by generating employment and new business ventures as household disposable income and business expenses circulate across the local economy. People's willingness to pay for the product, typically stated in monetary terms, is frequently used to calculate it. The economic worth of a good or service depends on the customer's preferences and the trade-offs they are willing to accept (Küfeoğlu 2022).

Jobs and Economic Growth	*Health, Wellbeing and the Environment*	*Strength of Community*
• Decent jobs for local people and hard to reach groups • Local people with the right skills for long-term employment • School leavers with aspirations of the industry • The local supply chain is supported and grown • Residents have comfortable homes which are affordable to operate • Thriving local businesses	• Good accessibility and sustainable transportation • Resilient buildings and infrastructure • High quality public and green spaces • Good mental health • Good physical health • Limit resource use and waste	• Strong local ownership of the development • Existing social fabric is protected from disruption • The new community is well integrated into the surrounding area • Thriving social networks • Vibrant diversity of building uses and tenures • Strong local identity and distinctive character

Fig. 2.13 Creating social value and its benefits (Crocker and Associates 2018; Küfeoğlu 2022)

Table 2.2 Social key metrics (GIIN 2024b)

Social metrics	Related SDGs	Metric definition
Anti-discrimination policy	5, 8, 10	Specifies whether the company has a formal anti-discrimination policy for its workers and whether a mechanism exists to ensure the policy is followed
Availability of basic services/facilities	3, 4, 6, 12, 13	Indicates if, as of the conclusion of the reporting period, essential services are available on-site and a mechanism is in place to maintain them
Bias-reducing investment practices	1, 2, 3, 4, 5, 8, 10, 11, 12, 16	This report provides an account of the steps taken by the investing organisation during the reporting period to lessen the impact of biases in investment criteria, rules, and procedures that disproportionately harm entrepreneurs from historically marginalised racial or ethnic groups
Board of Directors: Female and historically marginalised	5, 8, 10	Total number of women or members of historically marginalised racial/ethnic groups on the board of directors or other governing body as of the end of the reporting period
Community engagement strategy	1, 9, 11, 12, 13	Indicates whether the company plans to manage its interactions with the local communities impacted by its business
Conflict of interest policy		Indicates if there is a formal policy for the organisation to monitor and report any possible conflicts of interest between it and its owners, board members, or significant investors
Employee feedback and grievance system	8, 10	Indicates whether the organisation has a committee or established process to accept and handle employee feedback and whether there is a formal grievance system to receive and handle employee concerns and provide remedies
Training and transition programs offered and training hours	3, 4, 8, 10	This metric specifies the kind and extent of training programmes implemented, the help given to employees to improve their abilities, and the total number of hours full-time, part-time, and temporary employees received training throughout the reporting period
Employees receiving regular performance and career development reviews	8	The total number of temporary and permanent employees had regular performance and career development reviews throughout the reporting period
Employment benefits	3, 8, 10	This outlines the benefits offered to the organisation's full-time staff members throughout the reporting period: Medical insurance, life insurance, unemployment insurance, and paid time off
Founder demographics	5, 8, 10	This metric outlines the various demographic groupings that comprise the organisation's founders. Gender, disability status, and race/ethnicity/minority status are essential when determining whether an organisation was "founded" by a particular demographic group member
Gender ratio of promotions	5, 8, 10	The ratio of the number of internal promotions for female employees within the company during the reporting period to the number of internal promotions for non-female employees
Healthcare effective coverage	1, 3, 9, 10, 12	The proportion of the organisation's patients and clients who received the necessary medical intervention(s) during the reporting period
Historically marginalised wage equity	8, 10	The ratio of the average salary for a given position given by the company to workers who are members of historically marginalised racial or ethnic groups to the average salary offered to workers who are not members of these groups but hold the same position during the reporting period
Indigenous rights and stewardship practices	1, 2, 3, 6, 7, 8, 10, 11, 12, 13, 15	This document describes the organisation's steps to allow local indigenous communities to use biological resources on land, freshwater, and marine areas under direct or indirect management. These steps also recognise and support these communities' traditional knowledge, innovations, and practices

(continued)

Table 2.2 (continued)

Social metrics	Related SDGs	Metric definition
Length of road built or improved	9	Length of road that the organisation constructed or renovated throughout the reporting year
Number of employee grievances resolved	8	The number of official grievances resolved during the reporting period by company workers
Number of legal and regulatory complaints		The organisation was subject to the total number of official legal and regulatory complaints during the most recent reporting period
Public water point coverage	3, 6, 12, 13	The number of individuals who, as of the end of the reporting period, had primary water access through a public tap within a coverage territory (as defined by a municipality, district, or designated service area)
Racial equity advocacy strategy	1, 2, 3, 4, 5, 8, 10, 11, 12, 16	Indicates if, as of the conclusion of the reporting period, the organisation has an advocacy plan in place to promote local and national efforts towards racial equity and justice and whether the organisation completed a racial equity audit
Racial equity transparency practice	1, 2, 3, 4, 5, 8, 10, 11, 12, 16	Indicates whether the organisation has a formalised, continuous policy of publicly making its racial equity positions, practices, advancements, and advocacy available to foster trust and guarantee adherence
Service area population	3, 4, 6, 7	The number of people living inside the municipality, administrative region, or other geographic boundaries that make up the organisation's service area as of the conclusion of the reporting period
Number of service hours provided, interrupted, and number of interruptions	3, 6, 7, 12, 13	The number of service hours provided and interrupted, and the number of service interruptions experienced by clients during the reporting period
Social and Environmental Board Committee and performance incentives	1, 8, 9, 10, 11, 12	It states whether the company established any employee incentive programmes related to social and environmental performance goals during the reporting period and whether it has a formal board of directors or governing body committee dedicated to social and environmental performance
Social and environmental performance reporting	1, 8, 9, 11, 12, 13	It shows whether the company informs the appropriate stakeholders about its social and environmental performance
Value of investments in communities historically marginalised due to race and ethnicity	1, 2, 3, 4, 5, 8, 10, 11, 12, 16	Value of the investing organisation's managed assets as of the end of the reporting period allocated to communities primarily of historically marginalised racial and ethnic groups
Wage equity	1, 8, 10	The ratio between the highest-paid full-time employee's earnings (bonus included, benefits excluded) and the lowest-paid full-time employee's earnings during the reporting period
WASH facilities type	3, 4, 6, 12, 13	Describes the types of WASH (water, sanitation, and hygiene) facilities available on the spot due to the organisation's operations throughout the reporting period
Worker freedom of association and safety policy	1, 3, 8, 10	Denotes if the company has a documented policy to oversee, assess, and guarantee its employees' safety and freedom of association

The price at which something would be transferred to a foreign entity indicates its monetary value. Prices are based on various factors, such as the monetary value of labour, physical assets, intangible assets, and commodities. Using predicted monetary value, the financial impact of risk is calculated (Küfeoğlu 2022). Turning social and biophysical effect measurements into monetary units is known as monetary valuation, and it is used to calculate the worth of non-market products—i.e., items for which there is no market (Pizzol et al. 2015).

The word "business value" in management refers to any value that affects a company's long-term health and survival. Business value is the total value of all tangible and intangible

Table 2.3 Environmental key metrics (GIIN 2024b)

Environmental metric	Related SDGs	Metric definition
Greenhouse gas (GHG) emissions	3, 13	This metric measures direct and indirect GHG emissions to assess the environmental impact
GHG emissions sequestration	3, 13	This metric quantifies the amount of GHG emissions that have been sequestered
Water consumption	6, 8, 9, 11, 12	The metric examines the total volume of water used by the project or organisation for various purposes, including productive operations, human consumption, and agriculture
Wastewater treatment	6, 8, 9, 11, 12	This metric assesses the volume of wastewater treated by the organisation. The goal is to improve wastewater quality before it is released into the environment
Energy consumption	7, 8, 9, 11, 12	Total energy consumed by the organisation, including both generated and purchased energy. This metric also helps determine energy efficiency and strategies to reduce energy use
Land use	2, 6, 8, 12, 15	The organisation's total land area encompasses residential, industrial, commercial, farming, forestry, recreational, and conservation
Sustainable management of natural resources	2, 6, 8, 12, 15	The land area is managed sustainably by globally accepted standards, incorporating sustainable biodiversity and environmental services practices
Biodiversity	2, 9,13, 14, 15	This metric aims to quantify the company's or project's biodiversity footprint, measuring its impact on ecosystem integrity. This is typically represented in comparison to an undisturbed baseline condition. The biodiversity footprint can be assessed at various levels, including specific projects, locations, products, or the entire enterprise
Renewable energy usage	7, 8, 9, 11, 12	The amount or proportion of renewable energy the organisation uses, produces, or purchases. Renewable energy sources include biomass, solar, wind, geothermal, hydro, and hydropower
Waste management	6, 8, 9, 11, 12	This metric assesses waste collection, processing, recycling, and disposal efficiency and effectiveness

components of the business. Monetary assets, furniture, stockholder equity, and utilities are a few examples of tangible elements. Beyond its "economic worth", a company's "business value" includes many additional values. These are a few examples of trademarks, public benefit, goodwill, and brand recognition (Project Management Institute 2013). Many times, different kinds of value cannot be valued. While no universally accepted definition of company value exists, some economists contend that organisations cannot make decisions based only on financial indicators like profit and shareholder value (Küfeoğlu 2022).

Input-output models, which quantify how changes in demand by consumers, governments, and industry sectors affect other industry sectors, are the foundation for impact analysis and multiplier estimation (Market Street Services, Inc 2004). Depending on the impact assessment's scope, different metrics can be used. This difference helps people evaluate the project's economic impact according to their initial target. Calculating total value added by projecting changes in employment, income, business activity, and property value is more suitable for seeing results on different people in the affected area. Conversely, calculating the total economic effect by projecting direct, indirect, and induced impact is more suitable for showing the directness of the effects. Table 2.4 shows economic metrics for assessing economic impacts.

2.2.2 Challenges in Executing Impact Assessment

Assessing impact is a challenging task. Numerous potential challenges could impede the evaluation

Table 2.4 Economic key metrics (GIIN 2024b)

Economic metrics	Related SDGs	Metric definition
Business innovation	1, 2, 3, 4, 5, 6, 7, 8, 9, 10, 11, 12	This metric explains whether, during the reporting period, the company implemented or adopted a technology, internal procedure, product, or finance structure that was novel or not commonly utilised in the operating region and industry
Average agricultural yield	2, 8, 10, 12	The metric examines the average agricultural yield per hectare inside the set, which another metric indicates
Average insurance premium	1, 5, 8, 9, 10, 11	The organisation's average annualised insurance premium for a particular insurance product sold during the reporting period
Employee voluntary turnover rate	8, 10	The ratio between the average number of full- and part-time permanent workers at the company throughout the reporting period and the number of voluntarily departing permanent employees
Assets under management	1, 2, 3, 4, 5, 8, 10, 11, 12, 16	Value of all assets under management for the investing organisation after the reporting period
Voluntary savings accounts	1, 5, 8, 9, 10, 11	The value of demand and time deposits held in the organisation's voluntary savings accounts as of the conclusion of the reporting period
Female full-time wages	1, 2, 5, 8, 10	The value of pay (bonuses and benefits excluded) is given to each full-time female employee of the company during the reporting period
Employment	1, 2, 8, 9, 10	The employment component is the number of jobs (in full-time equivalents) produced per physical unit
Income	1, 2, 5, 8, 10	The rise in income levels is a statistic used to evaluate the effects of a project or investment
Changes in business activity	1, 9, 10	Any attempt to turn a profit in business is considered a business activity. The three categories of business activity are operating, investing, and funding
Property values	1, 8, 9, 11	Property values indicate wealth and generate income, while the subject policy is implemented
Return on assets	1, 8, 9	The entire amount of an organisation's net profit after taxes earned per dollar invested in total assets as of the end of the reporting period
Payments to government	8, 16	The total value of all transfers made by the organisation to governments throughout the reporting period

process and result in poor decision-making. Figure 2.14 shows the significant challenges and obstacles during IA.

Figure 2.13 shows the significant challenges of IA. In this part, these challenges are introduced.

- *Aligning Incentives*

  Aligning incentives around IA is essential to initiating an impact evaluation. It is the process of ensuring that all project stakeholders' goals and motivations are compatible. This alignment helps prevent implementers from resisting IAs (Gugerty et al. 2016).
- *Cost of Assessing Impact*
  Organisations must weigh the resources spent on assessment against the intervention (Gugerty et al. 2016). The resources available to non-profits, social entrepreneurs, and donors for program evaluation are all finite. Data collection costs can be prohibitively high when data must be frequently collected to remain relevant (Maurice et al. 2000).
- *Spillover Effect*
  Benefits from the program also spill over to non-participants, which creates another challenge. Spillover effects are beneficial for programs, but they might lead to contamination in the control group, making the program seem less impactful than it was (Creevey 2008). Crucially, before the initiative is implemented, the assessment must be well-planned to consider spillover effects (Angelucci and Di Maro 2015).

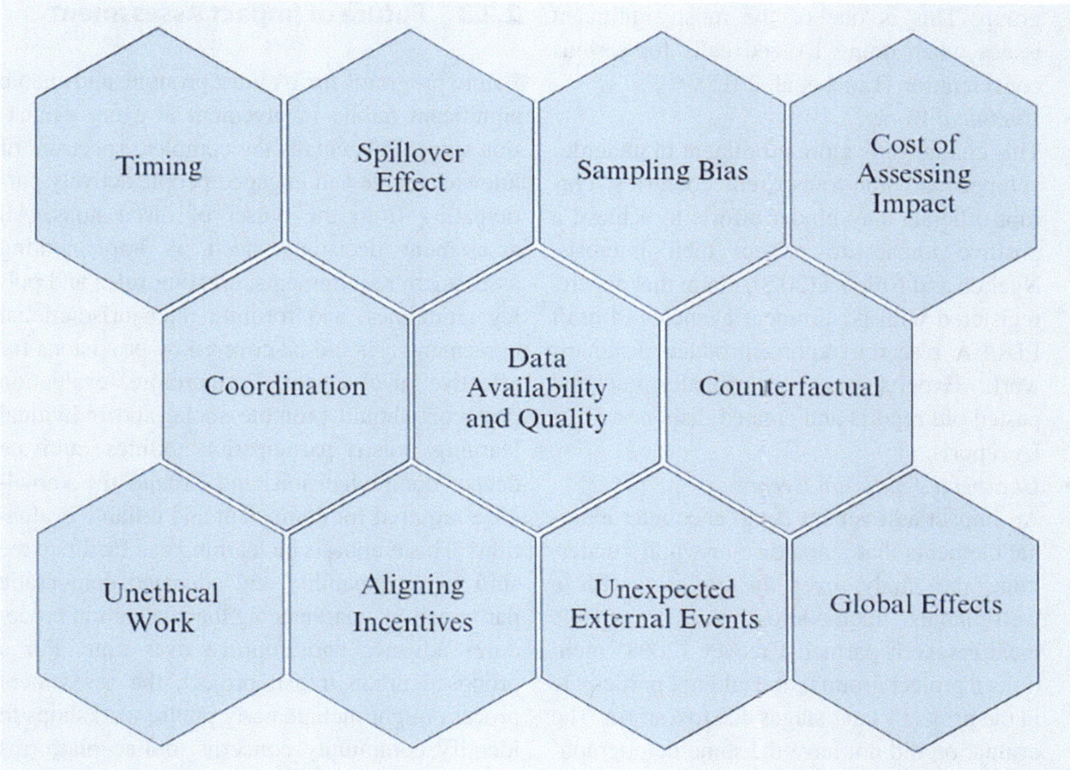

Fig. 2.14 Challenges to assessing impact

- *Timing*
  Timing of IA is another crucial challenge to consider. One advantage of finishing IA early is that more changes can be made to improve its efficacy and benefits. However, some program effects take time to manifest; rapid evaluation may overestimate or miss impacts (Lima et al. 2015). The time between the program's launch and the baseline study increases the likelihood that the IA will miss some program effects (Creevey 2008).
- *Sampling Bias*
  Collecting samples that do not fairly represent the total group is known as sampling bias. Selecting the treatment and control groups with a high level of statistical validity is a challenging aspect of performing a sound IA (Creevey 2008). Impact assessments should compare project participants with individuals with comparable social and economic characteristics who were not involved in assessing if the project had any effect (Lesa and Egger 2019). Randomised research approaches are susceptible to selection bias (Creevey 2008).
- *Coordination*
  One of the critical phases in the impact assessment process is choosing the local research partner and establishing coordination, even though it is not frequently stressed. Poor collaboration may lead to lower-quality work, dubious outcomes, and, in worst circumstances, the research's complete collapse (Creevey 2008). Full cooperation and ongoing communication between evaluators and program administrators is necessary for a practical IA (Lima et al. 2015).
- *Counterfactual*
  Determining what would have happened and what type of products or services beneficiaries would have access to instead of those provided by the intervention is crucial to the effectiveness of an evaluation. As there is no intervention-free area, the counterfactual needs to be represented by a comparison

group. This is one of the most significant issues when doing IA and calls for serious consideration (Lima et al. 2015).

- *Unethical Work*
  This challenge is more prominent in underdeveloped and non-transparent countries; corrupt officials may hinder efforts to achieve a positive impact to protect their interests. Ngetich and Ndiema (2020) claim that experts registered with government agencies to draft EIA/EA research reports produce dishonest work. Experts have occasionally cut and pasted old reports and created dishonest EIA/IA reports.
- *Unexpected External Events*
  An impact assessment could encounter external elements that jeopardise or wholly undermine the study, even in projects with a well-planned methodology and a capable local research partner. Creevey (2008) mentions a project group that could not participate in the project's later stages due to a storm. The evaluation did not have the same demographics as the original project group. Various types of external events may disrupt IA and lead to inconclusive results.
- *Data Availability and Quality*
  Finding reliable and consistent data is another difficulty when implementing IA. Poor data quality can significantly impact how conclusions are applied in decision-making, but this problem hasn't gotten enough attention yet (Curran 2014). Due to insufficient data, significant differences may arise between the suggested impact model and the related environmental mechanism.
- *Global Effects*

  Global effects play a significant role in IA, as they can dominate the impact of individual projects. These global megatrends related to resource usage, biodiversity, and climate change operate at such vast dimensions that they are challenging to comprehend (Bond and Dusík 2019). IA is often used to determine how planned projects affect local environmental problems (Ortolano and Shepherd 1995).

2.2.3 Future of Impact Assessment

Future programs for IA must promote and enable significant public involvement at every evaluation stage. This entails the complete spectrum of knowledgeable and engaged people actively participating from the outset of discussions. All assessment decisions, such as implementing assessment requirements, creating rules and policy guidelines, and forming inter-jurisdictional agreements, should be covered by provisions for effective involvement. Furthermore, evaluation protocols should promote social and individual learning, bolster participation abilities, cultivate deeper comprehension, and expand the knowledge required for competent and reliable evaluations. This emphasis on learning can facilitate the shift to sustainability and educated democratic participation, guaranteeing that evaluation procedures advance and improve over time. For a proposed urban transit project, the assessment process might include early public workshops to identify community concerns, online platforms for ongoing dialogue, public hearings at crucial decision points, and financial support for community groups to hire independent experts to review technical documents (Sinclair et al. 2021).

AI is being investigated globally for its potential application in many aspects of environmental monitoring. Many AI approaches are currently in use, and new ones are constantly being created. Artificial intelligence is utilised in various fields such as agriculture, sanitation and health, energy efficiency, environmental monitoring, fish stock prediction, water quality modelling, water treatment, seawater desalination, energy efficiency, energy conservation, and wildlife protection (Khan et al. 2024).

In a recent study by Khan et al. (2024) qualitatively addressing the concerns of stakeholders regarding how the advent of open-access chatbots such as ChatGPT may impact IA in the future, IA practitioners around the world experimented with various steps of IA. Twenty-five practitioners with diverse backgrounds and field experience levels were confidentially interviewed during the International Association for Impact Assessment (IAIA) conference held in Malaysia from 8 to 11 May 2023. The interview questions were created with the study's goal: to determine what stakeholders thought of ChatGPT's potential to strengthen or undermine the IA process (Khan et al. 2024).

Most interviewees expressed concerns that ChatGPT would be more dangerous for IA than helpful. Stakeholder interviewees expressed optimism that ChatGPT will speed up gathering information for IA reports, enhance their structure and grammar, and help collect data. However, they also voiced concern over the possibility of a decline in the reports' data quality. Together with one consultant, all regulators thought ChatGPT would speed up review and decision-making. But, they were also concerned that if regulators began depending too much on ChatGPT, the quality of reviews could decline. While academics hoped that the quality of assignments and papers would improve, they were equally concerned that the quality of IA practitioners may decrease. Consultants expressed optimism that ChatGPT could raise public awareness and encourage public involvement. Nevertheless, all the stakeholder groups were worried that public participation might be compromised, especially in the case of social surveys. All stakeholder groups' interviewees stated that legislation governing the use of ChatGPT in IA is necessary, as is the need to strengthen IA agencies' capacity. Changes in teaching practices are also required, according to academics and consultants. One of the study's main advantages is offering perspectives from international IA stakeholders. Therefore, the findings are not exclusive to any particular nation or area. But because AI is developing so quickly, it may be necessary to do more research in a few years to observe how things develop, assess the steps regulators have taken to guarantee quality control and determine what adjustments need to be made. Furthermore, country-specific research must be conducted to determine how the relevant parties have begun utilising ChatGPT or generative artificial intelligence in IA and the rules or regulations enacted by various nations so that other IA systems can benefit from their experiences. While the current study focused on ChatGPT, comparing ChatGPT, Bard, and other chatbots in subsequent research would be fascinating (Khan et al. 2024).

2.3 Monitoring, Evaluation, and Reporting for IA

Effective impact assessment requires rigorous monitoring, evaluation, and reporting processes. These three pillars ensure that green finance initiatives remain accountable, transparent, and aligned with financial and environmental goals. This section will focus on establishing baselines and targets, data collection techniques, reporting on IA, and possible failures and solutions.

2.3.1 Establishing Baselines and Targets

A baseline study establishes the reference point for tracking and evaluating the impact of a project or intervention. It provides essential data to assess the effectiveness of the intervention, identify gaps, and guide the development of appropriate actions and indicators for monitoring progress towards the desired goals (EvalCommunity 2024). A clear snapshot of the current situation is a benchmark to measure progress throughout the project cycle. Insufficient baseline data complicates decision-making and evaluation, making a comprehensive impact assessment unattainable (USAID 2010). Figure 2.15 illustrates the key purposes of a baseline study.

Researchers generally describe a target as the intended outcome of an action that is anticipated to be accomplished at a given point in the future and that all necessary resources and effort are dedicated to achieving. Every successful program and organisation in the humanities sets targets, and it is standard practice to write concise, well-organised statements that articulate objectives in a precise, quantifiable, and achievable manner (Day and Tosey 2011).

One of the developed frameworks for setting targets is the SMART framework, a popular method among others. A SMART target has five essential components or characteristics that

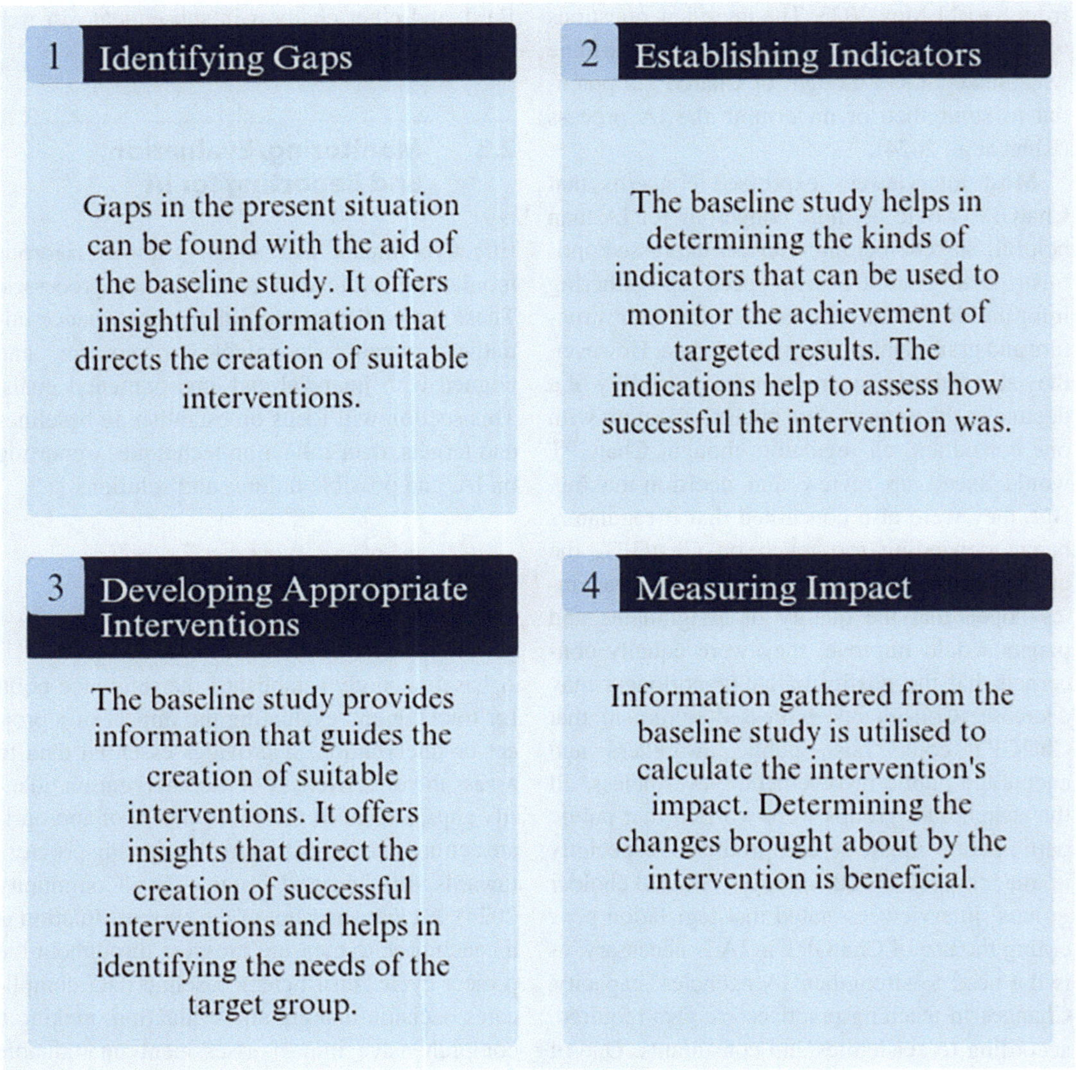

Fig. 2.15 Specific purposes of baseline study (EvalCommunity 2024)

describe it. Leaving out any one of the components may set goals without developing a successful plan (Leonard and Watts 2024). Table 2.5 details the five steps of the SMART framework.

2.3.2 Data Collection Techniques

Data collection is an integral part of the impact assessment process. All data gathering aims to obtain high-quality evidence, which enables the correct measure and analysis of the project's impact and success. Table 2.6 lists data collection techniques for IA.

2.3.3 Reporting on Impact Assessment

An IA report is a comprehensive document that provides a thorough evaluation, using both quantitative and qualitative methods, of the probable environmental, social, and economic effects, including those on small- and medium-sized businesses and competitiveness. It must be made

Table 2.5 SMART framework for target setting

1. Specific	Targets must be explicitly defined, providing an unambiguous understanding of the intended outcome. This ensures all stakeholders share a common direction and understand the precise steps required to achieve the objective
2. Measurable	Targets should include measurable indicators for effective progress evaluation. Quantifiable metrics allow for systematic performance tracking and ensure that progress can be objectively assessed
3. Achievable	Targets must be realistic and attainable within the given constraints while presenting a challenge. Dividing broader objectives into achievable tasks ensures that progress is motivating and feasible, supporting sustained effort towards completion
4. Relevant	Objectives must align with the organisation's broader mission and strategic priorities. The relevance of the target ensures that it contributes meaningfully to the organisation's long-term success and overall vision
5. Time-bound	A well-defined timeline for achieving the target is essential. Establishing a specific deadline creates a sense of urgency and facilitates consistent progress, ensuring the objective is pursued within a realistic and effective time frame

Table 2.6 Data collection techniques

Methods	Explanations
Interviews	The purpose of an interview is to learn more about the interviewee's motivations, emotions, and beliefs. As a result, the interviewer would be more eager to inquire about intriguing developments and allow the interviewee to go into further detail on a range of topics (Dörnyei 2007)
Focus groups	Focus groups are a method of data collection where a moderator leads a discussion with a group of participants on topics related to the research question. This approach allows researchers to gather multiple opinions and insights from several individuals simultaneously (Barrett and Twycross 2018)
Document analysis	This approach involves examining a range of document formats, such as books, newspaper stories, institutional reports, and academic journal articles (Morgan 2022)
Surveys	Surveys are among the most often used non-experimental data-gathering techniques. Various methods can collect survey data, including questionnaires, interviews, and observational surveys (Zheng 2021)
Experimental research	Experimental research seeks to determine the relationships between variables. It necessitates examining and evaluating the cause-and-effect relationship between two or more variables (Rana et al. 2021)
Longitudinal studies	Continuous or repeated measures are used in longitudinal studies to track specific individuals over extended periods, frequently years or decades (Caruana et al. 2015). These studies concentrate on measurable factors but do not involve researcher manipulation or random sampling (Ortega and Iberri-Shea 2005)
Sensors	Sensors help gather data in various fields, including energy, water, agriculture, transportation, and health. They can measure multiple parameters, including temperature, proximity, pressure, illumination, moisture, and contact. Because of their tiny size, sensors are rather intrusive, which is suited to monitoring individual and community behaviour (IPA 2016).
Macro data	Macro data can be described as an observational dataset obtained by deliberately combining statistical microdata that complies with statistical methods (OECD 2004). Various datasets can serve as valuable calibration input for assessing impact. However, this method is also inadequate for comprehending the implications at the individual level (Simler 2019). Macgillivray et al. (2017) state that macro data can also be used for more complex effect measurement activities like estimating indirect jobs created throughout a portfolio
Satellites	Satellites for data collection have various advantages, including eliminating the bias inherent in self-reporting and the capacity to collect objective data on both one-time occurrences. They enable remote, large-scale, and high-frequency data collection. However, using satellite data to infer outcomes demands high-accuracy ground data to calibrate and validate models (Simler 2019)
Company data	According to Rawlings (2013), utilising pre-existing data to manage a company's operations can offer insights into the company's impact without additional costs or time commitments. Understanding the effects of a project or business may be aided by using already-existing data sources, such as sales figures, retail prices, and product lifetime values (Simler 2019)

clear if any of these effects are considered insignificant. Additionally, the report must specify who will be impacted by the initiative and how, as well as the consultation strategy and its outcomes (European Commission 2024b).

2.3.4 Possible Failures and Solutions

IA serves as crucial for policies and programmes. However, seeing that many IAs have little value and some are useless is concerning. Regrettably, a lot of IAs are not able to halt risky or otherwise erratically advised projects (Laurance 2022). For example, an extraordinarily cursory EIA approved a massive housing development that would clear a vast tract of tropical forest in Panama since the research found just 12 bird species in the region. Thankfully, a local bird specialist conducted a second bird survey and saw 121 bird species in just 2 h of looking, including numerous rare and vulnerable species, proving how incredibly insufficient the IA was (Laurance and Salt 2018). Although IA is primarily significant and offers several advantages, many IAs, as seen, are failing to function for specific reasons.

2.3.4.1 Inadequate Investment

Conducting IAs requires substantial financial investments, time, and expertise, which forms an elusive question. Any project that earmarks only a tiny proportion of its funds for this critical process will result in incomplete or superficial assessments (Laurance and Salt 2018). For example, identifying vulnerable species is expensive, as it calls for field surveys and considerable investments in expert analysis; therefore, it is often ignored due to cost implications.

The inadequate investment may also cause an insufficiency in the necessary data collection, analysis, and reporting. Consequently, there will be a failure in identifying and mitigating potential adverse impacts satisfactorily. For instance, environmental impact assessments of mega infrastructure projects are mostly too narrow and shallow to allow adequate predictions about long-term environmental implications effectively because of funding insufficiency (Noble 2006).

Funding organisations and regulatory agencies should ensure that appropriate financial provisions are made for IAs. This must be secured for the initial assessment phase and its periodic monitoring and evaluation during a project cycle. In addition, making the cost of IA a part of the total project cost as a necessary element in the budget may help procure funds for this purpose (Geneletti 2002).

2.3.4.2 Insufficient Scope

An IA's scope must be imperative in deciding its effectiveness. However, most of the assessments lack the inclusion of the whole environment and social impacts; instead, they look at the area proposed for the project. This leads to ignoring the significant adverse consequences. For example, clear-cutting for large-scale mining in the Amazon has caused deforestation and habitat loss far from the mine sites, which the original assessments failed to capture (Sonter et al. 2017). Insufficient scope in interdisciplinary environmental assessments (IAs) can result from inadequate guidance and norms, leading to missed impacts and insufficient representation of competent experts in complex environmental interactions (Geneletti 2002).

Any improvement in the scope of IAs requires holistic and integrative approaches that consider the full range of potential impacts, including cumulative effects, indirect impacts, and interactions between environmental components. Improving guidelines and standards of IAs to consider these broader considerations can significantly enhance their utility. Moreover, interdisciplinary cooperation and stakeholder participation are means to ensure and reduce the wide scope of impacts involved with a large project (Cashmore et al. 2004).

2.3.4.3 Vested Interests

Vested interests can relentlessly erode the effectiveness of impact assessments. When stakeholders have a financial or political stake in a project, the pressures to produce congenial outcomes, regardless of the potential impacts, are too compelling to resist. The apparent implication of such a conflict of interest is biased assessment reports that ignore or downplay negative consequences. One example is data manipulation to favour side effects. This may involve selective data reporting, presenting only positive impacts and omitting or minimising negative impacts. Timing of releasing IA findings strategically with political cycles or decision-making processes also happens, further tainting any objectivity in the assessment (Noble 2006).

The IA process, characterised by transparency and regulatory accountability, is crucial in resolving vested interests, ensuring that reviews are conducted based on independent expert testimony. Moreover, in terms of maintaining openness and transparency with public participation and stakeholder involvement, it is essential to remove certain counterbalances in the vested interests so that views from all directions find their representation within the assessment (Sadler 1996).

2.3.4.4 Poor Governance

Finally, another critical factor that makes the IA a failure is the poor governance of institutional arrangements—failure to enforce compliance when accountability and transparency are absent may make any well-designed framework under IA ineffective. Most often, the institutions responsible for regulating IAs do not have adequate resources and power to enforce compliance, contributing to the unsatisfactory monitoring and enforcement of environmental standards (Bina 2008).

Another way in which IAs can be scuttled is through corruption. For example, when regulatory officials are captured through bribes or other forms of corruption, they allow projects that sometimes go against thorough assessments or do nothing regarding significant environmental impacts. This issue is especially surmounted within weak legal systems, where corrupt practices are rarely detected and taken to court (Van Der Heijden 2014). Improvement in governance arrangements is necessary for conducting IAs more effectively. Improved governance arrangements, institutional capacity enhancement, legal instrument development, independence, regulatory resource sufficiency, transparency, and accountability can enhance the effectiveness of IAs (Kolhoff et al. 2009).

2.4 Case Studies in Impact Assessment

IA processes and principles can be altered depending on the organisation's needs. Thus, using examples to demonstrate IA systems and ideas would be convenient. The International Finance Corporation's (IFC) Anticipated Impact Measurement and Monitoring (AIMM) system will be examined first. Second, the IA guiding principles of the European Bank for Reconstruction and Development will be reviewed. The World Bank's third example compares and illustrates potential IA approaches. Fourth, by looking at Intel's corporate responsibility report, we will discuss our research on IA by a company.

2.4.1 International Finance Corporation (IFC)

To quantify the effects of its initiatives at the project and systemic levels, IFC created its own IA technique, known as the AIMM. It is also a tool to incentivise impact and an emerging model for impact investors. The two aspects of this system are the project and the market results. Three distinct effect categories are examined by project dimension: stakeholder effects, economy-wide effects, and environmental effects. The market outcomes dimension assesses how an intervention improves market structure and function by promoting inclusivity, resilience, competitiveness, and sustainability goals. IFC has developed sector frameworks to evaluate projects in each of its companies. To help with IA, sector frameworks assign ratings in four areas, as shown in Fig. 2.16 (IFC 2022; Küfeoğlu 2022).

The gap analysis assesses the corresponding size of the development challenge that each desired effect addresses in the dimension of proj-

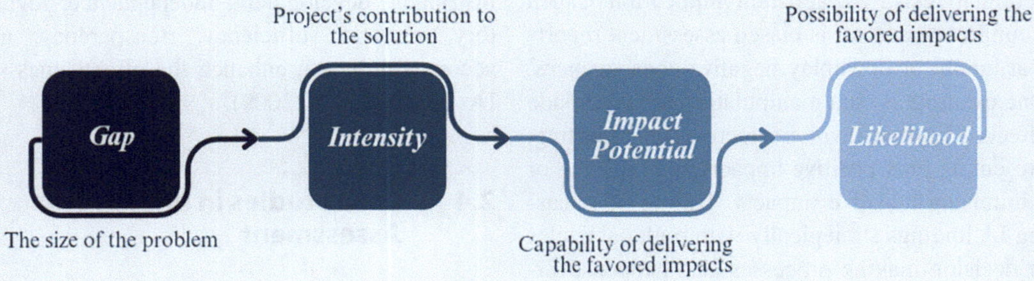

Fig. 2.16 Rating areas (IFC 2022; Küfeoğlu 2022)

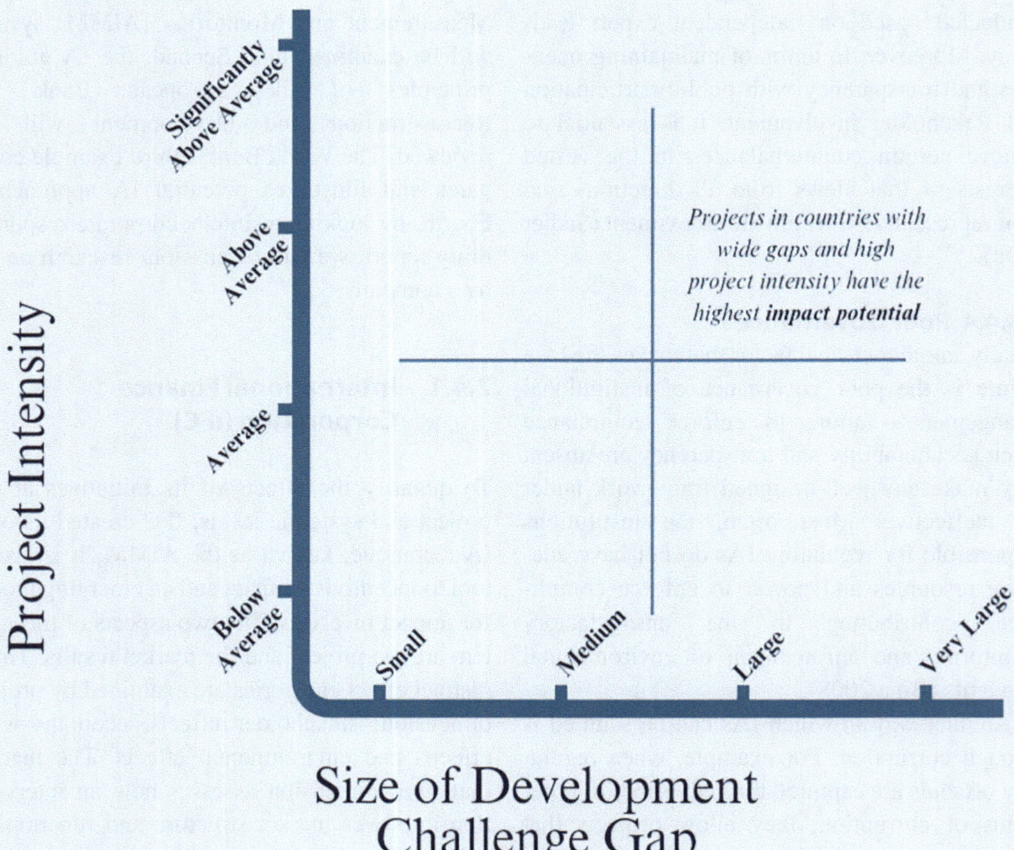

Fig. 2.17 Assessing project outcome (IFC 2022; Küfeoğlu 2022)

ect outcomes. The intensity analysis evaluates how much a project contributes to lessening development divergence. Sector-specific normalised criteria serve as the foundation for the intensity evaluation. An overall potential impact rating is assigned when an intensity assessment and a gap evaluation are combined, as seen in Fig. 2.17. The extent of the problem (the gap) and the effectiveness (the intensity) of the solution affect the likelihood that an effect will have an impact. Projects that address more significant development gaps and special or purpose-built programs that provide results quickly are given priority under this approach (IFC 2022; Küfeoğlu 2022).

For each of the five market characteristics, the market development dimension assigns market stages to market development. Conversely, market movement is employed to examine a project's attempts to modify the makeup and dynamics of markets. An overall prospective impact rating is authorised following the combination of the market stage and the market movement, as illustrated in Fig. 2.18. The market's development stage and the catalytic transformation that IFC expects its project to inspire indicate its capability for systemic change (IFC 2022; Küfeoğlu 2022).

The AIMM approach considers the uncertainty of attaining and maintaining the expected impacts over time for both the project- and market-level dimensions. Figure 2.19 shows that the likelihood assessment distinguishes between the risks that could prohibit a project's projected results from materialising and the results that could be produced (IFC 2022; Küfeoğlu 2022).

The IFC selects and assembles interventions with the greatest potential impact using AIMM evaluations. IFC accumulates AIMM ratings for several business areas to manage its interferences pipeline and create plans to address deficiencies. IFC will be able to pursue the portfolio strategy by balancing strategic aims with the help of portfolio ratings. Their goal is to provide sustainable and risk-adjusted financial returns while optimising the development effect. In addition, IFC creates information feedback loops, keeps track of and

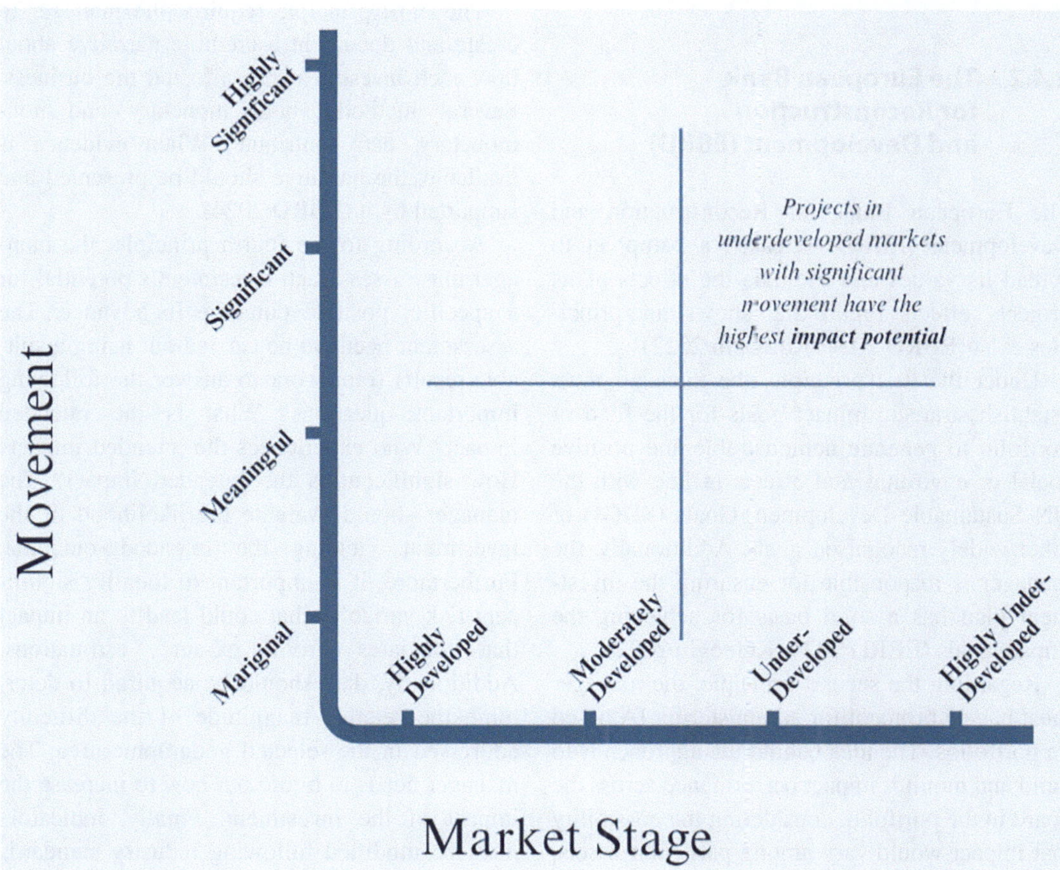

Fig. 2.18 Assessing market outcome (IFC 2022; Küfeoğlu 2022)

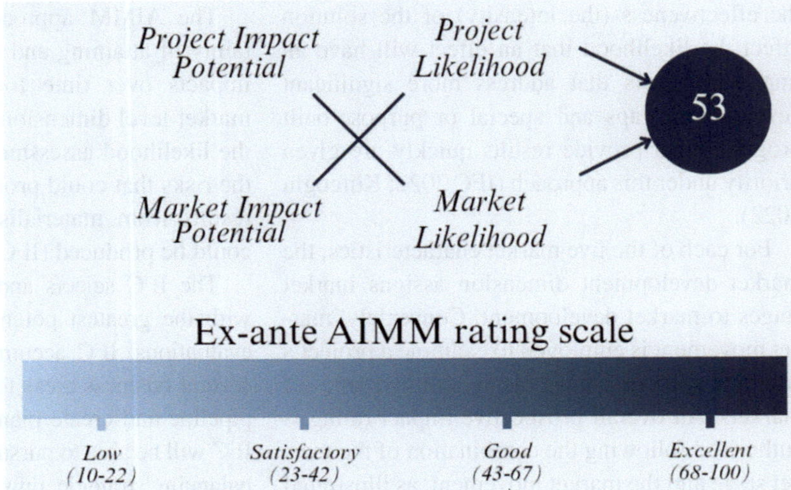

Fig. 2.19 Likelihood assessment (IFC 2022; Küfeoğlu 2022)

reports on ex-ante expectations, and creates an inventory of lessons (IFC 2022; Küfeoğlu 2022).

2.4.2 The European Bank for Reconstruction and Development (EBRD)

The European Bank for Reconstruction and Development (EBRD) released a pamphlet to spread its values and evaluate the effects of its projects' effects. Figure 2.20 shows nine principles exist (EBRD 2024; Küfeoğlu 2022).

Under the first principle, the manager must establish strategic impact goals for the fund or portfolio to generate demonstrable and positive social or environmental effects in line with the UN Sustainable Development Goals (SDGs) or other widely recognised goals. Additionally, the manager is responsible for ensuring the investment plan has a solid basis for achieving the impact goals (EBRD 2024; Küfeoğlu 2022).

Regarding the second principle, the manager must have a protocol for administering IA based on portfolios. The idea behind the approach is to build and monitor impact performance across the board in the portfolio, considering the possibility that impact would vary among particular assets. In addition to financial performance, the manager must acknowledge the relationship between impact achievement and staff incentive programs (EBRD 2024; Küfeoğlu 2022).

The third principle requires the manager to create and document a credible narrative about how each investment has affected the business. Several methods, both monetary and non-monetary, can contribute. When evidence is available, the narrative should be presented and supported by it (EBRD 2024).

According to the fourth principle, the manager must assess each investment's potential for a specific, positive outcome in advance. The assessment needs to be carried out using a suitable results framework to answer the following important questions: What is the intended impact? Who experiences the intended impact? How significant is the intended impact? The manager should evaluate the likelihood of the investment yielding the intended outcome. Furthermore, it is important to identify significant risk variables that could lead to an impact that deviates from ex-ante estimations. Additionally, data should be acquired to determine the relative magnitude of the difficulty addressed in the selected geographic area. The manager needs to figure out how to increase the impact of the investment. Finally, indicators must be modified following industry standards and best practices (EBRD 2024; Küfeoğlu 2022).

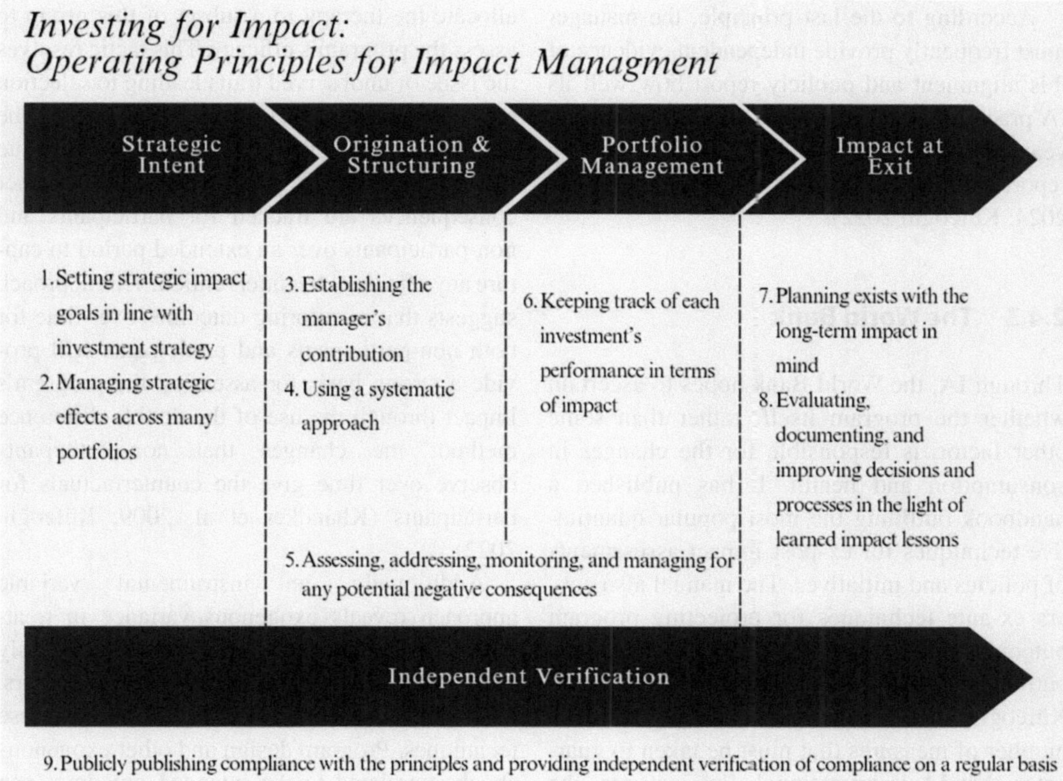

Fig. 2.20 Operating principles for impact management (Likhachova 2019)

Under the fifth principle, the manager is required to identify and prevent risks. If it is impossible to avoid investment, minimise the environmental, social, and governance (ESG) risks by implementing a systematic, documented approach for each investment (EBRD 2024; Küfeoğlu 2022). The manager must communicate with the investor to secure the investor's commitment to filling in the gaps in the current systems, protocols, and standards by utilising the finest global industry practices. Audits should also be conducted on investees' ESG risk and performance. If necessary, the management should interact with them to prevent gaps and unanticipated incidents (EBRD 2024; Küfeoğlu 2022).

The sixth principle requires that the manager use the results framework from principle four to track the process and produce positive outcomes for each investment. A pre-established procedure for sharing performance information with the investee should be followed to monitor progress. This should outline the design for data collection, data sources, data collection responsibilities, and data reporting procedures, in addition to the frequency of data collection. Furthermore, if monitoring indicates that the investment is no longer thought to have the anticipated results, the manager must make the appropriate efforts (EBRD 2024; Küfeoğlu 2022).

To adhere to the seventh principle, the management conducting an exit must assess how the timing, design, and exit process will affect the sustainability of the impact (EBRD 2024; Küfeoğlu 2022).

Regarding the eighth principle, management is in charge of updating and documenting the impact achieved on every investment and comparing the anticipated and actual effects and other outcomes. Furthermore, these conclusions should improve operational and strategic investment choices and administrative processes (EBRD 2024; Küfeoğlu 2022).

According to the last principle, the manager must frequently provide independent evidence of this alignment and publicly report how well its IA practices correspond with the principles each year. In addition, the results of this verification report ought to be released to the public (EBRD 2024; Küfeoğlu 2022).

2.4.3 The World Bank

Through IA, the World Bank hopes to ascertain whether the program itself, rather than some other factor, is responsible for the changes in consumption and health. It has published a handbook outlining the most popular quantitative techniques for ex-post impact assessments of policies and initiatives. The manual also covers ex-ante techniques for projecting program outcomes and strategies for quantifying distributional impacts (Khandker et al. 2009; Küfeoğlu 2022). The guidebook specifies a number of measures that must be taken to guarantee that IA is operational. For instance, the evaluation's purpose and significance must be clarified during project selection and preparation. There are also worries about the kind and promptness of evaluations. Plans for IA should be made to help program administrators evaluate and adjust their targeting during the intervention to isolate the program's impact on outcomes. A program's capability can also be determined by the quality and availability of its data. Hiring and training fieldwork personnel and developing a standardised data management and access approach are essential. Administratively speaking, the assessment team should be carefully selected during project implementation to ensure that it has the necessary technical and managerial expertise to guarantee accurate data and result reporting, as well as transparency in execution to enable precise data interpretation (Khandker et al. 2009; Küfeoğlu 2022).

Randomised evaluations, the first method in the book, identify a group of participants with comparable observed features and then randomly allocate the therapy to a subset of this group to assess the program's efficacy. This tactic resolves the issue of unobserved traits leading to selection bias (Khandker et al. 2009; Küfeoğlu 2022). The double-difference method is another technique that the World Bank uses to determine whether consequences are tracked for participants and non-participants over an extended period to capture any effects of the intervention. This approach suggests that monitoring outcomes over time for both non-participants and participants will provide a strong basis for assessing the program's impact through the use of the double-difference method; the changes that non-participants observe over time give the counterfactuals for participants (Khandker et al. 2009; Küfeoğlu 2022).

Additionally, an instrumental variable approach reveals exogenous variance in treatment by including a third variable that only impacts the therapy and not unobserved factors. Panel or cross-sectional data can be used in these techniques. Program design and other exogenous shocks unrelated to the intended outcomes can also be used to develop instruments. Instrumental variables and experimental approaches that employ exogenous program rules to compare participants and non-participants in a limited area are extended into regression discontinuity and pipeline techniques (Khandker et al. 2009; Küfeoğlu 2022). The World Bank asserts that only some assignments or assessment techniques are ideal. Therefore, it seems sensible to use alternative methods to confirm the results. Combining quantitative and qualitative methods with ex-ante and ex-post evaluation techniques is possible. It is necessary to employ particular methodologies. Key elements of intervention assessment include comprehending the goals and procedures of the program, the characteristics of both targeted and nontargeted regions, and the design and implementation of interventions. By leading enjoyable IA throughout the program and starting early in the project's planning and execution phases, one may also determine whether certain program elements can be altered to make it better (Khandker et al. 2009; Küfeoğlu 2022).

2.4.4 Intel Corporation

Intel's 2020 corporate responsibility report provides instances of IA from a business perspective. Intel argues in this paper that implementing and developing ethical business practices in their global operations and supply chain decreases risks and promotes human rights compliance. Intel's 2030 objectives include retaining and strengthening its focus on preserving and developing a solid safety culture as its organisation changes and grows, increasing the worldwide reach of its wellness initiatives. The targets also include a significant rise in the number of suppliers covered by their engagement programs, which will strengthen human rights responsibility throughout their global supply chain. Intel states it is also at the forefront of industry-wide initiatives to improve ethical mineral sourcing and responsible mobility (Intel 2021; Küfeoğlu 2022).

As we can see above, Intel makes an effort to show that it cares about people, the environment, and, in general, the welfare of all living things in the future. However, how does Intel evaluate the effects of its decisions in the real world? IA commissioners share their principles in their corporate responsibility reports and explore how they have evolved over the past year. While doing so, they stick to a single tactic that applies to their beliefs. Take, for instance, the importance of worker safety and well-being. In their 2021 report, they start by saying that they want to ensure that 50% of their employees participate in the company's worldwide corporate wellness program and that over 90% of its employees believe Intel has a solid safety culture. The next step is to establish a baseline. First, 22% of Intel employees participated in Intel wellness initiatives starting in 2020. Second, 37% of Intel employees completed the company's EHS Safety Culture Survey, with a baseline average of 79% on "safety is a value" measures. Intel then provides an overview of its achievements during the preceding year. Their health and wellness teams focused on early intervention and prevention programs and participation in the recently expanded virtual services offered by the Intel Vitality Program to raise employee awareness of and involvement in the company's initiatives throughout 2020. They look ahead to finish the assessment. The safety culture aim of Intel is to improve company-wide participation in the survey, involve employees and management more in safety activities, and reach 50% of workers by the end of 2021 (Intel 2021; Küfeoğlu 2022).

As a result, Intel, a technological behemoth, utilises these corporate responsibility reports to demonstrate its social and environmental decency to the public. It accomplishes this by applying several IA concepts. First, it defines the expected impact as ensuring that more than 90% of Intel's workers feel the company has a strong safety culture and that 50% engage in its global corporate wellness program. Second, it establishes baselines, such as the Safety Culture Survey, which has a baseline average of 79% on "safety is a value" measures, to enable comparisons of changes. Third, as we can see, it shows progress using experimental data. Finally, it uses the lessons from its study to create new targets, such as increasing the poll to 50% of workers by the end of 2021 (Intel 2021; Küfeoğlu 2022).

2.5 Sustainable Investing

Investment is the commitment of existing financial resources to achieve larger returns later (CMA 2024). Traditional investing only captures financial value in risk and return space since it is founded on the neoclassical paradigm of efficient markets and portfolio theory (Schoenmaker and Schramade 2019). Figure 2.21 demonstrates the relation between different investment types.

Sustainable investment encourages the growth of an economy's potential for production while advancing decarbonisation, protecting the environment, fostering the creation of jobs and skill development, and guaranteeing that everyone has equitable access to opportunities. Whether carried out by domestic or international businesses, it is essential to boost living standards and build a better, more sustainable future for everybody (OECD 2024). However, not all countries share

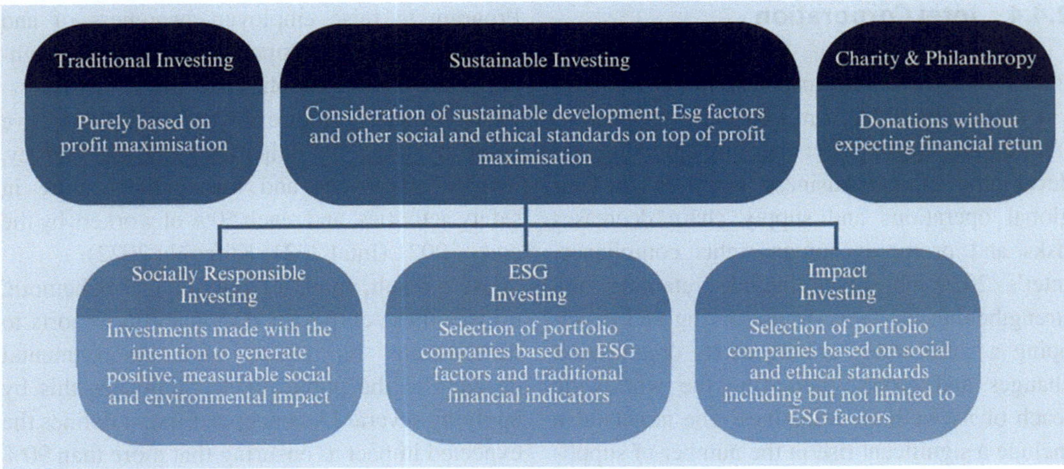

Fig. 2.21 Relationship between different investment approaches

Snapshot of global sustainable investing assests, 2016-2022 (USD billions)

Region	*2016*	*2018*	*2020*	*2022*
Europe	12,040	14,075	12,017	14,054
United States	8,723	11,995	17,081	8,400
Japan	474	2,180	2,874	4,289
Canada	1,086	1,699	2,423	2,358
Australia & New Zealand	516	734	906	1,220
Total (USD Billions)	22,838	30,683	35,301	30,321

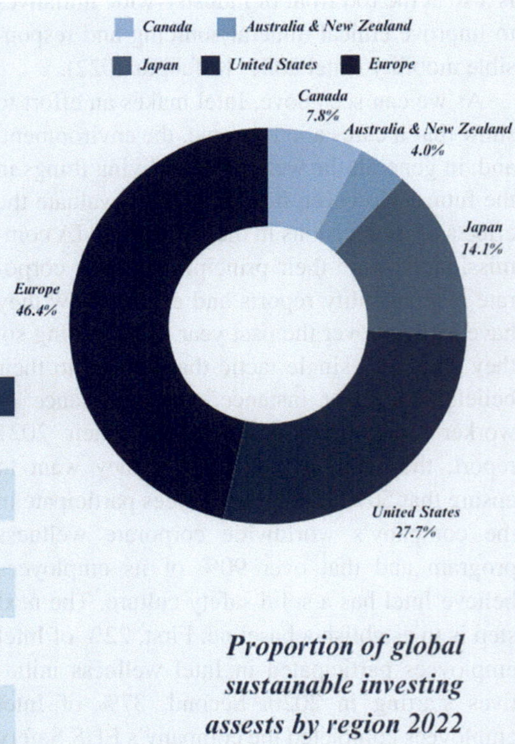

Fig. 2.22 Global sustainable investing assets between 2016 and 2022 and their distribution to world regions (GSIA 2023)

this vision, and strategies may vary. Figure 2.22 shows how sustainable investing assets have changed throughout 2016–2022 and their distribution between world regions.

In 2020, global sustainable investing assets reached USD 35 trillion in five major areas (the USA, Canada, Japan, Europe, and Australasia). Thus, 35.9% of total assets under management are used in sustainable investing (GSIA 2021). In 2022, the number decreased to USD 30 trillion in 2022. Major areas except Canada and the USA have grown their sustainable investing assets, while assets in Canada stayed neutral, and the USA had a major decrease in sustainable investing assets (GSIA 2023).

According to Bloomberg Intelligence (2024), the focus on socially responsible investing has steadily increased for several years, and the pandemic only fuelled it further. Assets are expected to rise to USD 40 trillion by 2030. Figure 2.23 demonstrates how sustainable and traditional investors differ in weighing their investments.

2.5.1 Socially Responsible Investing

Socially responsible investing (SRI) or responsible investing is an investment approach that aims to consider financial return in addition to ethical, social, or environmental goals (Sparkes 2003). The term "ethical investment" denotes SRI in its early stages. Social investing, socially responsible investing, socially aware investing, socially conscious investing, green investing, value-based investing, and mission-based or mission-related investing all refer to the same general process and are often used interchangeably (Schueth 2003).

Various strategies are used in SRI to match financial choices with moral principles. By selecting or excluding businesses based on ethical and environmental standards, screening helps investors meet their financial objectives and encourages beneficial effects. Investors can advocate for improved corporate practices by interacting with management on significant topics and presenting proxy motions. Also, community investment allocates funds to underprivileged areas, giving financial access to those who might not otherwise have access to standard financial services in low-income or high-risk locations.

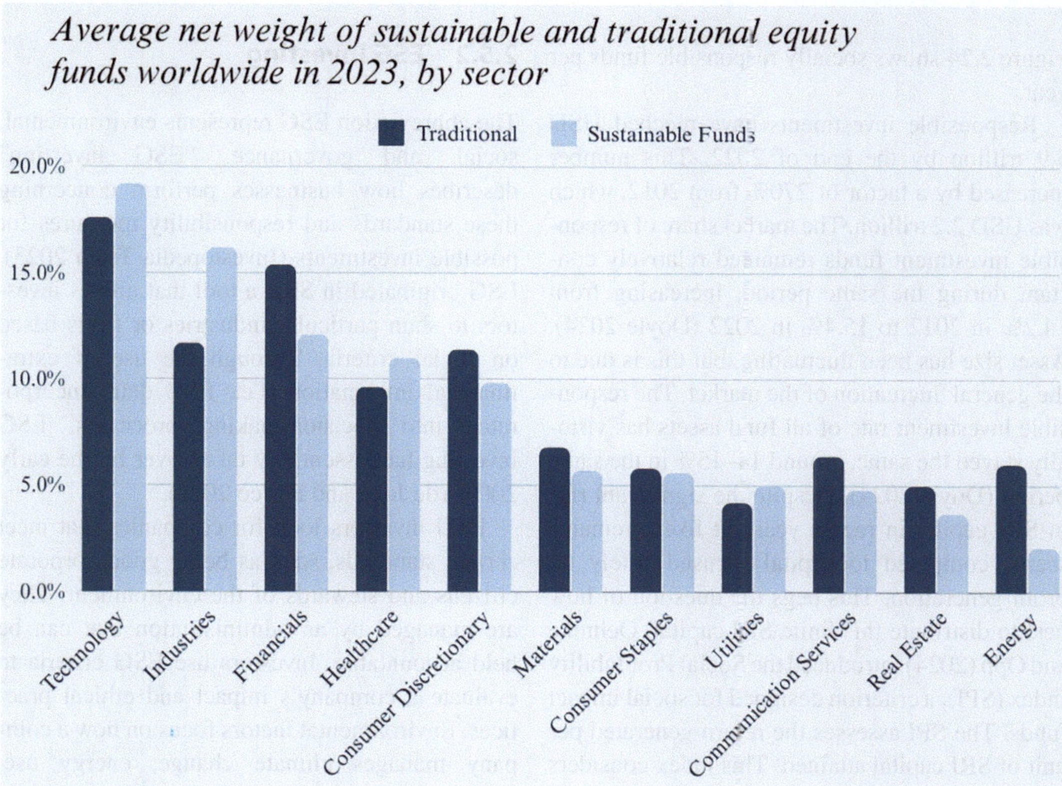

Fig. 2.23 Sustainable and traditional fund asset distribution worldwide in 2024 (Statista 2023)

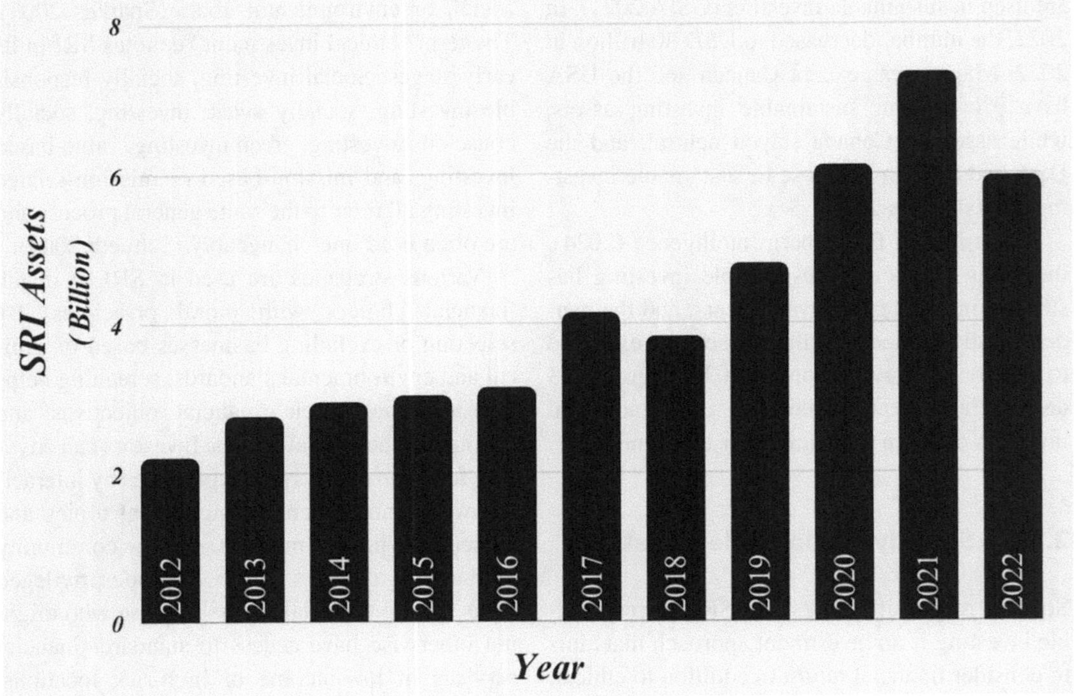

Fig. 2.24 Fund assets allocated to socially responsible investing between 2012 and 2022 (Doyle 2024)

Figure 2.24 shows socially responsible funds per year.

Responsible investments have reached USD 5.9 trillion by the end of 2022. This number increased by a factor of 270% from 2012, which was USD 2.2 trillion. The market share of responsible investment funds remained relatively constant during the same period, increasing from 14.2% in 2012 to 15.4% in 2022 (Doyle 2024). Asset size has been fluctuating, but this is due to the general fluctuation of the market. The responsible investment rate of all fund assets has virtually stayed the same, around 14–15% in the same period (Doyle 2024). Despite the significant rise in SRI capital in recent years, it likely remains scarce compared to capital focused solely on profit generation. This begs the question of how best to distribute this finite SRI capital. Oehmke and Opp (2024) introduced the Social Profitability Index (SPI), a criterion designed for social impact funds. The SPI assesses the return generated per unit of SRI capital attained. This index considers the financial return and extra -financial data in investment decision-making.

2.5.2 ESG Investing

The abbreviation ESG represents environmental, social, and governance. "ESG investing" describes how businesses perform concerning these standards and responsibility measures for possible investments (Investopedia Team 2023). ESG originated in SRI, a tool that allows investors to shun particular industries or firms based on social criteria. Through the use of extra-financial information (i.e., ESG data) incorporated into decision-making processes, ESG investing had essentially taken over by the early 2000s (de Jong and Rocco 2022).

ESG investors look for companies that meet certain standards, such as being good corporate citizens and stewards of the environment. They are managed by an administration that can be held accountable. Investors use ESG criteria to evaluate a company's impact and ethical practices. Environmental factors focus on how a company manages climate change, energy use, pollution, waste, natural resource conservation, and animal welfare, including handling hazardous materials and greenhouse gas emissions. Social factors examine a company's relationships with its employees, customers, and communities, assessing practices like employee volunteering, charitable contributions, and maintaining safe and healthy working conditions. Governance involves the structures and processes for corporate oversight, ensuring accountability, integrity, and transparency in leadership and operations. Investors focused on ESG issues seek companies that avoid conflicts of interest, do not misuse political contributions, and uphold ethical standards in all activities.

According to PwC (2022), asset managers globally are expected to increase their ESG-related assets under management (AUM) to USD 33.9 trillion by 2026 from USD 18.4 trillion in 2021. AUM or funds under management refers to the total market value of the securities financial institutions own or manage on behalf of their clients. Figure 2.25 shows the growth of ESG investing AUM and three possible scenarios for ESG investment assets in 2026.

2.5.3 Impact Investing

Impact investments are monetary investments made to businesses, funds, and organisations to produce a financial return and quantifiable social and environmental benefits. According to GIIN (2019), impact investing has four core characteristics: intentionality, using evidence and impact data in investment design, managing impact performance, and contributing to the industry's growth.

Impact investing is based on four key characteristics. The first is intentionality, where investors explicitly target positive social or environmental outcomes. The second is using data to design evidence-based investments to maximise impact. Thirdly, impact performance management includes ongoing monitoring and evaluation of social outcomes. Finally, impact performance management contributed to the sector's growth, where investors sought to positively influence the entire impact investment ecosystem (GIIN 2019). With 34% each, the USA, Canada,

	2015	*2020*	*2021*	*2026 Low*	*2026 Base*	*2026 Best*
Asia-Pasific	0.2	0.7	1.0	2.1	3.3	5.0
Europe	1.1	4.7	12.8	14.3	19.6	25.7
Latin America	0.0	0.0	0.0	0.1	0.2	0.3
Middle East & Africa	0.0	0.1	0.1	0.2	0.3	0.4
North America	0.8	3.8	4.5	7.7	10.5	16.3
Total	***2.2***	***9.4***	***18.4***	***24.4***	***33.9***	***47.6***

Fig. 2.25 ESG investing AUM and future scenarios (PwC 2022)

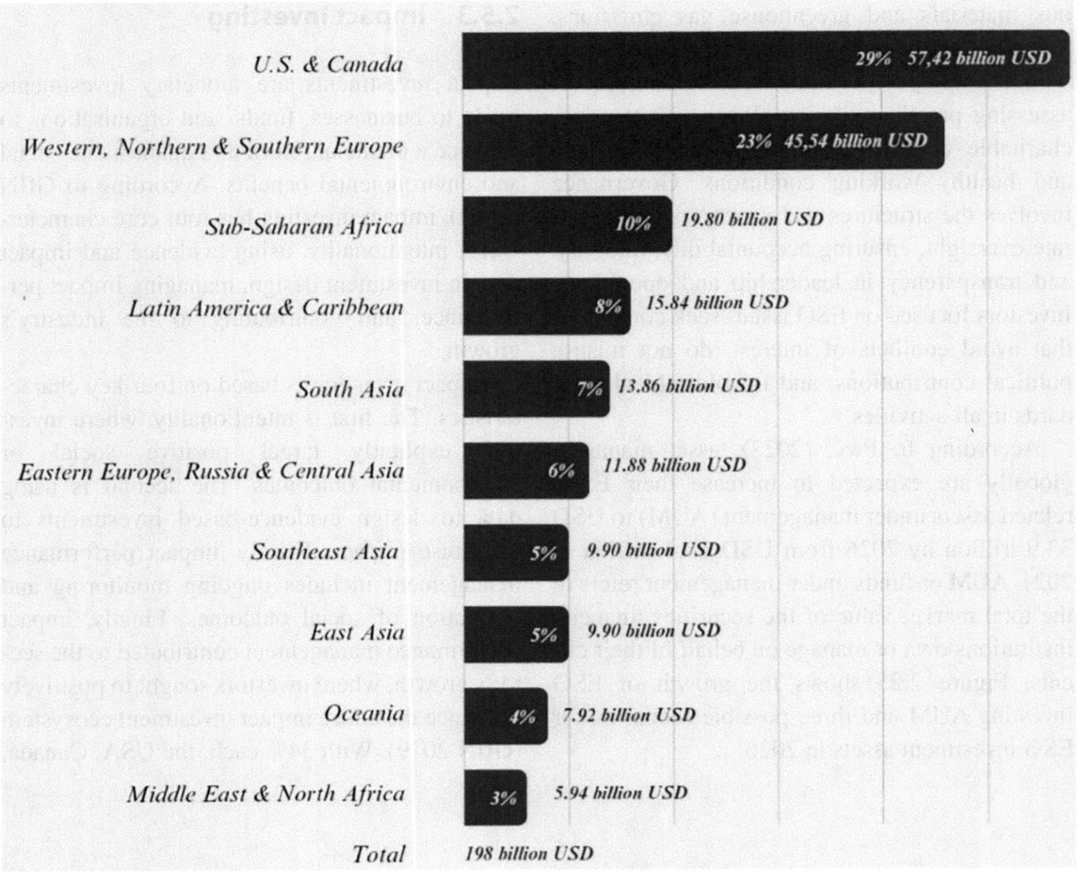

Fig. 2.26 The proportion of impact investments by region (Hand et al. 2022)

Western, Northern, and Southern Europe account for just over two-thirds of impact investors' headquarters (Hand et al. 2023). Figure 2.26 shows the proportion of impact investments in each world region.

In their report for the GIIN, Hand et al. (2022) anticipate that organisations responsible for managing assets in impact investing will be USD 1.164 trillion in 2021. In recent years, the financial markets have witnessed unprecedented commitments from asset owners and managers to allocate capital towards the USD 4.2 trillion funding gap necessary to achieve the SDGs as the 2030 target approaches (OECD 2020). Figure 2.27 shows the proportion of impact investments in each sector.

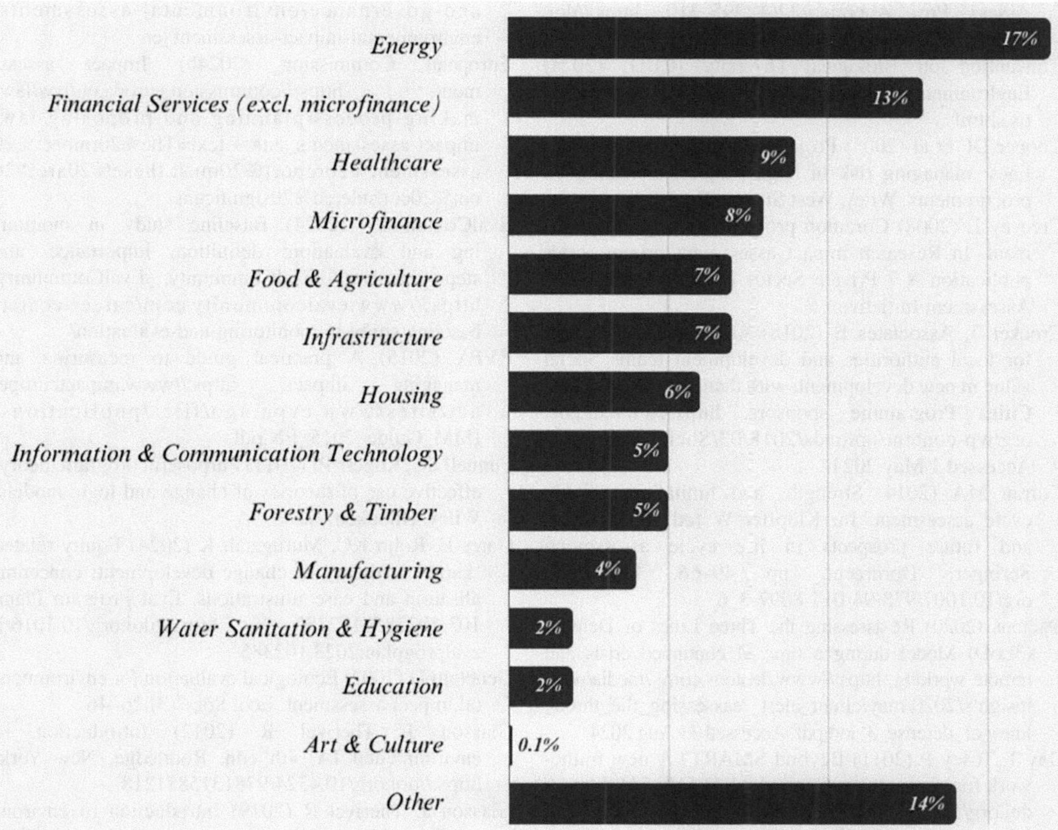

Fig. 2.27 The proportion of impact investments by sector (Hand et al. 2023)

References

Angelucci M, Di Maro V (2015) Programme evaluation and spillover effects. J Dev Eff 8(1):22–43. https://doi.org/10.1080/19439342.2015.1033441

Bank for International Settlements (2011) Principles for the sound management of operational risk. https://www.bis.org/publ/bcbs195.pdf

Banke-Thomas AO, Madaj B, Charles A, van den Broek N (2015) Social return on investment (SROI) methodology to account for value for money of public health interventions: a systematic review. BMC Public Health 15(1):582. https://doi.org/10.1186/s12889-015-1935-7

Barrett D, Twycross A (2018) Data collection in qualitative research. Evid Based Nurs 21(3):63–64

Bateman K (2018) Implementing a three lines of defence approach to risk management, information age. https://www.information-age.com/implementing-defence-risk-management-10049/. Accessed 31 Aug 2024

Bina O (2008) Strategic environmental assessment. Edward Elgar Publishing eBooks, Cheltenham. https://doi.org/10.4337/9781848445062.00019

Bloomberg Intelligence (2024) Global ESG Assets predicted to hit $40 Trillion by 2030, despite challenging environment. Forecasts Bloomberg Intelligence Press, Bloomberg LP. https://www.bloomberg.com/company/press/global-esg-assets-predicted-to-hit-40-trillion-by-2030-despite-challenging-environment-forecasts-bloomberg-intelligence/. Accessed 17 Aug 2024

Bond A, Dusík J (2019) Impact assessment for the twenty-first century—rising to the challenge. Impact Assess Proj Apprais 38:94–99. https://doi.org/10.1080/14615517.2019.1677083

Bose S, Dey A, & Sardashti H (2019) Financial inclusion and its determinants: Evidence from emerging economies. Journal of Economic Studies, 46(5), 1130–1144. https://doi.org/10.1108/JES-02-2018-0055

Burdge RJ (1990) The benefits of social IA in third world development. Environ IA Rev 10(1–2):123–134. https://doi.org/10.1016/0195-9255(90)90012-o

Caruana EJ, Roman M, Hernández-Sánchez J, Solli P (2015) Longitudinal Studies. J Thorac Dis 7(11):E537

Cashmore M et al (2004) The interminable issue of effectiveness: substantive purposes, outcomes and research challenges in the advancement of environmental impact assessment theory. Impact

Assess Proj Apprais 22(4):295–310. https://doi.org/10.3152/147154604781765860
Convention on Biological Diversity (CBD) (2024) Environmental IA. https://www.cbd.int/impact/whatis.shtml
Cooper DF et al (2005) Project risk management guidelines: managing risk in large projects and complex procurements. Wiley, West Sussex, England
Creevey L (2008) Common problems in impact assessment. In Research impact assessment primer series publication # 7 Private Sector Development Impact Assessment Initiative
Crocker T, Associates B (2018) An introductory guide for local authorities and development teams Social value in new development: with thanks to the UKGBC Cities Programme sponsors. https://www.ukgbc.org/wp-content/uploads/2018/03/Social-Value.pdf. Accessed 1 May 2024
Curran MA (2014) Strengths and limitations of life cycle assessment. In: Klöpffer W (ed) Background and future prospects in life cycle assessment. Springer, Dordrecht, pp 49–66. https://doi.org/10.1007/978-94-017-8697-3_6
Dantons (2020) Re-assessing the Three Lines of Defense (3LoD) Model during a time of continued crisis and remote working. https://www.dentons.com/-/media/pdfs/insights/2020/may/client_alert_reassessing_the_three_lines_of_defense_3_lod.pdf. Accessed 31 Aug 2024
Day T, Tosey P (2011) Beyond SMART? A new framework for goal setting. Curric J 22(4):515–534. https://doi.org/10.1080/09585176.2011.627213
de Jong M, Rocco S (2022) ESG and impact investing. J Asset Manag 23:547–549. https://doi.org/10.1057/s41260-022-00297-7
Dean M (2020) Multi-criteria analysis. Adv Transp Policy Plann 6:165–224. https://doi.org/10.1016/bs.atpp.2020.07.001
Desmond M (2007) Strategic environmental assessment (sea): a tool for environmental decision-making. Ir Geogr 40(1):63–78. https://doi.org/10.1080/00750770709555887
Dörnyei Z (2007) Research methods in applied linguistics: quantitative qualitative, and mixed methodologies. Oxford University Press, Oxford
Doyle J (2024) Responsible investment funds build consistent market presence. CFA Institute. https://rpc.cfainstitute.org/-/media/documents/article/industry-research/responsible-investment-funds.pdf. Accessed 4 Sept 2024
European Bank for Reconstruction and Development (EBRD) (2024) Operating principles for impact management. https://www.ebrd.com/documents/policy/operating-principles-for-impact-management.pdf. Accessed 17 Aug 2024
European Commission (2009) European Commission Impact Assessment Guidelines. https://ec.europa.eu/smart-regulation/impact/commission_guidelines/docs/iag_2009_en.pdf
European Commission (2024a) Environmental IA. https://environment.ec.europa.eu/law-and-governance/environmental-assessments/environmental-impact-assessment_en
European Commission (2024b) Impact assessment. https://commission.europa.eu/law/law-making-process/planning-and-proposing-law/impact-assessments_en#:~:text=The%20impact%20assessment%20report%20must,these%20are%20not%20considered%20significant
EvalCommunity (2024) Baseline study in monitoring and evaluation: definition, importance, and steps involved—EvalCommunity, EvalCommunity. https://www.evalcommunity.com/career-center/baseline-study-in-monitoring-and-evaluation/
EVPA (2015) A practical guide to measuring and managing impact. https://www.impacteurope.net/sites/www.evpa.ngo/files/publications/IMM_Guide_2015_EN.pdf
Funnell SC, Rogers PJ (2011) Purposeful program theory: effective use of theories of change and logic models. Wiley, Hoboken, NJ
Gates E, Rohn KC, Murugaiah K (2024) Equity-related 'knots' in theory of change development: conceptualisation and case illustrations. Eval Program Plann 103:102385–102385. https://doi.org/10.1016/j.evalprogplan.2023.102385
Geneletti D (2002) Ecological evaluation for environmental impact assessment. Ecol Soc 7(3):26–46
Glasson J, Therivel R (2012) Introduction to environmental IA, 4th edn. Routledge, New York. https://doi.org/10.4324/9781315881218
Glasson J, Therivel R (2019) Introduction to environmental IA, 5th edn. Routledge, New York. https://doi.org/10.4324/9780429470738
Global Impact Investing Network (GIIN) (2024a) IRIS+ System. [online] iris.thegiin.org. https://iris.thegiin.org/about/. Accessed 15 Apr 2024
Global Impact Investing Network (GIIN) (2024b) IRIS Catalog of Metrics | IRIS+ System. iris.thegiin.org. https://iris.thegiin.org/metrics/?search=&focus%5B0%5D=social&sortby. Accessed 8 May 2024
Government of South Australia (2016) Government Institutional Impact Statement Guideline 2016. https://www.premiers.qld.gov.au/publications/categories/policies-and-codes/handbooks/cabinet-handbook/assets/government-institutional-impact-statement-guidelines.pdf
GSIA (2021) Global Sustainable Investment Review 2020. https://www.gsi-alliance.org/wp-content/uploads/2021/08/GSIR-20201.pdf. Accessed 31 Aug 2024
GSIA (2023) Global Sustainable Investment Review 2022. https://www.gsi-alliance.org/wp-content/uploads/2023/12/GSIA-Report-2022.pdf. Accessed 1 Sept 2024
Gugerty MK, Karlan D, Welsh D (2016) In: Batcheck D et al (eds) Organisational challenges of impact evaluation. https://poverty-action.org/sites/default/files/publications/Goldilocks-Deep-Dive-Organizational-Challenges-of-Impact-Evaluation_0.pdf. Accessed 2 Jun 2024

Guijarro F, Tsinaslanidis P (2020) Analysis of academic literature on environmental valuation. Int J Environ Res Public Health 17(7):2386. https://doi.org/10.3390/ijerph17072386

Hand D, Ringel B, Danel A (2022) Sizing the Impact Investing Market: 2022. The Global Impact Investing Network (GIIN), New York. https://thegiin.org/assets/2022-Market%20Sizing%20Report-Final.pdf

Hand D, Sunderji S, Pardo N (2023) 2023 GIINsight: impact investing allocations, activity & performance. The Global Impact Investing Network (GIIN), New York. https://thegiin.org/assets/documents/pub/2023-GIINsight/2023%20GIINsight%20%E2%80%93%20Impact%20Investing%20Allocations,%20Activity%20&%20Performance.pdf

Hayes H, Parchman ML, Howard R (2011) A logic model framework for evaluation and planning in a primary care practice-based research network (PBRN). J Am Board Fam Med 24(5):576–582

Hubbard D (2009) The failure of risk management. Google Books. https://books.google.com.tr/books?hl=tr&lr=&id=fMbKDwAAQBAJ&oi=fnd&pg=PR11&dq=The+Failure+of+Risk+Management:+Why+It%27s+Broken+and+How+to+Fix+It.&ots=kia7ySdNuY&sig=Mo8-jtbr67KRQ_lCbQ0V8uleIF0&redir_esc=y#v=onepage&q&f=false. Accessed 31 Aug 2024

IIA (2013) IIA Position Paper: The Three Lines of Defense in effective risk management and control. https://theiia.fi/wp-content/uploads/2017/01/pp-the-three-lines-of-defense-in-effective-risk-management-and-control.pdf

Impact Canada (2020) Logic model and narrative IA of challenges under Impact Canada. https://impact.canada.ca/sites/default/files/2021-08/ENlogicmodel-Jul22.pdf

Intel (2021) 2020–21 Corporate Responsibility Report. http://csrreportbuilder.intel.com/pdfbuilder/pdfs/CSR-2020-21-Full-Report.pdf

International Association for IA (IAIA) (2003) International Principles for Social IA. https://www.iaia.org/pdf/Sections/SIA/IAIA%20SIA%20International%20Principles.pdf

International Association for IA (IAIA) (2023a) IA. https://www.iaia.org/wiki-details.php?ID=4

International Association for IA (IAIA) (2023b) Economic and fiscal assessment. https://www.iaia.org/wiki-details.php?ID=11

International Finance Corporation (2022) How IFC measures the development impact of its interventions. https://www.ifc.org/content/dam/ifc/doc/2022/202012-ifc-aimm-brochure.pdf. Accessed 16 Aug 2024

International Organization for Standardization (ISO) (2006a) ISO 14040 Environmental management—Life cycle assessment—Principles and framework. https://www.iso.org/obp/ui/fr/#iso:std:iso:14040:ed-2:v1:en

International Organization for Standardization (ISO) (2006b) ISO 14044 Environmental management—Life cycle assessment—Requirements and guidelines. https://www.iso.org/obp/ui/fr/#iso:std:iso:14044:ed-1:v1:en

Investopedia Team (2023) What is environmental, social, and governance (ESG) investing? In: Investopedia. https://www.investopedia.com/terms/e/environmental-social-and-governance-esg-criteria.asp

IPA (2016) Sensing impacts: remote monitoring using sensors. IPA. poverty-action.org. https://poverty-action.org/publication/sensing-impacts-remote-monitoring-using-sensors. Accessed 1 Aug 2024

Islam G, Greenwood M (2022) The metrics of ethics and the ethics of metrics J Bus Ethics 175:1–5

Jolly D, Thompson-Fawcett M (2021) Enhancing indigenous IA: lessons from indigenous planning theory. Environ IA Rev 87:106541. https://doi.org/10.1016/j.eiar.2020.106541

Jonkers K, Tijssen R, Karvounaraki A, Goenaga X (2018) A regional innovation IA framework for universities. [online]. https://doi.org/10.2760/623825

Joseph C, Gunton T, Knowler D, Broadbent S (2020) The role of cost-benefit analysis and economic impact analysis in environmental assessment: The case for reform. IA Proj Apprais 38(6):491–501. https://doi.org/10.1080/14615517.2020.1767954

Khan M et al (2024) ChatGPT and the future of impact assessment. Environ Sci Policy 157:103779–103779. https://doi.org/10.1016/j.envsci.2024.103779

Khandker SR, Koolwal GB, Samad HA (2009) Handbook on impact evaluation: quantitative methods and practices. World Bank Publications

Kingdom of Saudi Arabia Capital Market Authority (CMA) (2024) Investor Awareness Guidebooks: Investment. https://cma.org.sa/en/Awareness/Publications/booklets/Booklet_1.pdf

Knoll MS (2002) Ethical screening in modern financial markets: the conflicting claims underlying socially responsible investment. Bus Law 57:681

Kolhoff AJ, Runhaar HAC, Driessen PPJ (2009) The contribution of capacities and context to EIA system performance and effectiveness in developing countries: towards a better understanding. Impact Assess Proj Apprais 27(4):271–282. https://doi.org/10.3152/146155109x479459

Küfeoğlu S (2022) Emerging technologies: value creation for sustainable development. Springer Nature

Laurance WF (2022) Why environmental impact assessments often fail. Therya 13(1):67–72. https://doi.org/10.12933/therya-22-1181

Laurance W, Salt D (2018) Environmental impact assessments aren't protecting the environment. Ensia. https://ensia.com/voices/environmental-impact-assessment/

Leonard K, Watts R (2024) The ultimate guide to S.M.a.R.T. Goals, Forbes Advisor. Forbes. https://www.forbes.com/advisor/business/smart-goals/

Lesa A, Egger EM (2019) The challenges of conducting impact assessments of community driven development projects. IFAD. https://www.ifad.org/en/web/latest/-/blog/the-challenges-of-conducting-impact-assessments-of-community-driven-development-projects

Li F (2009) Documenting accountability: environmental IA in a Peruvian mining project. PoLAR Polit Leg Anthropol Rev 32(2):218–236. https://doi.org/10.1111/j.1555-2934.2009.01042.x
Likhachova I (2019) Operating principles for impact management. Harvard.edu. https://corpgov.law.harvard.edu/2019/05/05/operating-principles-for-impact-management/
Lima L et al (2015) Impact evaluation of development. In: eVALUAtion Matters, p 18
Macgillivray A et al (2017) Measuring total employment effects: a lean data methodology for a portfolio of investments in developing countries. https://assets.cdcgroup.com/wp-content/uploads/2018/06/25150849/Methodology-for-measuring-total-employment-effects.pdf. Accessed 21 Jul 2024
Mackay M et al (2024) Integrated IA for land and water management. Environ IA Rev 105:107397. https://doi.org/10.1016/j.eiar.2023.107397
Market Street Services, Inc (2004) Economic impact analysis Carolina North. Market Street Services. https://facilities.unc.edu/wp-content/uploads/sites/256/2015/12/economic_impact.pdf
Marshall R (2005) Environmental IA follow-up and its benefits for industry. IA Proj Apprais 23(3):191–196. https://doi.org/10.3152/147154605781765571
Martinez DE, Cooper DJ (2020) Seeing through the logical framework. VOLUNTAS 6:1239. https://doi.org/10.1007/s11266-020-00223-8
Maurice B et al (2000) Uncertainty analysis in life cycle inventory. Application to the production of electricity with French coal power plants. J Clean Prod 8:95–108
Miller C (2019) IA framework. [online]. https://wgu.yourlearningportal.com/wgu/resources/courses/wgu_courses/hcm/ihcm_masters/C409_IHCM5110_innovation_project/resources/impact_assessment_framework.pdf
Morgan H (2022) Conducting a qualitative document analysis. Qual Rep 27(1):64–77
Morrison-Saunders A et al (2014) Strengthening IA: a call for integration and focus. IA Proj Apprais 32(1):2–8. https://doi.org/10.1080/14615517.2013.872841
Ngetich JK, Ndiema AC (2020) Challenges and opportunities in environmental impact assessment and environmental audit practice in Kenya. J Geogr Reg Plan 13(3):70–76. https://doi.org/10.5897/jgrp2019.0753
Noble BF (2006) Introduction to environmental impact assessment: a guide to principles and practice. Oxford University Press, Toronto, ON
O'Flynn M (2010) M&E Paper 7 IA: understanding and assessing our contributions to change. [online]. https://www.intrac.org/wpcms/wp-content/uploads/2016/09/Impact-Assessment-Understanding-and-Assessing-our-Contributions-to-Change-1.pdf
OECD (2004) OECD Glossary of Statistical Terms. https://www.paris21.org/sites/default/files/1300.pdf. Accessed 21 Jul 2024
OECD (2020) Global outlook on financing for sustainable development 2021: a new way to invest for people and planet. OECD Publishing, Paris. https://doi.org/10.1787/e3c30a9a-en
OECD (2024) Sustainable investment. https://www.oecd.org/en/topics/sub-issues/sustainable-investment.html
Oehmke M, Opp MM (2024) A theory of socially responsible investment. Rev Econ Stud:rdae048. https://doi.org/10.1093/restud/rdae048
Office of Internal Audit (2015) The three lines of defense, internalaudit.Olemiss.Edu. The University of Mississippi. https://internalaudit.olemiss.edu/the-three-lines-of-defense/
Okediji OO (1965) The socio-cultural problems in the Western Nigeria land settlement scheme: a case study. Niger J Econ Soc Stud 7:301–310
Ortega L, Iberri-Shea G (2005) Longitudinal research in second LANGUAGE acquisition: recent TRENDS and future directions. Annu Rev Appl Linguist 25:26–45. https://doi.org/10.1017/S0267190505000024
Ortolano L, Shepherd A (1995) Environmental impact assessment: challenges and opportunities. Impact Assess 13(1):3–30. https://doi.org/10.1080/07349165.1995.9726076
Pizzol M et al (2015) Monetary valuation in life cycle assessment: a review. J Clean Prod 86:170–179. https://doi.org/10.1016/j.jclepro.2014.08.007
Project Management Institute (2013) A guide to the Project Management body of knowledge: PMBOK guide, 5th edn. Project Management Institute, Newtown Square, PA
PwC (2022) Asset and wealth management revolution 2022: exponential expectations for ESG. https://www.pwc.com/gx/en/industries/financial-services/asset-management/publications/asset-and-wealth-management-revolution-2022.html?WT.mc_id=CT11-PL1000-DM2-TR2-LS4-ND1-TTA9-CN_gx-fy22-xlos-esg-awm-esg-revolution-pressrelease
PWC and The Forest Products Association of Canada (FPAC) (2010) Life Cycle Assessment and Forest Products: a White Paper. https://www.pwc.com/gx/en/forest-paper-packaging/pdf/fpac-lca-white-paper.pdf
Queensland Department of State Development (2017) Economic IA Guideline. https://www.statedevelopment.qld.gov.au/__data/assets/pdf_file/0012/33420/economic-impact-assessment-guideline.pdf
Rana J, Gutierrez PL, Oldroyd JC (2021) Quantitative methods. In: Global Encyclopedia of Public Administration, Public Policy, and Governance, pp 1–6
Rawlings L (2013) Who's afraid of administrative data? why administrative data can be faster, cheaper and sometimes better. World Bank Blogs. https://blogs.worldbank.org/en/impactevaluations/whos-afraid-administrative-data-why-administrative-data-can-be-faster-cheaper-and-sometimes-better. Accessed 1 Aug 2024
Sadler B (1996) Environmental assessment in a changing world. Evaluating practice to improve performance—final report. EN106-37/1996E. [online]. https://trid.trb.org/view/653262
Schoenmaker D, Schramade W (2019) Investing for long-term value creation. J Sustain Financ Invest

9(4):356–377. https://doi.org/10.1080/20430795.2019.1625012

Schueth S (2003) Socially responsible investing in the United States. J Bus Ethics 43(3):189–194. https://doi.org/10.1023/a:1022981828869

Schueth S, Gravitz A, Larsen T (1999) 1999 report on socially responsible investing trends in the United States. [online]. https://www.griequity.com/resources/InvestmentIndustry/Trends/SIF%20Research%201999%20Trends%20Report.htm. Accessed 7 Apr 2024

Scudder T (1968) Social anthropology, man-made lakes and population relocation in Africa. Anthropol Q 41:168–176

Simfoni (2023) IA. https://simfoni.com/impact-assessment-101/

Simler C (2019) Impact measurement: a practical guide to data collection https://assets.cdcgroup.com/wp-content/uploads/2019/10/29172105/CDC_ImpactMeasurementHandbook.pdf. Accessed 14 Jul 2024

Sinclair AJ, Doelle M, Gibson RB (2021) Next generation impact assessment: exploring the key components. Impact Assess Proj Apprais 40:1–17. https://doi.org/10.1080/14615517.2021.1945891

Slootweg R, Mollinga P (2009) The IA framework. Cambridge University Press eBooks, pp 87–124. https://doi.org/10.1017/cbo9781139195775.008

Slootweg R, Vanclay F, van Schooten M (2001) Function evaluation as a framework for the integration of social and environmental IA. IA Proj Apprais 19(1):19–28. https://doi.org/10.3152/147154601781767186

Sonter LJ et al (2017) Mining drives extensive deforestation in the Brazilian Amazon. Nat Commun 8(1). https://doi.org/10.1038/s41467-017-00557-w

Spangenberg JH (2004) Reconciling sustainability and growth: criteria, indicators, policies. Sustain Dev 12(2):74–86. https://doi.org/10.1002/sd.229

Sparkes R (2003) Socially responsible investment. Google Books. https://books.google.com.tr/books?id=JgnqYLa13EQC&redir_esc=y. Accessed 17 Aug 2024

Statista (2023) Global: equity funds average net weight by sector, 2023. Statista. Statista. https://www.statista.com/statistics/1454508/average-net-sector-weight-of-sustainable-and-traditional-equity-funds/. Accessed 4 Sept 2024

Stein P, Ardic OP, & Hommes M (2012) Closing the Credit Gap for Formal and Informal Micro, Small, and Medium Enterprises. International Finance Corporation (IFC), World Bank Group. https://documents.worldbank.org/en/publication/documents-reports/documentdetail/113671468154509368

The GIIN (2019) Core characteristics of impact investing. https://thegiin.org/publication/post/core-characteristics-of-impact-investing/

UN (2023) Core indicators for sustainability and SDG impact reporting: training manual. https://unctad.org/publication/core-indicators-sustainability-and-sdg-impact-reporting-training-manual

UN Department of Economic and Social Affairs (UN DESA) (2024) Core Publications Agenda 21. https://www.un.org/esa/dsd/agenda21/

UNESCO (2024) Environmental IA. https://whc.unesco.org/fr/glossaire/239/

United Nations Environment Programme (UNEP) (2024) Green Financing. https://www.unep.org/regions/asia-and-pacific/regional-initiatives/supporting-resource-efficiency/green-financing

USAID (2010) Performance monitoring and evaluation tips: baselines and targets. https://www.ngoconnect.net/sites/default/files/resources/Performance%20Monitoring%20%26%20Evaluation%20TIPS-%20Baselines%20and%20Targets.pdf

Van Der Heijden J (2014) Governance for urban sustainability and resilience. https://doi.org/10.4337/9781782548133

van Es M, Guijt I, Vogel I (2015) Theory of change thinking in practice: a stepwise approach. Hivos

Vanclay F (2002) Conceptualising social impacts. Environ IA Rev 22(3):183–211. https://doi.org/10.1016/s0195-9255(01)00105-6

Vanclay F (2003) International principles for social IA. IA Proj Apprais 21(1):5–12. https://doi.org/10.3152/147154603781766491

Wang L, Juslin H (2009) The impact of Chinese culture on corporate social responsibility: the harmony approach. J Bus Ethics 88(Suppl 3):433–451. https://doi.org/10.1007/s10551-009-0306-7

Zheng X (2021) Data collection in quantitative research. In: Research methods for student radiographers. CRC Press, Boca Raton, FL, pp 79–92

3 Measuring Sustainability: Metrics and Standardisation

Abstract

This chapter explores the important role of measurement and standardisation in improving sustainability efforts in finance and policy. It starts by emphasising the significance of sustainability metrics and discusses how these metrics are developed, integrated, and applied to evaluate progress in economic, social, environmental, and governance areas. This chapter covers the challenges of creating reliable indicators and addresses issues like greenwashing and inconsistent data. It also explores the landscape of data collection, focusing on methods, ethical concerns, and the need for strong, clear data. A significant emphasis is placed on standardisation, explaining key global and regional frameworks—such as CSRD, ESRS, EU Taxonomy, GRI, ISSB, SFDR, SASB, and others—that promote consistent sustainability reporting. This chapter concludes with a detailed overview of greenhouse gas (GHG) and carbon accounting, presenting the frameworks necessary for trustworthy climate-related disclosures and performance tracking.

3.1 Measuring Sustainability and Its Importance

The concept of sustainable development is based on the idea that industrial progress leads to the depletion of natural resources and extensive environmental harm. Experts argue that human consumption patterns have exceeded the Earth's limitations, creating more resource-efficient technology to sustain future generations (Kim et al. 2015). This sense of urgency has led to sustainability measurement, aiming to mitigate the unsustainable path of current human activities by improving system performance across environmental, economic, and social domains. The role of sustainability measurement cannot be overstated because it serves as a guidepost for decision-makers to make more accountable and resilient choices (Jones et al. 2019). In this matter, the choice of indicators is of great importance. Indicators are quantitative or qualitative tools used to track, monitor, and evaluate changes over time. Over the years, numerous indicators have been proposed to capture the various components of sustainability, ranging from carbon footprint to economic viability measures. However, substantial debate still needs to be made over the most appropriate measurement scale and methodologies. Bell and Morse (2003) emphasised numerous critical characteristics of successful indicators, including specificity, measurability, usefulness, sensitivity, availability, and cost-effectiveness The selection of indicators acts as a frame through which system performance is evaluated, thereby influencing decisions. Once selected, these indicators require robust data collection and accurate calculations, which form the basis for informed decision-making procedures (Bell and Morse 2003).

S. Küfeoğlu, *Financing the Green Transition*, Sustainable Development Goals Series,
https://doi.org/10.1007/978-3-032-10440-3_3

Moreover, synthesising sustainability's environmental, economic, and social components into a comprehensive evaluation is daunting, yet it is vital for incorporating scientific knowledge into decision-making processes. Environmental consequences, often multidimensional and linked with societal and economic factors, are the principal pressures leading to departures from sustainability. Given their complexity, environmental indicators demand a thorough investigation, offering insights into the systematic approaches required for handling intricate data (Sikdar et al. 2017). The challenge of promoting sustainable development through environmental policies also requires rapid socio-technical transitions at a systemic level. Indicators are essential tools for providing feedback to policymakers, identifying and analysing unsustainable transitions, and exploring pathways towards achieving desired social change (Lehtonen et al. 2016).

When discussing sustainability measurement, it is essential to note the difference between indicators and metrics. Figure 3.1 illustrates these differences across four areas.

As shown, indicators are measures or signals used to monitor qualitative or quantitative changes. They function as guideposts, highlighting areas that require focused attention. Conversely, metrics are precise measures or calculations used to quantify the performance associated with each indicator. In the context of sustainability measurement, they are tools that enable measuring, comparing, and documenting progress in sustainable growth (Pertceva and Zyulyaeva 2019). For instance, greenhouse gas emissions may be an indicator, showing the level of emissions that affect climate change. Then, a metric could be tonnes of CO_2 emitted per year, as it provides the exact amount of greenhouse gases released. Organisations must translate indicators into specific, quantifiable metrics to effec-

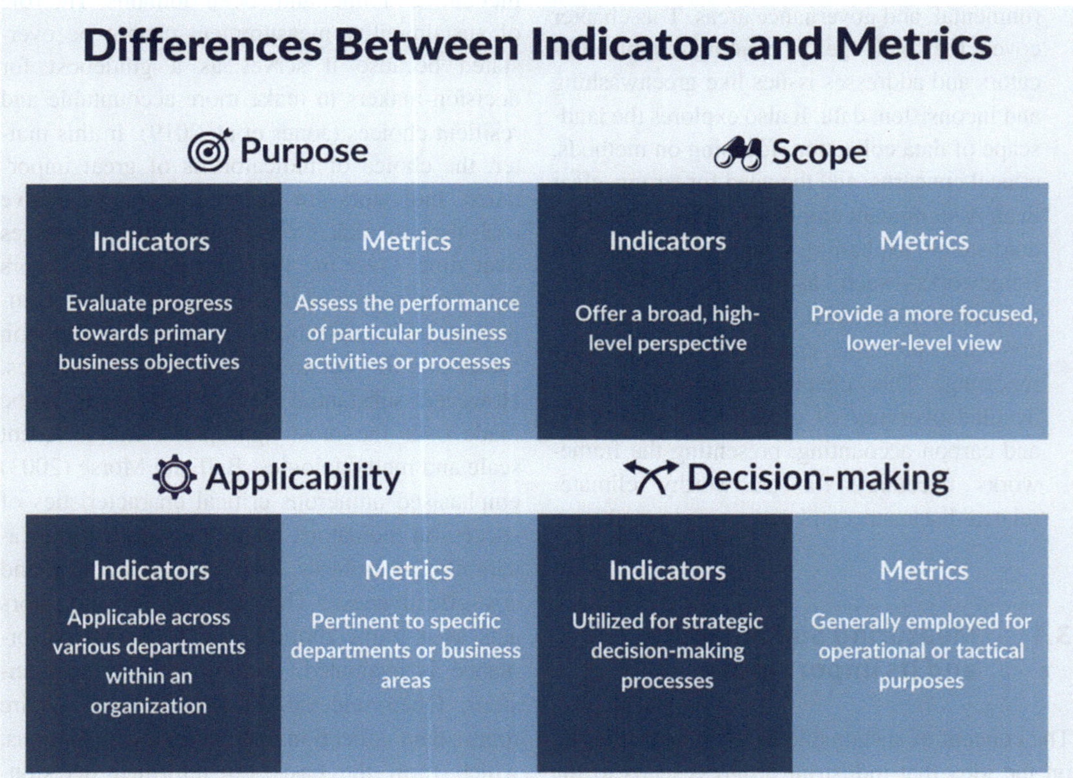

Fig. 3.1 Differences between indicators and metrics. (RIB Software 2024)

tively monitor progress towards sustainability goals. The following section provides a more detailed analysis of indicator and metric definitions.

The pursuit of measuring and quantifying sustainability has evolved remarkably over the past five decades. Figure 3.2 offers a chronological summary of the critical turning points in developing sustainable assessment measures. It draws attention to how our knowledge of sustainability is changing, as evidenced by essential studies, global accords, and the creation of institutions and frameworks devoted to incorporating environmental factors into decisions about the economy and politics. The timeline follows the development of increasingly complex and inclusive methods for quantifying and overseeing sustainable development, starting with the early conversations on sustainability in economic literature.

3.2 Understanding Sustainability Metrics

3.2.1 Defining Metrics for Sustainability Development

Sustainable development has become widespread over time and has been subject to various interpretations and analyses. It is a common misconception to understand sustainability and sustainable development as interchangeable or synonymous, but these concepts differ (Hopwood et al. 2005). Sustainability represents the final objective or endpoint of a process known as sustainable development (Mebratu 1998). The international community has recognised the environmental, social, and economic dimensions of sustainable development at Agenda 21 and the UN World Summit on Sustainable Development (WSSD) (UNCED 1992). Sustainable development ensures a harmonious balance between environmental integrity, social well-being, and economic prosperity. Therefore, achieving sustainable development requires the integration of dimensions into decision-making frameworks (OECD 2004; Hák et al. 2012). The environmental perspective mainly emphasises biodiversity and ecological health. The economic perspective aims for financial and natural accumulation and sustainable use (Acemoglu and Robinson 2013). The social perspective seeks to enhance cultural, political, religious, health, and education systems to ensure the dignity and welfare of individuals. This assertion suggests that sustainable development intrinsically upholds the principle of intergenerational justice, considering both immediate and enduring sustainability outcomes (Acemoglu and Robinson 2013).

3.2.2 Integrating Indicators: Measuring Progress in Key Dimensions

A framework that integrates various indicators is vital for effectively monitoring and tracking sustainable development. The triple bottom line (TBL) sustainability concept, introduced by Elkington (1994), is a framework that aims to evaluate sustainability performance by investigating environmental, social, and economic dimensions. The TBL framework was accepted and widely used across various sectors, from businesses to governments and non-profit organisations. Its popularity stemmed from its comprehensive approach to sustainability, which went beyond the traditional focus on financial performance. By incorporating environmental and social aspects alongside economic considerations, the TBL framework provided a more holistic view of an organisation's impact and performance. Hoque et al. (2020) conducted a survey and examined four environmental, four social, and three economic indicators to assess their importance. They gathered 30 responses, evenly distributed among academia, industry, and government (10 each). Indicators deemed important by at least 50% of experts were selected, with an option for suggesting new ones. Table 3.1 illustrates these indicators, their corresponding units of measurement, and the percentage of survey respondents who rated each indicator as important.

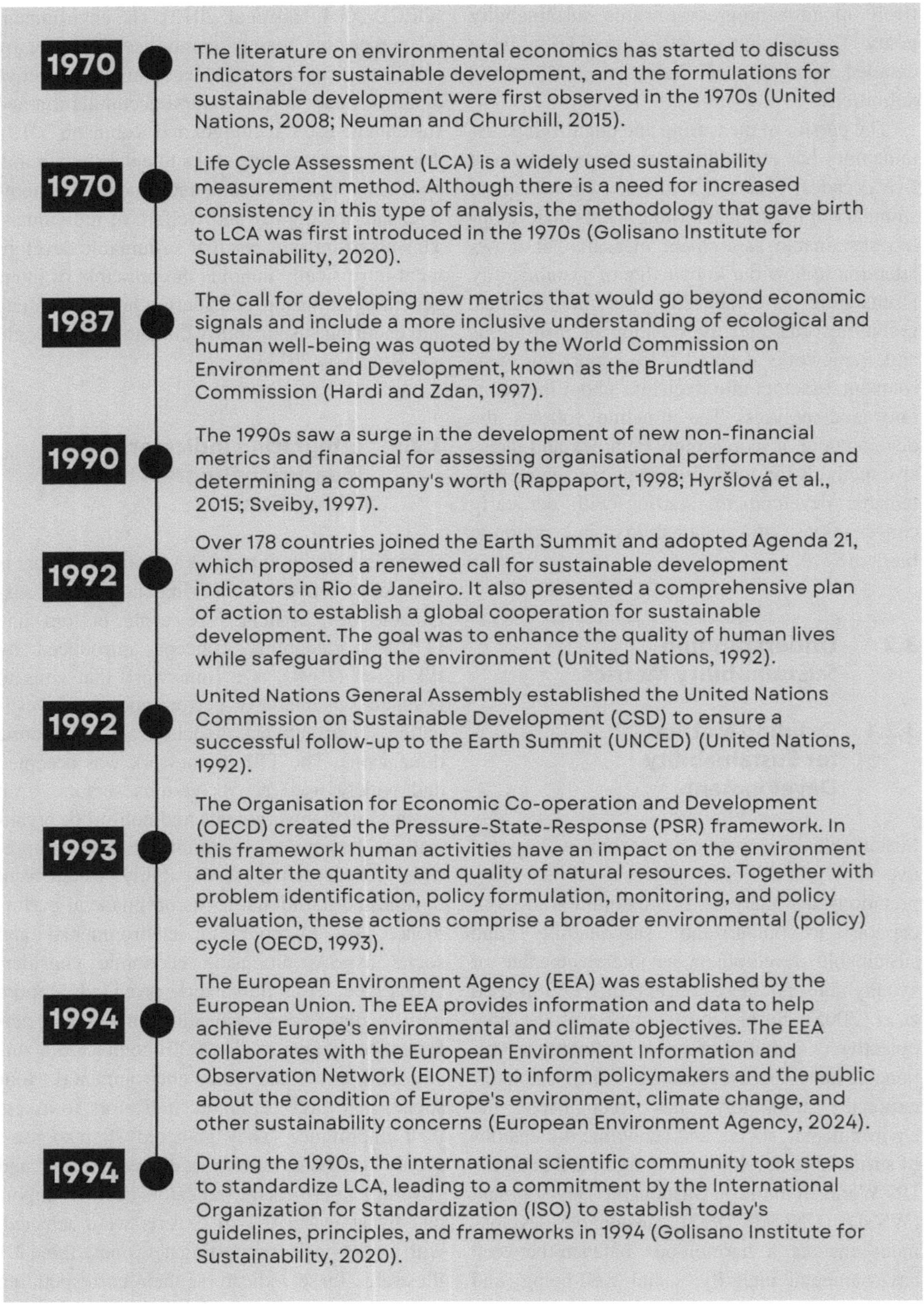

Fig. 3.2 History of sustainable measurement

- **1996** The International Organization for Standardization (ISO) published ISO 14001, the first standard in this series, which lays out the specifications for an environmental management system (EMS). The goal of ISO 14001 is to assist businesses in handling their environmental obligations in a methodical way that advances the environmental pillar of sustainability (ISO, 1996)
- **1997** The Global Reporting Initiative (GRI) was founded in Boston by the Tellus Institute, the Coalition for Environmentally Responsible Economies (CERES), and the United Nations Environment Program (UNEP, 2014; GRI, 2022).
- **1997** The Bellagio Principles were created by a global group of measurement and assessment experts to provide direction and encourage best practices towards sustainable development (The Global Development Research Center, 2024).
- **1999** The DJSI (Dow Jones Sustainability Index), was launched to track the performance of world's largest corporations in terms of sustainability (S&P Global, 2024)
- **1999** European Environment Agency developed the Driver-Pressure-State-Impact-Response framework to report environmental issues (European Environment Agency, 1999).
- **2000** Carbon Disclosure Project (CDP) was founded seeking disclosure from companies of their climate impact (CDP, 2024).
- **2000** The initial edition of the GRI Guidelines (G1), which later became known as the first global framework for sustainability reporting, was published (GRI, 2000).
- **2002** The World Summit on Sustainable Development, held in Johannesburg, South Africa, strongly endorsed the complete implementation of Agenda 21 (United Nations, 2008).
- **2002** To encourage the application of life cycle assessment (LCA), the United Nations Environment Programme (UNEP) and the Society for Environmental Toxicology and Chemistry (SETAC) established the Life Cycle Initiative (Golisano Institute for Sustainability, 2020).
- **2006** Committee on Sustainable Assessment was founded aiming to thoroughly evaluate the environmental, economic, and social aspects of sustainability (Committee on Sustainable Assessment, 2024).
- **2009** The Conference of European Statisticians (CES) established a collective comprising the United Nations Economic Commission for Europe (UNECE), European Commission (Eurostat) and OECD. This collective seeks to standardize the various metrics used to quantify sustainable development (United Nations, 2014).
- **2010** The International Integrated Reporting Council (IIRC) was founded with global alliance of corporations, legislators, entrepreneurs and accounting professionals (IIRC, 2024).

Fig. 3.2 (continued)

Fig. 3.2 (continued)

Table 3.1 Selected TBL indicators and their significance (Hoque et al. 2020)

Sustainability dimension	Indicator	Unit	Percentage of participants deeming it significant
Environmental	Global Warming Potential (GWP)	Kg CO_2/VKT	87%
	Fossil Fuel Depletion (FFD)	MJ/VKT	67%
	Water Consumption (WC)	m^3/VKT	63%
	Land Use (LU)	Ha.a/VKT	53%
Social	Local job creation	Man.Hour/VKT	67%
	Conservation of Fossil Fuel (CFF)	MJ/VKT	50%
	Occupational Health and Safety (OHAS)	Qualitative	50%
	Human Health Based on Vehicle Exhaust Emission (HHver)	Gm/VKT	86%
Economic	Life Cycle Costing (LCC)	AUD/VKT	83%
	Carbon Reduction Credit (CRC)	AUD/VKT	57%
	Net benefit	AUD/VKT	70%

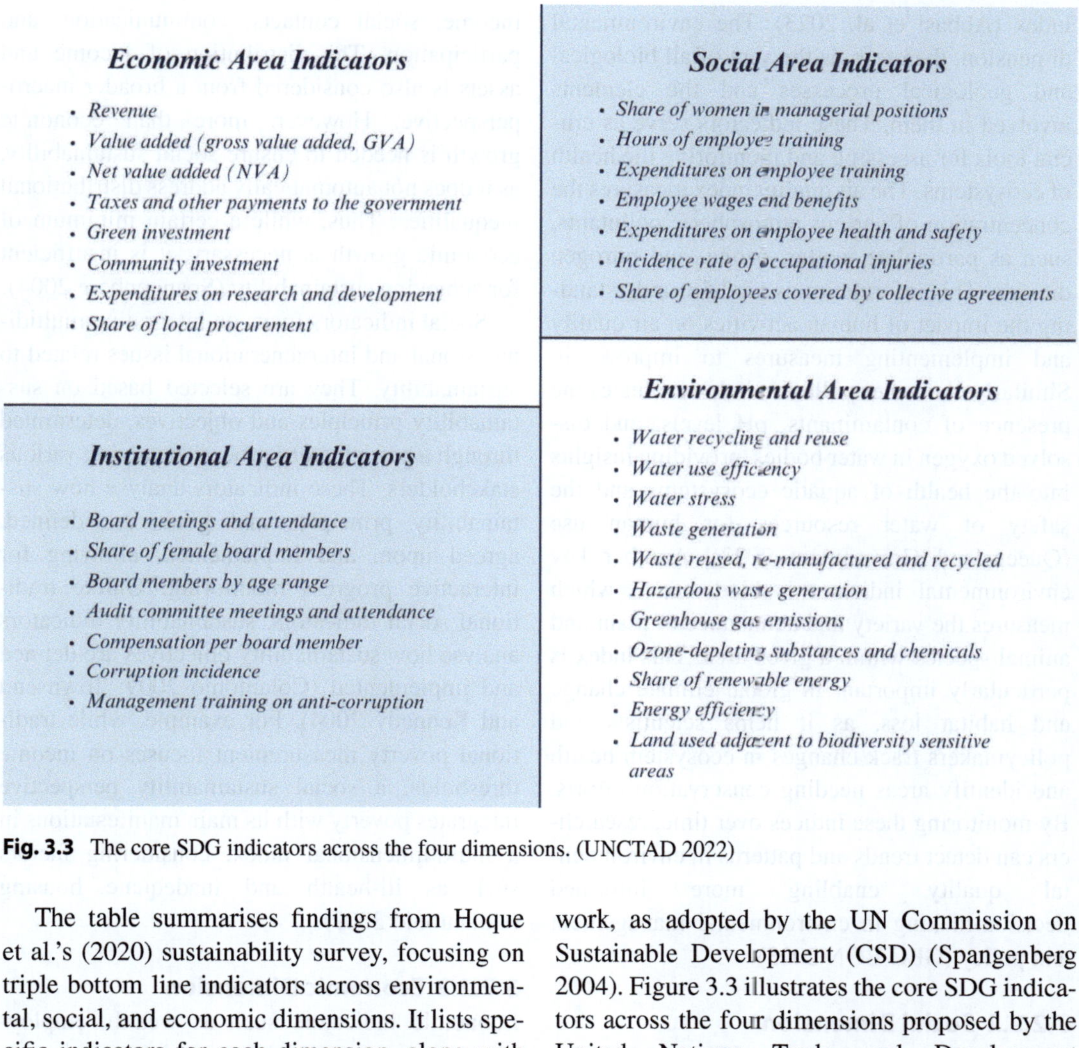

Fig. 3.3 The core SDG indicators across the four dimensions. (UNCTAD 2022)

The table summarises findings from Hoque et al.'s (2020) sustainability survey, focusing on triple bottom line indicators across environmental, social, and economic dimensions. It lists specific indicators for each dimension, along with their units of measurement and the percentage of survey participants who deemed them significant. Key environmental indicators include Global Warming Potential and Fossil Fuel Depletion. Social indicators cover aspects like Local Job Creation and Human Health, while economic indicators include Life Cycle Costing and Carbon Reduction Credit. The percentages reveal which indicators experts considered most important, with Global Warming Potential (87%), Human Health Based on Vehicle Exhaust Emission (86%), and Life Cycle Costing (83%) ranking highest in their respective categories.

The TBL framework was later expanded to include the institutional dimension, the fourth component of the sustainability indicators framework, as adopted by the UN Commission on Sustainable Development (CSD) (Spangenberg 2004). Figure 3.3 illustrates the core SDG indicators across the four dimensions proposed by the United Nations Trade and Development (UNCTAD):

Figure 3.3 shows a comprehensive framework for evaluating performance across four key dimensions: Economic, Social, Institutional, and Environmental. Each dimension includes specific indicators that can be used to measure success and identify areas for improvement. A more detailed explanation of these dimensions is given in the paragraphs below:

3.2.2.1 Environmental Dimension

Environmental indicators, which are measurements used to quantify different types of pollution affecting locally or globally and their effects on the environment, can include the air quality index, water pollution index, and biodiversity index (Abbasi et al. 2023). The environmental dimension, therefore, is the sum of all biological and geological processes and the elements involved in them. These indicators serve as crucial tools for assessing and monitoring the health of ecosystems. The air quality index measures the concentration of various atmospheric pollutants, such as particulate matter, ozone, and nitrogen dioxide. This information is vital for understanding the impact of human activities on air quality and implementing measures to improve it. Similarly, the water pollution index evaluates the presence of contaminants, pH levels, and dissolved oxygen in water bodies, providing insights into the health of aquatic ecosystems and the safety of water resources for human use (Queensland Government 2022). Another key environmental indicator is biodiversity, which measures the variety and abundance of plant and animal species within a given area. This index is particularly important in global climate change and habitat loss, as it helps scientists and policymakers track changes in ecosystem health and identify areas needing conservation efforts. By monitoring these indices over time, researchers can detect trends and patterns in environmental quality, enabling more informed decision-making in environmental management and policy (GEO BON 2023).

3.2.2.2 Social Dimension

Social indicators assess the effects on human well-being, fairness, community liveability, historical and cultural heritage, and aesthetics (Ajmal et al. 2017). This includes access to knowledge, information, and experience, such as fair education, opportunities to work and participate in social processes, access to technology, and the ability to select and transform information into relevant knowledge. Human capital is made up of human beings' personal qualities, such as their skills, dedication, and experiences (Hák et al. 2007). These indicators help organisations address important aspects of social sustainability, ensuring fairness, well-being, and inclusivity in their operations (UNCTAD 2022).

At the micro-level, the core elements of social sustainability include education, training, income, social contacts, communication, and participation. The distribution of income and assets is also considered from a broader macro-perspective. However, more than economic growth is needed to ensure social sustainability, as it does not automatically address distributional inequalities. Thus, while a certain minimum of economic growth is necessary, it is insufficient for achieving sustainability (Spangenberg 2004).

Social indicators focus on integrating multidimensional and intergenerational issues related to sustainability. They are selected based on sustainability principles and objectives, determined through a participatory process involving various stakeholders. These indicators analyse how sustainability principles and goals are defined, agreed upon, and implemented, allowing for interactive progress monitoring. Unlike traditional social indicators, sustainability indicators analyse how sustainability objectives are defined and implemented (Colantonio 2009; Townsend and Kennedy 2004). For example, while traditional poverty measurement focuses on income thresholds, a social sustainability perspective integrates poverty with its main manifestations in a multidimensional index, considering factors such as ill-health and inadequate housing (Colantonio 2009).

3.2.2.3 Economic Dimension

Economic indicators such as wealth, employment, income, and high productivity are important in helping societies assess economic activities' impact and performance. These include individual well-being, a secure minimum income, and access to the sociocultural system that allows people to participate actively in society (Clune and Zehnder 2018). These measurements guide efforts to ensure economic growth and contribute positively to sustainable goals.

Furthermore, the economic dimension includes measures that ensure equitable distribution of resources and opportunities, fostering inclusive growth. Every individual must have an equal opportunity to access the benefits of the economy and contribute towards wealth generation through different forms of work at a national level. This opportunity should be open to all, with access to finance without barriers to entry and on non-discriminatory credit terms. Participation in the national and international economies should be accessible to all, with the removal of any barriers that may exist (Hák et al. 2007). Indicators that assess income distribution, employment rates, and access to essential services are important. Ensuring that economic benefits are distributed fairly across different segments of society helps reduce inequalities and promotes social cohesion (Al-Thani and Koç 2024).

Economic indicators provide an understanding of the stability and sustainability of an entity. Revenue measures the total income generated from the sale of goods or services, which reflects economic activity and growth potential. In contrast, value-added and net value-added represent the wealth created by an entity. The metrics assessing wealth creation indicate contributions to the economy and potential wealth distribution among stakeholders (UNCTAD 2022). Understanding an entity's public finance contributions by assessing taxes and other government payments helps support public services and infrastructure. Green investment indicates an entity's commitment to environmentally sustainable practices, reflecting how economic activities contribute to environmental preservation. Community investment reflects the extent to which an entity contributes to social initiatives, illustrating its role in promoting community development and social well-being (UNCTAD 2022).

3.2.2.4 Institutional Dimension

The institutional dimension refers to the role of institutions in providing the necessary mechanisms to influence action and change in any pillar. A participatory political system has several components that ensure equal access and participation. These include the right to participate in political systems, fair access to justice, non-discriminatory social security systems, and legal provisions that guarantee access to environmental, social, and economic resources (Hák et al. 2007). Institutions are formed through interpersonal processes such as cooperation and communication. These processes lead to the interaction of individuals in a society, which includes knowledge and rules (Spangenberg 2004). A political institution's role gains importance by governing decision-making and implementation. They include mechanisms for decision-making and rule enforcement but not societal values. Fundamental institutional objectives must complement sustainability characteristics, such as effective justice, compliance, or participation (Spangenberg 2004).

Institutional indicators are essential for evaluating the institutional effectiveness in promoting sustainable development. Key institutional indicators include governance and policy frameworks, institutional capacity and resources, social capital and community engagement, and legal and institutional mechanisms. For example, having national sustainable development strategies and spending on research and development (R&D) as a percentage of gross domestic product (GDP) are important institutional indicators. These metrics show how well institutions integrate sustainability into their policies and practices, ensuring transparent, inclusive, and accountable decision-making processes (Spangenberg 2002). These indicators reflect a commitment to ethical practices and high standards of governance. By systematically evaluating these indicators, institutions can identify strengths and areas for improvement, enhancing their contributions to sustainable development.

3.2.3 Interlinkages Between the Four Dimensions

While an indicator for a single dimension could be an important tool for tracking sustainable development, some important conceptual deficiencies still need to be resolved. For example, there needs to be more certainty about whether the social dimension alone ensures or hinders sustainable social development due to its complex structure (Moldan et al. 2007). Some concepts, such as social capital, can be noted as an aspect of the social dimension. However, there has yet to be an explicit representation of the indicators. For the institutional dimension, only a few indicators have been found in this sector (Hák et al. 2012). Economic indicators, however, do not include measurements regarding long-term sustainability in the economic system (Moldan et al. 2007). All these one-dimensional approaches are inadequate because they address sustainability in isolation and do not fully address sustainable development. Identifying and analysing the trade-offs between achieving high levels of human welfare and ecosystem protection and restoration more holistically is important. However, there are also studies in the literature that have investigated the interlinkages between dimensions and the effects of these interlinkages.

According to Nondo (2009), the relationship between environmental regulation and economic growth was revealed by analysing simultaneous relationships between population, employment, per capita income, and changes in environmental regulations for 410 counties in the Appalachian region of the USA. In this regard, a four-equation regional growth model was used. The study found that initial conditions for environmental regulation had a negative impact on regional growth of population, per capita income, and total employment. Therefore, when firms initially invest in abatement capital, productivity (including labour demand) tends to fall, but a gradual increase in environmental quality will offset this (Nondo 2009). It is concluded that shifting resources from production and investment activities to pollution abatement is inadvertently transferred to other sectors of the economy, thus causing regional growth to slow down. This indicates a trade-off between environmental quality and economic growth in the short run. However, the study also found robust evidence that, over the longer run, changes in environmental regulations positively influenced changes in population, total employment, and per capita income, offering a promising outlook for the future (Nondo 2009).

In their study, Borhan and Elsadig (2010) investigate the relationship between economic growth and different air and water pollution indicators in Malaysia. They analyse air pollution using indicators such as nitrogen dioxide (NO_2), sulfur dioxide (SO_2), carbon monoxide (CO), ozone (O_3), and particulate matter (PM10). For water pollution, they considered biochemical oxygen demand (BOD), cadmium, and arsenic levels. They measured GDP per capita quarterly from 1996 to 2006 and formulated a four-equation simultaneous model. As a result, the study's findings supported the Environmental Kuznets Curve (EKC) hypothesis for SO_2 and PM10. The EKC hypothesis suggests that environmental degradation increases as the economy grows until a certain point is reached (Kostakis et al. 2023). According to Grossman and Krueger (1995), the relationship between economic growth and environmental degradation is not linear but follows an inverted U-shaped curve. This means that environmental degradation worsens during the early stages of a country's per capita income, but as economic growth continues, environmental degradation decreases.

According to Gangadharan and Valenzuela (2001), integrated health outcomes with economic growth and environmental quality are investigated in an extended EKC framework for a cross-section of countries widely dispersed on the economic development spectrum. Their results showed that environmental stress significantly negatively affects health status, while gross national product varies positively with health status. The emission of air pollutants, organic water pollutants, commercial energy use, and deforestation represented environmental quality or stress in the model. At the same time, health outcomes were measured through life expectancy and infant mortality rates. In their model, environmental stress and the population's health status are the only two endogenous variables. In contrast, economic growth, non-income factors affecting the environment, and health factors are treated as exogenous variables (Gangadharan and Valenzuela 2001).

Sustainable development is an integrated system with multiple linkages and feedback loops. Decision-makers need indicators to build mutually reinforcing links between dimensions, but defining and quantifying these links remains challenging (Moldan et al. 2007). An integrated and holistic approach or "whole systems" approach is required, encompassing complex dimensions' emergent properties and interactions. In Fig. 3.4, some selected indicators are examples of dimensional indicators used for measurements of dimensional interconnections:

3.2.3.1 Dow Jones Sustainability World Index (DJSI)

Sustainability techniques affect accounting and performance indicators, requiring adjustments to resource consumption standards (López et al. 2007). A sustainability index called the Dow Jones Sustainability World Index (DJSI) ranks the most sustainable businesses in each industry by assessing corporations according to social, environmental, and economic standards (Gond et al. 2011). Companies listed in sustainability indices include data in their reports that reflect sustainability standards (Durand et al. 2019). The DJSI overviews a corporation's sustainability tasks (Hawn et al. 2018). It examines environmental control, resource performance, social impact, and governance techniques (Friede et al. 2015). For instance, agencies are assessed based on their usual environmental performance, emission discounts, waste control, and biodiversity safety (Hawn et al. 2018). They are also evaluated on their environmental initiatives and how they deal with sustainability issues (Dong et al. 2017). The DJSI additionally examines how corporations engage with their stakeholders, such as personnel, groups, clients, and providers (Denuwara et al. 2022). Labour practices, human rights, network participation, and product duty are all considered. This ensures that businesses are assessed not simply on their monetary success but also on their societal impact (Friede et al. 2015). The DJSI governance component evaluates corporations' control structures, board variety, transparency, and ethical company

Interlinkage Indicators	
Environmental-economic	• Resource productivity (GDP/TMI) • Transport intensity
Socio-economic	• Labour productivity (production per capita) • Income distribution per decentile
Socio-environmental	• Environmental health problems • Access to common goods (to be specified regionally)
Economic-institutional	• Corruption rate (Transparency International Index) • Share of taxes on labour, capital and the environment in total tax revenues
Socio-institutional	• Co-decision rights of workers • Reliability of the healthcare and social security system
Environmental-institutional	• NGO right to file suit • Freedom of information

Fig. 3.4 The interlinkage indicators

behaviour. Good company governance ensures duty, integrity, and long-term viability (Durand et al. 2019). By analysing the group's governance structures, the DJSI assists investors in identifying agencies that adhere to robust company ethics and integrity (Durand et al. 2019). The DJSI is a useful resource for traders, giving insights into firms' sustainability, overall performance, and positioning (Garcia-Blandon et al. 2023).

3.2.3.2 Morgan Stanley Capital International Sustainability Index

MSCI (Morgan Stanley Capital International) Inc. produced the MSCI Sustainability Index, a crucial instrument for sustainable investment (Widyawati 2019). It combines advanced analytics and data-driven techniques to assist investors in incorporating sustainability concerns into their investing plans (Jain et al. 2019). The index evaluates companies based on environmental, social, and governance parameters (Friede et al. 2015). Carbon emissions, labour rights, board diversity, and supply chain management are all considered (Hawn et al. 2018). It examines how businesses handle these areas and how they comply with international standards. The MSCI Sustainability Index fosters openness and accountability, encouraging companies to improve their sustainability performance (Garcia-Blandon et al. 2023). These frameworks help investors with their decision-making process and represent a company's commitment to creating long-term value by following ESG principles (Jain et al. 2019).

3.2.4 Metric Construction

Metrics created by the combination of indicators allow investors, policymakers, and companies to measure and report sustainability and help them make informed decisions on this issue. Therefore, metrics are the language of measurements to be made and facilitate communication by simplifying a complex reality (Smeets and Weterings 1999). Constructing a sustainability metric diverges from the conventional trajectory of development, which follows a linear path and ends with a developed product, as mentioned in Energy Efficiency Indicators: Fundamentals on Statistics (IEA 2014). Developing indicators should not be seen as an end in itself but more as a beginning for further use. It's an ongoing process, a feedback loop that constantly revises and refines itself. First, the metric is formulated, followed by establishing a feedback loop that continually refines the metric by taking qualitative and quantitative inputs, even changing it as a whole whenever a better-performing, better-measuring metric is available for the defined goal. Figure 3.5 illustrates the procedure to construct a sustainability metric.

The construction of sustainability metrics involves a systematic approach beginning with determining the organisation's most important social and environmental effects, as shown in the figure above. This initial step requires comprehensive assessments to identify key sustainability-related impacts, including waste management, resource use, emissions, labour practices, inclusivity, biodiversity impact, and community well-being (Williams 2023).

The selection of metrics is crucial, as they must be easily measured, sensitive to system stresses, respond predictably, be anticipatory, and integrative (Dale et al. 2015; Cairns et al. 1993). The process continues with implementing data collection mechanisms, conducting data analysis to identify patterns and areas for improvement (Williams 2023), and consulting with stakeholders. Stakeholder involvement is critical for maintaining the metrics' relevance and progress, ensuring alignment of objectives and public expectations (Dale et al. 2015). This process is iterative, with stakeholder input being used to refine the metrics by returning to determining sustainability objectives, creating a continuous cycle of improvement and adaptation.

3.3 Challenges in Measuring Sustainability

Measuring sustainability is a complicated process fraught with political, social, and technical challenges. This section explores the various bar-

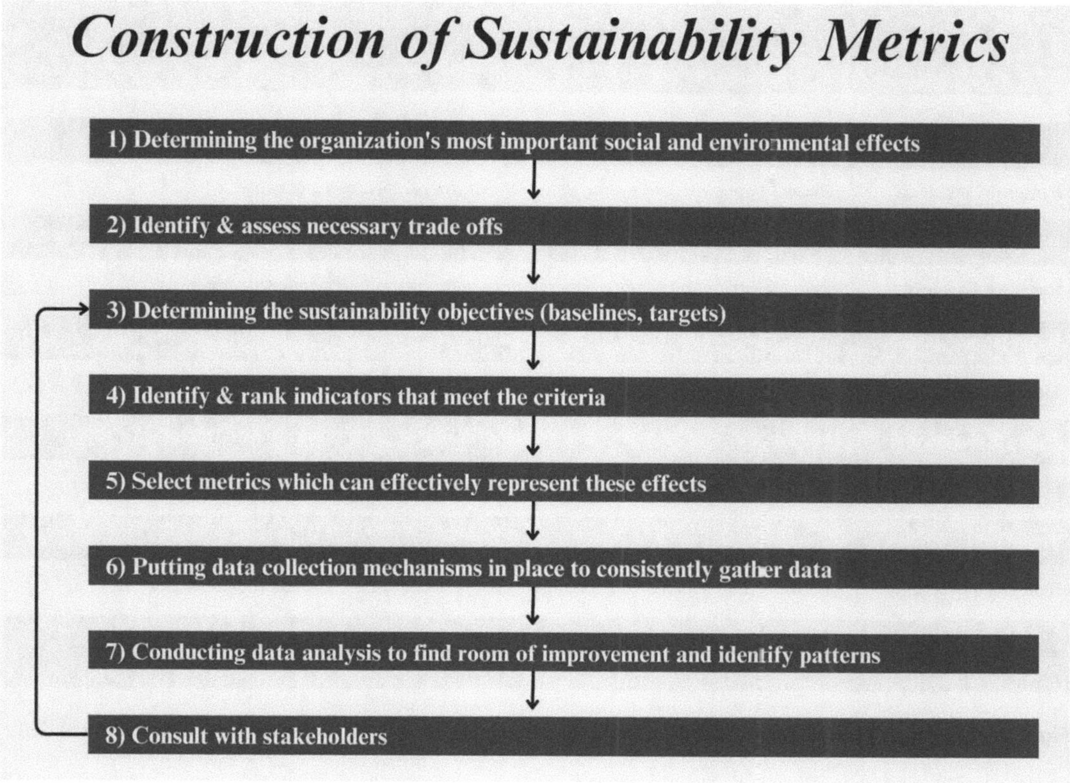

Fig. 3.5 Process of constructing a sustainability metric

riers to successfully tracking sustainability measures and highlights the significance of these issues for research, practice, and policy.

3.3.1 Overview of Sustainability Indicators Complexities

The interpretation of indicators used to assess sustainability is challenging due to the concept's complexity. One challenge is that effective policymaking and planning sometimes overlook the interconnectivity of the components of sustainable development, even though most indicators address each component separately. Consequently, developing indicators that accurately depict these interconnections presents a substantial challenge.

Another issue stems from determining the longevity and efficiency of human-environmental systems, a process complicated by political, philosophical, and cultural variations. To address this complexity, combining systematic scientific approaches that allow for multiple models reflecting different human systems and cultures is essential. These models assess the sensitivity, robustness, and transitions of complex systems. Sustainability indicators must include fundamental characteristics and demonstrate scientifically validated methods for maintaining or improving system functionality. The goal is to identify irreversible thresholds beyond which recovery is difficult, considering variables such as loss criticality, substitution feasibility, compensatory methods, and uncertainties that necessitate cautious responses (Hák et al. 2012). Aside from what is mentioned above, there are many different challenges in measuring sustainability. An overview of these challenges is given in Fig. 3.6.

Challenges in Measuring Sustainability

Tool Overload:	*The wide range of sustainability assessment methods used globally*
Cultural Complexity:	*Acknowledging and integrating different cultural perspectives and values*
Interdependence:	*The complexity and interdependence of factors in measuring sustainable innovation performance*
Resource Fairness:	*The fair distribution of resources and opportunities through democratic processes*
Inconsistent Metrics:	*Discrepancies in the evaluation of business performances*
Data Gaps:	*The factor for creating reliable indicators, determining their quality and effectiveness*
Policy Relevance:	*The importance of indicators in informing decision-makers throughout the policy life cycle*
Scales Variability:	*The different timescales and conceptual structures used to measure sustainability*
Temporal Dynamics:	*The realisation and the evaluation of sustainable efforts and innovations are determined by time*
Indicator Diversity:	*Quantitative and qualitative measurements, including proxy indicators and indexes*
Stakeholder Engagement:	*Tailoring indicators to target audiences and involving them in the design process ensures relevance*

Fig. 3.6 Challenges in measuring sustainability. (World Economic Forum 2017; Dempsey et al. 2011; Hák et al. 2012)

3.3.2 Greenwashing

Since environmental contamination has worsened, many stakeholders and enterprises around the world have become more aware of environmental concerns as environmental issues and public awareness have grown (Chen and Chang 2012). As a result of this growing knowledge, stakeholders such as governments, investors, consumers, and corporate clients have put more pressure on businesses during the last decade to disclose their environmental performance and produce environmentally friendly products. Under this pressure, for example, enterprises in the energy industry face increasing demand from many stakeholders to deliver clean energy and sustainable products (Vollero et al. 2016). Companies must create green marketing strategies in response to the growing consumer demand to demonstrate their social responsibility and positive business image (Zhang et al. 2018; de Freitas Netto et al. 2020). According to Delmas and Burbano (2011), the green market is expanding daily. The phenomenon of "greenwashing" appears in tandem with the growth of green markets (Majláth 2017).

Making up information or misrepresenting a product's or behaviour's environmental benefits is known as greenwashing. Environmentalist Jay Westervelt first used the phrase "greenwashing" in 1986. He wrote a piece for publication discussing the hospitality sector's initiatives to encourage towel reuse (de Freitas Netto et al. 2020). Companies may use it to abuse the system or make money off of well-meaning, environmentally conscious customers while expanding or maintaining their damaging and polluting practices (Lindwall 2023).

Several businesses spend money on green marketing materials to be seen as socially conscious and environmentally conscious. To improve purchasing intentions and brand perceptions, they use advertising and corporate social responsibility (CSR) (Nyilasy et al. 2013). The truth about corporate environmentalism, meanwhile, can sometimes be disheartening. In Canada and the USA, 95% of products labelled as green did at least one of the so-called sins of greenwashing, which range from the iniquity of worshipping false labels to the iniquity of concealed trade-off (de Freitas Netto et al. 2020).

Fig. 3.7 Signs of greenwashing. (Moran and O'Neill 2022; Blesserholt 2021; Faber and Sick 2022)

Advertising agency Ogilvy & Mather claims that greenwashing methods have become widespread in the last few decades (Hsu 2011). Customers need help distinguishing between genuine green claims due to the rise in green marketplaces and greenwashing, which has created a trust issue (Nyilasy et al. 2013). Strähle and Müller (2016) have devised "The Seven Sins of Greenwashing", a list of seven specific allegations businesses can be charged with. The word "sin" has a strong connotation that draws attention to the unethical or unlawful behaviour of companies that engage in greenwashing. This definition can be unsettling because of the word's harshness. The following list of "signs"'in Fig. 3.7 might be used to start conversations about environmental pollution and labour exploitation (Strähle and Müller 2016; Blesserholt 2021).

This figure presents 13 key signs of greenwashing, providing a comprehensive guide for identifying deceptive environmental claims. These signs range from hidden trade-offs and vague language to more blatant tactics like fraudulent claims and fibbing. "Fibbing" in this context refers to companies presenting false or misleading information about their environmental practices or product benefits. The figure also highlights subtler forms of greenwashing, such as highlighting irrelevant features or presenting a "lesser evil" option within an unsustainable industry. "Scare tactics" is another noteworthy sign where companies exaggerate environmental threats to promote their supposedly green products or services. Other important indicators include unfounded expectations, concealed risks, and the prioritisation of profit over genuine environmental protection. By understanding these signs, consumers, investors, and regulators can better distinguish between authentic environmental initiatives and deceptive marketing strategies, ultimately promoting true corporate sustainability and accountability. Building on this awareness, there are effective strategies to counter greenwashing, as illustrated in Fig. 3.8, which outlines approaches for both consumers and businesses to combat this issue.

By following these guidelines, both consumers and businesses can contribute to a more hon-

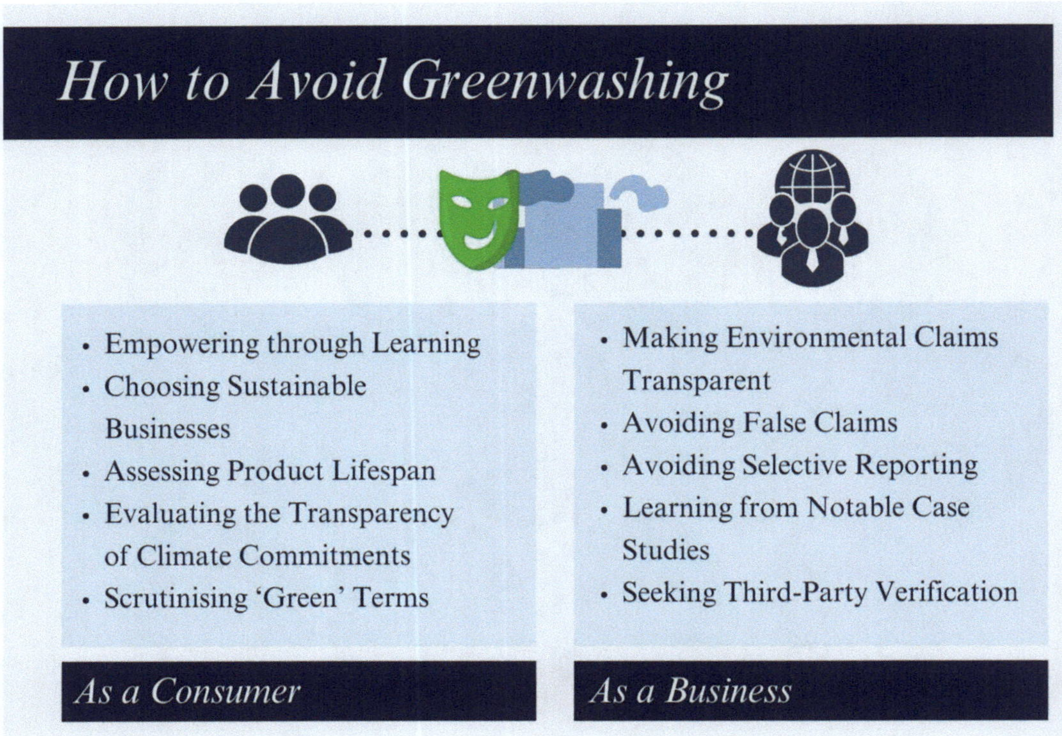

Fig. 3.8 How to avoid greenwashing. (Koons 2023; UN 2024; Mlaba 2021; IFC 2023)

est and effective approach to sustainability, reducing the prevalence of greenwashing in the marketplace. On the policymaker and regulators' side, there is an urgent need for more robust methods to scrutinise and verify corporate environmental claims. As sustainability becomes increasingly critical in business and investment, regulatory bodies must ensure that companies' green initiatives are genuine and impactful. This need has spurred research into developing more comprehensive and reliable evaluation methods.

3.4 Sustainability-Related Data Collection

3.4.1 The Definition and the Impact of Data Collection

Data collection is the process of gathering, measuring, and analysing information on a specific topic or area of interest (Case Western Reserve University 2023). In the context of sustainability, this refers to the collection and analysis of data related to environmental, social, and economic aspects of sustainable practices and outcomes. It is a category of data collection that involves measuring and monitoring indicators that express the impact of human activity on the planet and society and progress towards goals of sustainable development in general. Collecting this type of data facilitates the creation of more sustainable policies, business strategies, or individual choices related to sustainability. Many studies have observed that companies integrate sustainability-related data into their investment decision-making processes (Rahman and Homayoun 2020; Khan et al. 2016; Social Investment Forum 2012; Kim et al. 2012). Collecting sustainability-related data is crucial since it combats significant global challenges towards a more sustainable future. There are some key reasons why it is essential. Figure 3.9 illustrates the integration of sustainability-related data and why it is significant.

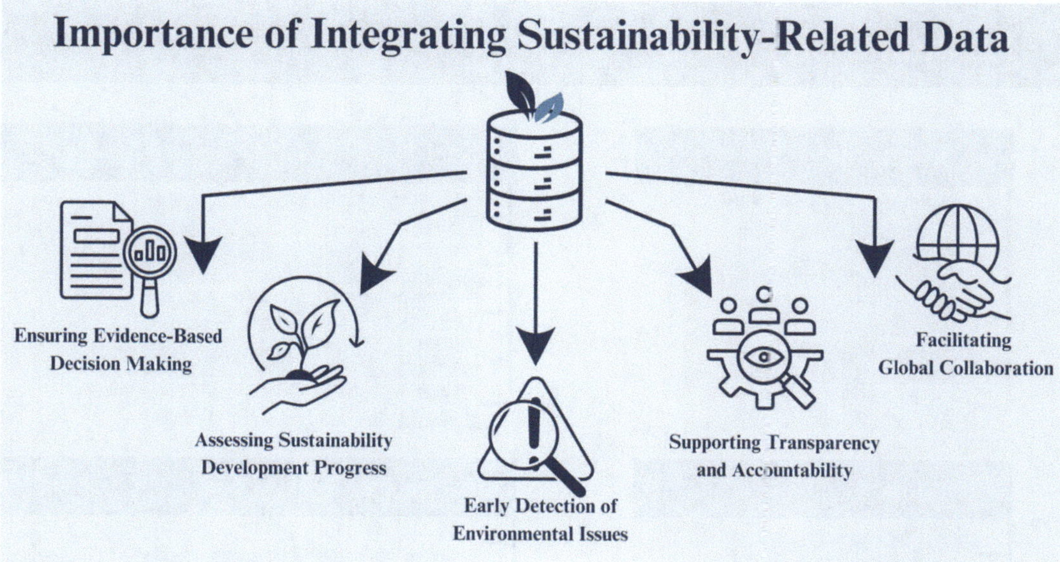

Fig. 3.9 Importance of integrating sustainability-related data. (Dissanayake et al. 2016; Zumente and Lāce 2021; Doni et al. 2022)

3.4.2 Data Collection Landscape

3.4.2.1 Types of Data

Data categorisation is a fundamental component of data management and analysis. It involves organising data into distinct groups based on specific criteria or characteristics. Categorising data is preferred because it enhances efficiency through better organisation (Thurmond 2023). There are numerous methods to categorise general data and also sustainability-related data. For example, researchers can organise general data into categories based on the intended type of use, such as operational, strategic, or analytical purposes. Additionally, data can be grouped according to user preferences, such as by topic, region, or time, depending on the specific needs of the analysis. Sustainability-related data spans a wide range of categories, reflecting the dimensions of sustainability itself. Figure 3.10 displays the categories of sustainability-related data and provides examples:

3.4.2.2 Volume of Data

Data collection has become a vital component of the digital landscape. It is growing exponentially, reaching an unprecedented scale. However, the amount of data created is accelerating even faster, making it nearly impossible to estimate the total amount of data created in human history (Duarte 2024). Some estimations on global data generation are given in Fig. 3.11:

These figures do not represent non-digital data, such as books and other physical documents. Those contribute a small portion of the global volume and tend to grow much slower than digital data. Therefore, this rapid growth can be mainly attributed to digital factors, such as:

- Increased Internet usage worldwide.
- Proliferation of Internet of Things (IoT) devices.
- Growth in social media usage and created online content.
- Advancements in data storage and processing capabilities.

The role of IoT devices in data creation and collection is particularly significant and will be explained in more detail further (Rivery 2023).

The volume of sustainability data has grown exponentially in recent years, reflecting the increasing global focus on environmental, social, and governance issues. This surge in data produc-

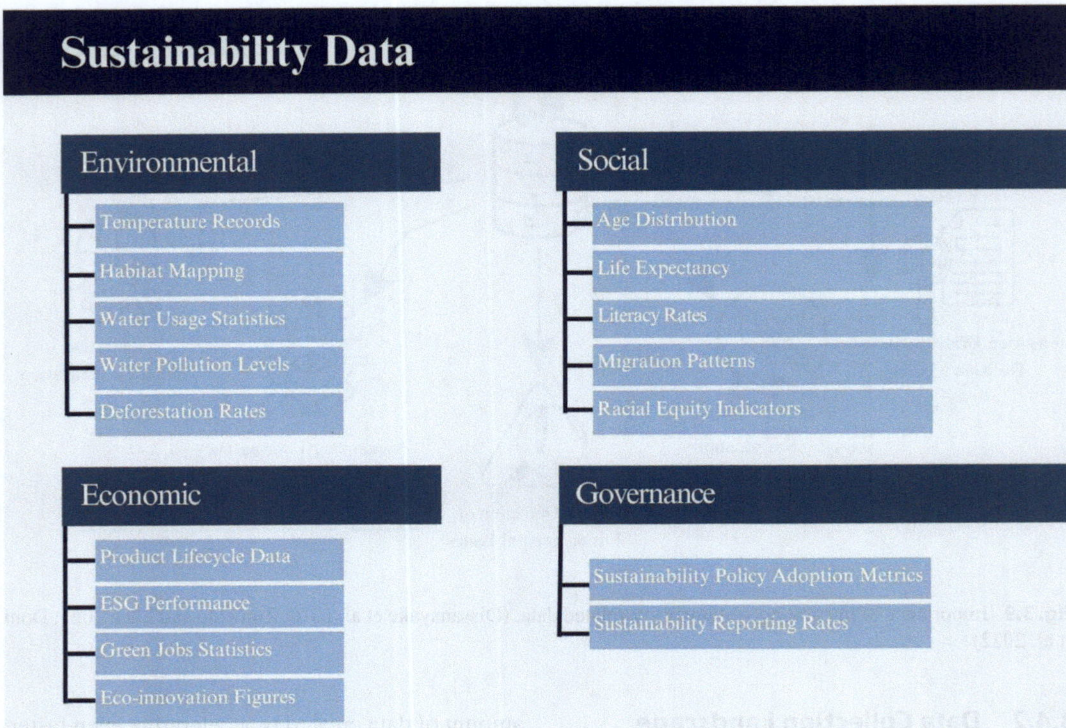

Fig. 3.10 Types of sustainability-related data

Fig. 3.11 Global data generation. (Duarte 2024; Marr 2024; Statista 2024)

tion is driven by heightened awareness, regulatory requirements, and technological advancements in data collection and analysis. Figure 3.12 illustrates this trend by showcasing the cumulative number of sustainability publications over time and their geographical distribution.

This sharp upward trend on the left shows the rapidly growing interest and investment in sustainability research and reporting across various sectors. On the right-hand side, this distribution highlights the global nature of sustainability efforts, with significant contributions coming from both developed and developing nations, though there is still a notable concentration in Western countries.

3.4.2.3 Internet of Things (IoT) in Data Collection

The Internet of Things (IoT) is a network of connected devices that may gather and share data without requiring human interaction (Atzori et al. 2010). IoT devices include sensors that collect data from their surroundings and send it to a central system for analysis and decision-making (Gubbi et al. 2013). These devices can range from modest home appliances to complex industrial equipment, all Internet-connected and communicate with one another (Borgia 2014). IoT includes a wide range of devices designed to collect data and automate various processes. These include smart home appliances that optimise domestic functions, wearable devices that monitor health and fitness, and industrial sensors that improve manufacturing processes. IoT technology also extends to environmental monitoring systems for conservation efforts, transportation devices that enhance logistics and supply chain management, and agricultural devices that enable precision farming (Küfeoğlu 2021a; Küfeoğlu 2021b). Given the vast array of applications across diverse sectors, the amount of data IoT

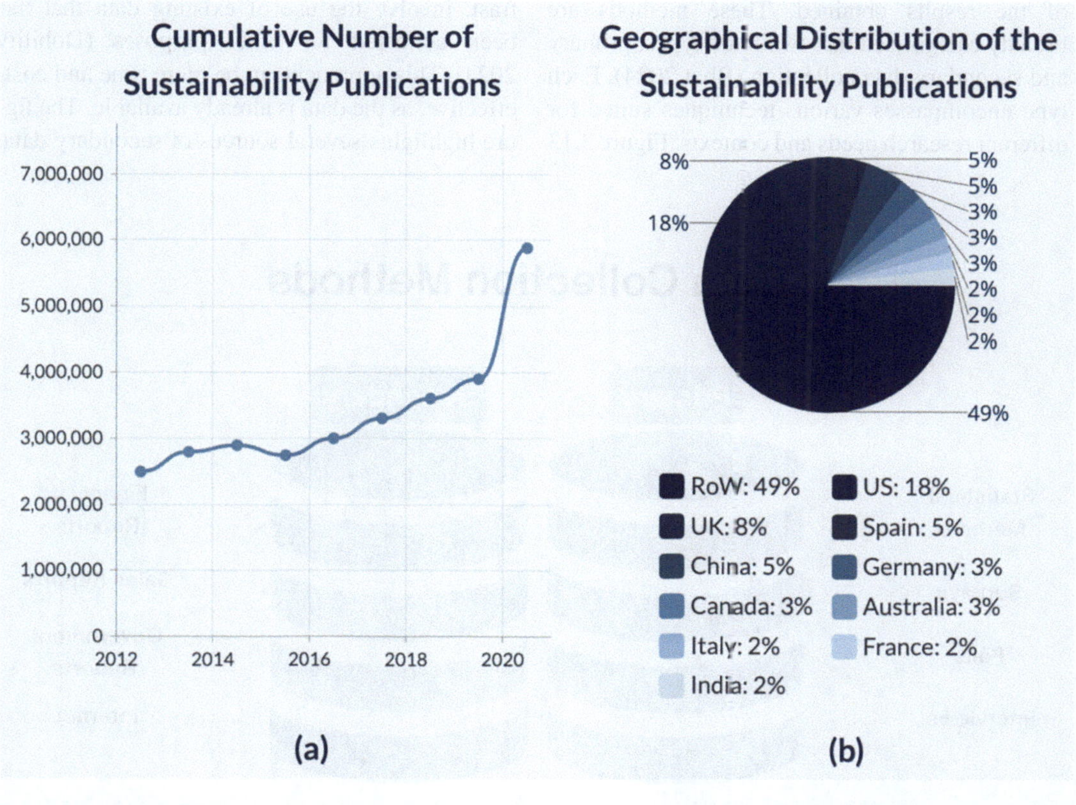

Fig. 3.12 (**a**) Cumulative number of sustainability publications and (**b**) geographical distribution of the sustainability publications. (Mazzoni et al. 2022)

devices create is massive and continues to expand dramatically. It is expected that by 2025, there will be over 75 billion IoT devices worldwide, creating over 79.4 zettabytes of data (Statista 2022). To extract valuable insights from such enormous amounts of data, efficient storage and processing capabilities are required (Sivarajah et al. 2017).

As IoT devices play a pivotal role in collecting vast amounts of data, they are also becoming increasingly crucial in sustainability monitoring and environmental conservation efforts. These interconnected sensors and devices offer real-time insights into various ecological parameters, enabling more effective resource management and informed decision-making.

3.4.3 Data Collection Methods

The methods used to collect data can significantly impact the quality, reliability, and validity of the results obtained. These methods are broadly categorised into two main types: primary and secondary data collection (Bhat 2024). Each type encompasses various techniques suited for different research needs and contexts. Figure 3.13 illustrates the primary and secondary data collection methods, showcasing the diverse approaches available to researchers and analysts.

Primary data collection methods involve gathering new data directly from first-hand sources. These methods are particularly useful when specific information is required that is not available from existing sources (Dobility 2024). As shown in Fig. 3.12, primary data collection includes statistical methods, surveys, polls, and interviews. Statistical methods involve using mathematical techniques to collect and analyse numerical data, often used in experimental research or large-scale studies. Surveys are widely used to gather information from a large number of respondents, typically through questionnaires. Polls are similar to surveys but are often shorter and focused on specific topics or opinions (Simplilearn 2024). Interviews, however, involve direct interaction with respondents, allowing for in-depth exploration of topics and the ability to clarify responses.

Secondary data collection methods, in contrast, involve the use of existing data that has been collected for other purposes (Dobility 2024). This approach can be more time and cost-effective, as the data is already available. The figure highlights several sources of secondary data,

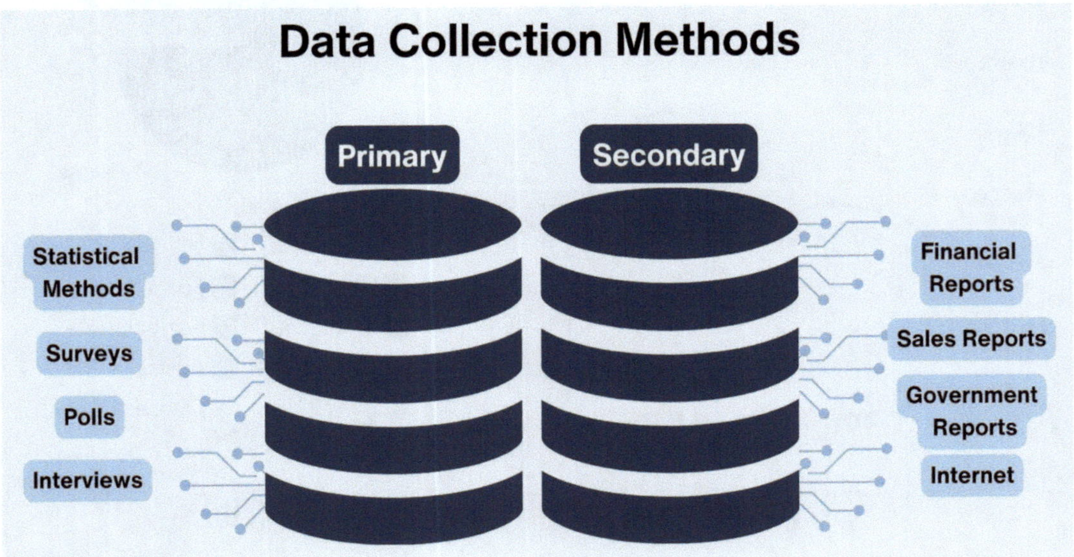

Fig. 3.13 Data collection methods. (Bhat 2024; Dobility 2024)

including financial reports, sales reports, government reports, and the Internet. Financial reports provide valuable information about companies' economic performance and can be used for market analysis or investment decisions. Sales reports offer insights into consumer behaviour and market trends. Government reports are a rich source of demographic, economic, and social data collected by official agencies (Indeed 2023). The Internet has become an increasingly important source of secondary data, providing access to a vast array of information from academic publications to social media trends. Each of these methods has its strengths and limitations, and the choice of method depends on the research objectives, resources available, and the nature of the information required.

3.4.4 Hard Data Points

Hard data refers to quantifiable information that can be statistically analysed and compared with other data points. The counterpart of hard data is soft data, which, as opposed to hard data, is qualitative and descriptive. Both hard data and soft data are important in the decision-making process (QuestionPro 2023). Hard data is characterised by its measurability, objectivity, and consistency. Being numeric and quantifiable, hard data allows for easy comparison between different sets of information (Saheed 2021). Its pure numerical nature ensures objectivity, eliminating biases that might be present in other forms of data. Additionally, the consistency of hard data means it can be reliably reproduced, making it a valuable resource for analysis and decision-making.

3.4.4.1 Hard Data Collection Methods and Tools

Accurate and reliable hard data collection is critical for any quantitative research or analytical project. Data cleaning is an essential step in this process since it involves identifying and correcting errors, inconsistencies, and gaps in numerical and categorical datasets. The following table outlines several common data-cleaning techniques that can be employed for hard data (Table 3.2).

Applying these data cleaning methods ensures that researchers' hard data is accurate, consistent, and ready for further quantitative analysis. The appropriate arrangement of numerical and categorical data is critical for efficient data management and statistical investigation.

Implementing efficient data organising strategies enables researchers to streamline their quantitative data management procedures, making numerical and categorical data more accessible and analysable.

Several ways can be used to organise hard data, and also proper hard data storage and management methods are given together in Table 3.3.

Researchers may ensure that their data is secure, properly organised, and conveniently accessible for research and reporting by carefully implementing these data storage and management procedures. To recapitulate, the efficient

Table 3.2 How to clean hard data

Data-cleaning techniques	Description
Duplicate record removal	Duplicate entries might affect analysis and generate incorrect results. Duplicate removal tools are useful for detecting and deleting duplicates (Wickham et al. 2017)
Missing data management	Missing data is a typical problem in data collection. This difficulty can be mitigated by methods such as maximum likelihood (ML) and Bayesian multiple imputation (MI), which estimate missing values based on current data (Schäfer et al. 2002)
Correction of formatting and typographical errors	Variations in data formatting, such as conflicting date or time formats, might cause analysis problems. Having data formats examined and standardised can help to ensure data integrity (Wickham et al. 2017)
Outlier detection and removal	Outliers, or data points that differ significantly from the norm, can impact data analysis accuracy. These issues should be thoroughly identified and addressed (Osborne 2013)

Table 3.3 Storage and management methods for hard data

Data management type	Description
File naming conventions	Adopting a uniform approach to naming files, such as incorporating descriptive titles and versioning, significantly aids in organising and navigating your data efficiently (Wickham et al. 2017)
Folder structure	Establishing a clear folder hierarchy is essential for data organisation. To ensure easy access and management, utilise separate folders for distinct data stages, including raw data, cleaned data, analysis files, and final results (Wickham et al. 2017)
Metadata documentation	Maintaining detailed metadata records, including data origins, collection techniques, and variable definitions, is critical for providing context and clarity about your dataset (Ravat et al. 2019)
Data dictionaries	Creating a comprehensive data dictionary that outlines each variable in your dataset can promote consistent interpretation and understanding of the data (Kahn et al. 1983)
Data storage	Choosing a suitable data storage option, such as a secure server, cloud storage, or a local hard drive, is critical to protecting your data from loss or unwanted access (Hsien et al. 2016)
Data backup	Regular data backups, whether manual or automated, are required to prevent data loss caused by hardware breakdowns, natural disasters, or other unforeseen events (Ramesh et al. 2023)
Data security	Adopting strong data security policies, such as access controls, encryption, and secure data transfer techniques, is critical for preventing unwanted access or misuse of data (Inukollu et al. 2014)
Data handling protocols	Creating and adhering to clear data handling procedures, including stating who has access to the data, how it should be used, and how it can be shared, is critical for guaranteeing ethical and responsible data use (Inukollu et al. 2014)

collection, cleansing, organising, and storage of substantial data is critical for ensuring the quality and reliability of investigation and evaluation. By implementing the solutions outlined, professionals can improve their data management workflows, increase the precision and quality of their data, and generate more solid and significant findings.

3.4.5 Data Privacy and Ethics

Data privacy is safeguarding private information against abuse, disclosure, or illegal access (Solove 2006). Personal data protection is important to ensure the confidentiality and security of individuals. Various procedures and measures should be implemented throughout the data life cycle to ensure appropriate data management and prevent unauthorised use and disclosure (United Nations Development Group 2021). Collecting and storing data securely is a key element of data privacy. Secure data storage methods protect data from unauthorised access (Garfinkel 2015).

Today, there are many regulations for data privacy and management. Some of these include the General Data Protection Regulation (GDPR) adopted by the European Union and the California Consumer Privacy Act (CCPA) adopted by California (European Commission 2018; California Office of the Attorney General 2020). GDPR is a regulation that regulates the rights of individuals whose personal data is processed, the obligations of those engaged in data processing, methods of compliance with the rules, and the sanctions to be imposed on those who do not comply with the regulations. Today, when more and more data is being processed, the GDPR, which came into force in 2018, regulates companies that process EU citizens' data regarding data privacy and security. These regulations include data minimisation, which requires that personal data be collected and processed only in the minimum amount necessary for the intended purpose. Accuracy is another vital aspect, mandating that personal data must be kept accurate and up to date. Additionally, retention limitations stipulate that data should be retained only for as long as necessary, preventing unnecessary long-term storage of personal information. Finally, the principles of confidentiality and security are paramount, requiring that data must be protected against unauthorised access and disclosure. It also gives individuals the right to access their data, delete their data, edit their data, and move their data (Voigt and Von dem Bussche 2017).

CCPA is one of the most comprehensive data privacy laws in the USA. Similar to the GDPR, the CCPA subjects organisations that collect data to various regulations. It also gives individuals multiple rights, such as deletion of their data. The CCPA also provides various penalties for violations.

3.5 The Role of Standardisation in Sustainability Metrics

3.5.1 Importance of Standardisation and Major Frameworks

Standardisation in sustainability metrics involves establishing consistent criteria, regulations, and methods for measuring, evaluating, and reporting sustainability indicators across various sectors and regions. This process aims to create a cohesive framework based on established principles, moving away from diverse unregulated methods. Table 3.4 lists major standardisation bodies and their respective years of establishment.

Table 3.4 Major standardisation bodies and frameworks

Abbreviation	Name	Founded
FAO	Food and Agriculture Organization	1945
ISO	International Organization for Standardization	1947
–	Rainforest Alliance	1987
–	Fairtrade Foundation	1992
BSR	Business for Social Responsibility	1992
CSR	Corporate Sustainability and Responsibility	1995
GRI	Global Reporting Initiative	1997
CDP	Carbon Disclosure Project	2000
PRI	Principles for Responsible Investment	2006
A4S	Accounting for Sustainability	2004
–	Carbon Trust	2001
–	B Lab	2006
CDSB	Climate Disclosure Standards Board	2007
SASB	Sustainability Accounting Standards Board	2011
NFRD	Non-financial Reporting Directive	2014
SDG	Sustainable Development Goals	2015
SECR	Streamlined Energy and Carbon Reporting	2019
CSRD	Corporate Sustainability Reporting Directive	2021
ISSB	International Sustainability Standards Board	2021
VRF	Value Reporting Foundation	2021

Table 3.4 presents a comprehensive list of major standardisation bodies and frameworks in the field of sustainability, showcasing the evolution of sustainability standardisation efforts over the past several decades. These organisations and initiatives range from long-established entities like the Food and Agriculture Organization (FAO), founded in 1945, to more recent additions such as the International Sustainability Standards Board (ISSB), established in 2021.

The table illustrates the diverse focus areas within sustainability, including environmental concerns (e.g., Carbon Disclosure Project), social responsibility (e.g., Fairtrade Foundation), and corporate governance (e.g., Business for Social Responsibility). It also highlights the increasing global recognition of the need for standardised sustainability reporting, as evidenced by the creation of bodies like the Global Reporting Initiative (GRI) in 1997 and the more recent Corporate Sustainability Reporting Directive (CSRD) in 2021.

The need for standardisation becomes evident when considering the multitude of methodologies, measurements, and terminologies currently in use. This fragmentation can create mistrust among stakeholders and make it difficult to identify important trends and patterns. By establishing consistent standards and procedures, standardisation allows accountability, comparability, interoperability, and transparency for diverse organisations, industries, and regions.

3.5.2 Corporate Sustainability Reporting Directive (CSRD)

The Corporate Sustainability Reporting Directive (CSRD) became operative for businesses on January 5, 2023. In January 2024, it formally replaced the Non-Financial Reporting Directive (NFRD), which was proposed by the European Union in 2014 and implemented in 2016 (Plana Earth 2024a). The CSRD is significant because it adopts a broad multi-stakeholder approach that addresses potential information users, including nongovernmental organisations, customers, policymakers, and the general public. By ensuring that investors and other stakeholders have access to comprehensive and relevant data, the CSRD enhances transparency and enables informed decision-making regarding companies' impacts on people and the environment, as well as the financial risks and opportunities associated with climate change and other sustainability issues (European Commission 2024a, b). This all-encompassing approach is embodied in the double materiality perspective of the directive, which requires companies to report on their effects on the environment and society, including any externalities they generate and how ESG issues affect them (Christensen et al. 2021).

One of the most creative aspects of the CSRD is that it mandates that businesses submit data in line with the European Sustainability Reporting Standards (ESRSs), which define the structure and substance of sustainability disclosures. When compared to the prior legislation, this is a significant novelty. The CSRD goes one step further by creating formal standards for reporting information. In contrast, the Non-Financial Reporting Directive (NFRD) sought to reach a certain level of comparability and dependability of the information given (Iozzelli and Sandoval Velasco 2023).

All medium-sized and small trading companies (SMEs) and major European trading organisations with more than 500 employees must adhere to the CSRD, except microenterprises whose securities are listed for trade on a Union-regulated market. Proportionate requirements may be voluntarily applied by small- and medium-sized businesses whose stocks are not listed for trading. The directive additionally applies to non-EU companies that produce more than EUR 150 million in net income annually within the Union and have at least one subsidiary that makes more than EUR 40 million in revenue or a subsidiary that falls within the parameters defining the business as large, medium, or small based on the securities listed for trading on the Union market. The number of companies that will have to submit sustainability reports will significantly rise as a result of this enlargement of the circle of obligees, from about 11,700 to about 49,000 businesses and organisations throughout the Union (Odobaša and Marošević 2023).

Depending on their size and categorisation, organisations must meet the Corporate Sustainability Reporting Directive (CSRD) deadlines for sustainability reporting. Beginning with the 2024 fiscal year and for reports issued in 2025, the first corporations must implement the new regulations (European Commission 2024a, b). Figure 3.14 illustrates the timeline for the implementation of the Corporate Sustainability Reporting Directive (CSRD) by various types of organisations.

The timeline depicted in Fig. 3.17 outlines the phased approach to implementing the CSRD and ESRSs from 2022 to 2028. It begins with the provisional political agreement on the CSRD proposal in June 2022, followed by the adoption of the first set of ESRSs by the EU Commission in June 2023. The figure then shows the gradual rollout of reporting requirements for different categories of companies. Entities already subject to the Non-Financial Reporting Directive (NFRD) are the first to start gathering data in 2024 for reporting in 2025 (Duvernay 2024). Subsequently, large entities not previously subject to NFRD, listed SMEs, and companies with non-EU parent companies are brought into the reporting framework in successive years. This staggered implementation allows for a smooth

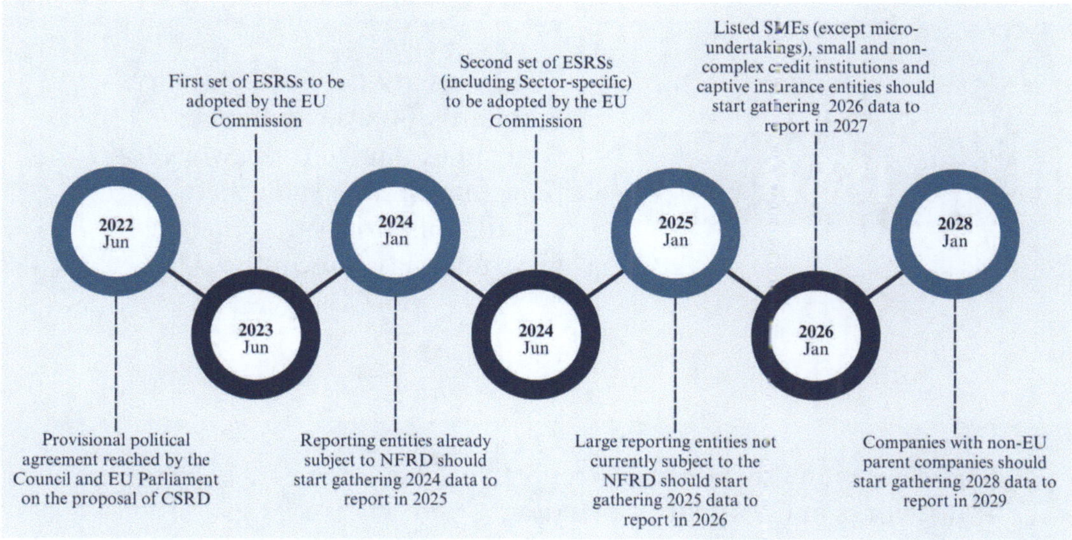

Fig. 3.14 Timeline for CSRD implementation and ESRS adoption in the EU. (Duvernay 2024)

transition to the new sustainability reporting standards across the EU business landscape.

While the timeline outlines the implementation schedule for the CSRD and ESRSs, organisations need a structured approach to meet these new reporting requirements. Table 3.5 presents a framework for implementing the CSRD, breaking down the process into four key steps.

Table 3.5 presents a comprehensive framework for implementing the Corporate Sustainability Reporting Directive (CSRD). This four-step approach guides organisations through the process of integrating sustainability into their corporate governance and reporting practices, from identifying material themes to producing sustainability reports that comply with the European Sustainability Reporting Standards (ESRSs).

3.5.2.1 Challenges

The process of transforming the European economy and society into one that is sustainable may be slowed down by several difficulties with the use of CSRD.

Clarity A significant barrier for businesses aligning their reporting processes with the new law is navigating the complex reporting framework, necessitating a detailed understanding of the precise data points, metrics, and formats defined in the CSRD (Clifft 2023).

Double Materiality The relationship between the aspects of double materiality—that is, the effects of the company's activities on people and the environment and the impact of sustainability issues on the company—and the implementation of CSRD will be complicated because this relationship is still not fully developed (Odobaša and Marošević 2023).

Data It can be difficult and time-consuming to coordinate data gathering from various sources and guarantee its accuracy and consistency throughout the reporting process (Branquart 2023).

Costs The increased reporting requirements would probably result in higher labour, administrative, and other costs for them; this will be especially painful for medium-sized and small enterprises. These costs may result in higher manufacturing and product prices, which would lower their market competitiveness within the global framework (Odobaša and Marošević 2023).

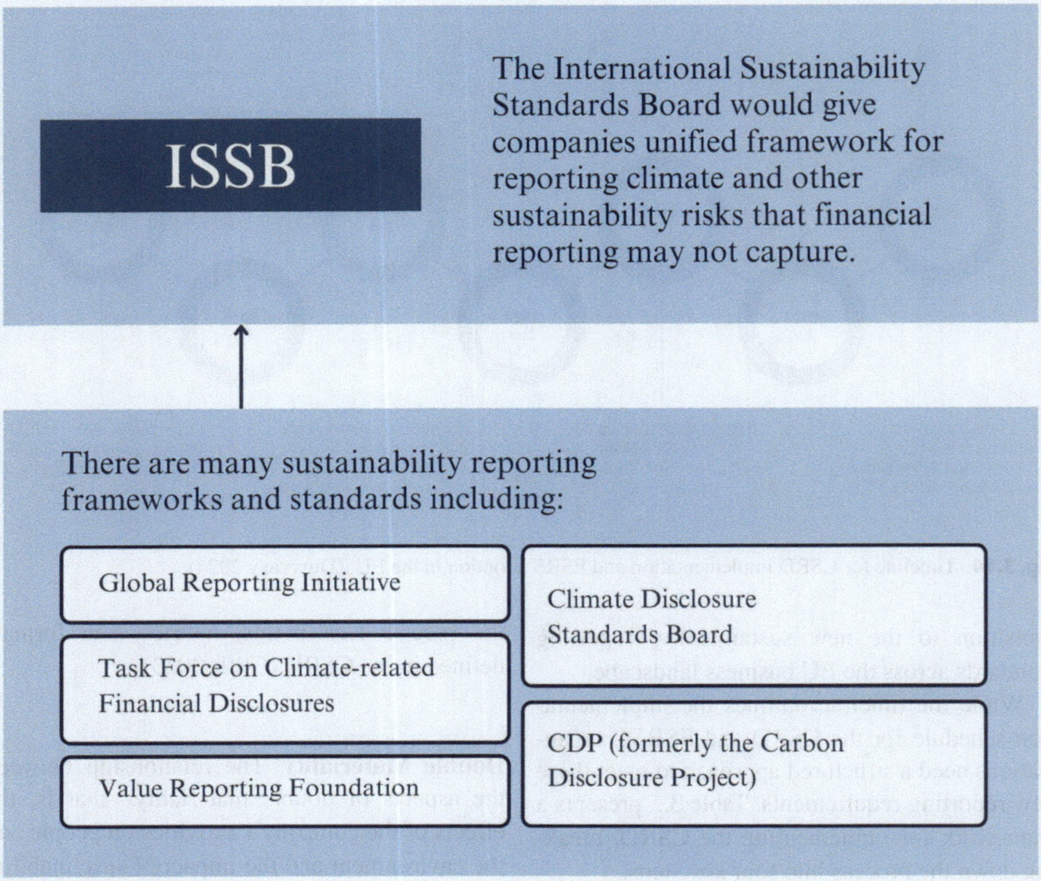

Fig. 3.17 The ISSB and the landscape of sustainability reporting frameworks

3.5.2.2 Benefits

While ESG reporting is going mainstream, keeping track of and adapting to evolving sustainability regulations has become a critical strategic priority. Leveraging our approach to CSRD can help you in many ways.

Reputation Businesses can improve their reputation and establish credibility with suppliers, employees, consumers, and regulators by taking proactive measures to manage risks associated with climate change (Walsh 2024).

Innovations Creating a CSRD report can provide insightful information about a business's operations and point up areas for efficiency and innovation. Long-term benefits include more operational effectiveness and enhanced resistance to changes in the market and environment (Thecsrdcompass 2023).

Competitive Differentiation Businesses can differentiate themselves from rivals and position themselves as sustainability leaders by adhering to CSRD. A company's dedication to sustainable business operations is demonstrated by this compliance, which may boost market share and customer loyalty (Thecsrdcompass 2023).

Motivating Staff Motivating, upholding, and increasing productivity are all enhanced when workers participate in sustainability projects. A strong workplace culture based on shared values

Table 3.5 Framework for implementing the CSRD

Step 1 Identifying material themes	The materiality themes for companies and their stakeholders are identified at the outset of the framework. Integrating sustainability with corporate governance is contingent upon the extent to which the organisation holds itself accountable for its actions, including identifying and managing risks and opportunities, which makes the activity all the more critical. A bottom-up (i.e., stakeholder viewpoints) and top-down (i.e., management and board perspectives) approach should inform material themes (Asif et al. 2013). After consulting with the company's managers and key stakeholders, a graphical materiality matrix can be used to determine the most critical topics
Step 2 Initial assessment	Planning the best approaches to integrate sustainability with corporate governance requires a critical, introspective, and analytical initial examination. This preliminary evaluation stage satisfies the intent and parameters stated during the materiality analysis stage. The primary evaluation procedure is to gauge the organisation's engagement with sustainability. A straightforward method is proper when gauging company involvement in sustainability issues
Step 3 Strategic formulation and action	It is necessary to make precise connections between the materiality themes identified during the initial evaluation and the company's strategy based on its corporate governance structure (Monks and Minow 2011). Creating the strategic goals should involve deciding which ESRS-compliant actions to take, how long it will take to accomplish the goals, and how to handle the projects. This method establishes long-term sustainability goals, strategies, and resource deployment by addressing the CSRD criteria early in the planning phase
Step 4 Sustainability reporting	The company's sustainability reporting should disclose, by applicable European Sustainability Reporting Standards (ESRSs), all the material information regarding impacts, risks, and opportunities in environmental, social, and governance matters' once the plan has been developed and implemented. According to the CSRD, double materiality assessment should be the main emphasis of sustainability reporting (EFRAG 2022). The 12 ESRSs provide comprehensive guidelines for preparing the company's sustainability reporting (Hristov and Searcy 2024)

and teamwork is fostered by this inclusive approach, which is terrific for your employer brand (Cahill 2024).

3.5.3 European Sustainability Reporting Standards (ESRS)

The European Sustainability Reporting Standards (ESRS) form the backbone of the Corporate Sustainability Reporting Directive (CSRD), providing a comprehensive framework for sustainability reporting in the EU. Developed by the European Financial Reporting Advisory Group (EFRAG), an independent organisation that convenes diverse stakeholders, these standards aim to standardise and enhance the quality of sustainability disclosures (UNEP FI 2024).

This historic project requires companies covered by the CSRD to implement these standards to improve the rigour and transparency of non-financial reporting to the level of financial reporting. The ESRS framework seeks to present a clear, accurate, and comparable picture of an organisation's opportunities, risks, and ESG impacts. As a reflection of the European Union's commitment to a sustainable economic future, this is not just a statutory requirement but a move towards integrating sustainability into the core of business reporting (Plana Earth 2024a).

The standards address environmental, social, and governance concerns, including biodiversity, human rights, and climate change. They offer data so that investors may comprehend how their investments will affect the sustainability of the businesses they invest in. In addition, they consider the conversations that have taken place with the Global Reporting Initiative (GRI) and the International Sustainability Standards Board (ISSB) to guarantee a high level of compatibility between EU and international standards and to avoid needless double reporting by businesses (European Commission 2023).

Subject to the CSRD, all companies must report by the 12 standards created by the European Financial Reporting Advisory Group (EFRAG). They outline the details an organisation must provide regarding the opportunities, dangers, and material impacts on sustainability issues (ClimatePartner 2024) (Fig. 3.15).

The Delegated Act of the Commission includes the two Cross-Cutting Standards (ESRS 1 and ESRS 2) and the ten particular sustainability standards for reporting on environmental, social, and governance issues (Topical Standards), which are included in the ESRS package. ESG dimensions are reflected in these standards. Every standard has a letter and number (ESRS E1 and ESRS S4). Sustainability themes, subtopics, and sub-subtopics of sustainability are how they are organised (Generation Impact Global 2024).

A company's impacts, risks, and opportunities regarding environmental, social, and governance (ESG) aspects are outlined in ESRS 1, "General Requirements", which establishes broad guidelines for sustainability reporting. It requires that information about your employees be disclosed, such as organisational information, goals, risk management, governance frameworks, and business plans. The reporting concept of double materiality is also made more evident by ESRS 1, specifying what data is required, optional, and subject to materiality assessment (Sweep 2024).

All businesses must comply with ESRS 2, which specifies the disclosure criteria for creating a sustainability statement, regardless of their operations or the significance of sustainability to their business. Additionally, details regarding

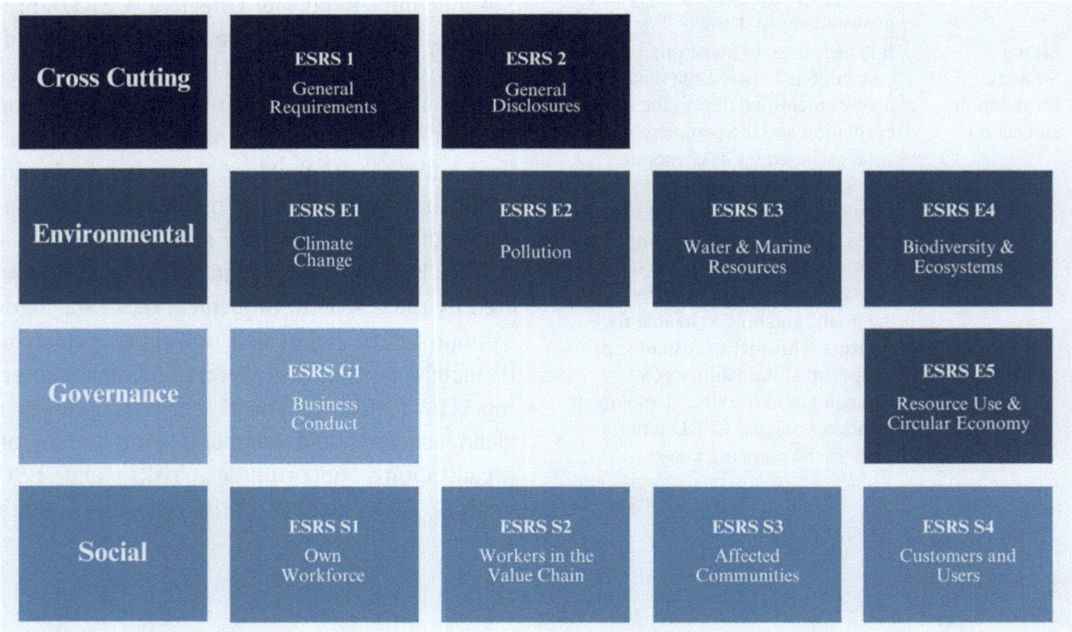

Fig. 3.15 Overview of European Sustainability Reporting Standards (ESRS). (ClimatePartner 2024)

how much of the upstream and downstream value chain are covered by the statement must be supplied. Furthermore, clarifications are needed regarding the following subjects: strategy (business model, stakeholders, risks, and opportunities), materiality assessment, governance (role of administrative/management/supervisory bodies, incentive systems, due diligence, and risk management), and key performance indicators and targets (Rödl and Partner 2024).

3.5.3.1 Challenges

Investing in Sustainability Teams The ESRS involves complicated reporting, necessitating personnel with specialised expertise and abilities. Increasing the size of your sustainability team or hiring new members demands financial outlays, which can be challenging for smaller businesses.

Standardisation Issues The ESRS still needs to be resolved. By June 2026, the European Commission is anticipated to adopt the second set of standards, which will contain industry-specific guidelines. Companies may need help to prepare their reporting strategy thoroughly due to incomplete conditions. Therefore, firms must stay flexible and adjust their operations once the final standards are released.

Scale and Complexity The ESRS and CSRD are comprehensive frameworks that address a variety of subjects. This can be highly daunting for businesses, especially those unfamiliar with such thorough reporting obligations (Coolset 2024).

3.5.3.2 Benefits

Improved Comparability Stakeholders can more readily and precisely evaluate the sustainability performance of various organisations according to the ESRS, which offers a standard set of reporting requirements. This evaluation allows stakeholders to make more informed decisions and motivates businesses to keep raising the bar on their sustainable performance (Esgpro 2024).

Complying with Sustainable Development Goals (SDGs) The ESRS complies with UN Sustainable Development Goals (SDGs) recommendations and other worldwide sustainability initiatives. Businesses can support these global sustainability goals by implementing ESRS, improving their standing as ethical corporations, and opening up new business opportunities (Thecsrdcompass 2024).

Simplified Reporting Process Companies' reporting processes are simplified by the ESRS, which unifies sustainability reporting standards. According to Esgpro (2024), businesses may ensure consistency and comparability of their reports while saving time and money by eliminating the need to navigate a complex terrain of numerous frameworks and norms.

Attraction of Investments Including environmental, social, and governance (ESG) considerations in investment decisions is a top priority for today's investors. Businesses may provide investors with the transparent and comparable data needed to assess their sustainability performance by implementing the ESRS. Investor interest may result from this, which could lower the cost of capital (Thecsrdcompass 2024).

3.5.4 EU Taxonomy Regulation

On June 22, 2020, the Taxonomy Regulation (Regulation (EU) 2020/852) was published in the European Union's Official Journal. It became operative on July 12, 2020. Based on six environmental objectives, the taxonomy is created. These six objectives are: climate change mitigation; climate change adaptation; the sustainable use and protection of water and marine resources; the transition to a circular economy; pollution prevention and control; and the protection and restoration of biodiversity and ecosystems (IIASA 2023).

The first thorough, scientifically based classification system to assist in determining whether an economic activity is sustainable is EU Taxonomy for Sustainable Activities. This framework aims to guide investors, companies, and policymakers in identifying and supporting activities that contribute to the EU's environmental objectives. By establishing performance thresholds, also known as technical screening criteria, for these activities, it gives a common definition for about eighty sustainable activities (Schütze and Stede 2021).

The content of the Taxonomy Regulation includes criteria for determining the environmental sustainability of economic activities. It includes specific technical screening criteria for various sectors. The technical screening criteria, which are detailed performance thresholds, help determine whether an activity can be considered sustainable. These criteria are developed based on scientific and technical input from various stakeholders, including industry representatives, academia, environmental experts, and public bodies. For example, the criteria for climate change mitigation may include specific greenhouse gas emission limits for different industries. Similarly, criteria for the sustainable use of water resources may include standards for water use efficiency and pollution control.

Making taxonomy-related disclosure mandatory for major companies in the EU is one of the objectives. Large companies must disclose the extent to which their activities align with the taxonomy in their annual reports. Financial intermediaries, such as investment funds, must disclose the proportion of their investments that meet the taxonomy criteria. This transparency is intended to help investors make informed decisions and support the shift towards a more sustainable economy. In the end, this is anticipated to support the EU's transition towards climate neutrality by

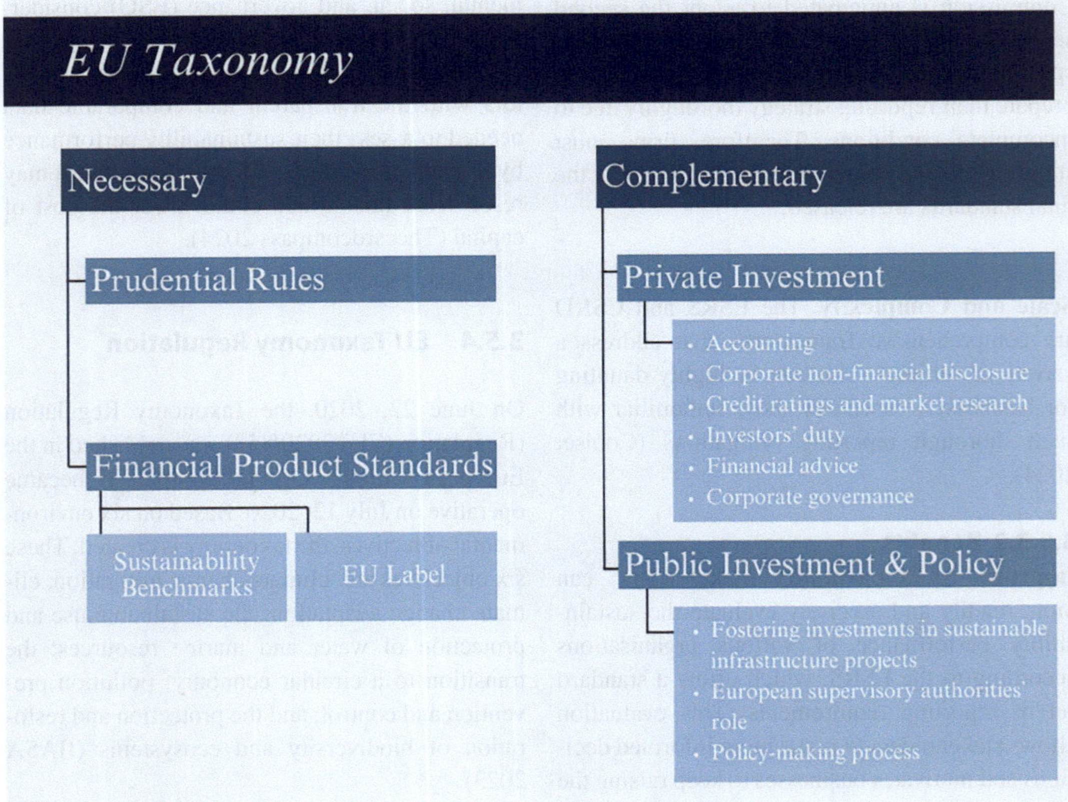

Fig. 3.16 Role of EU taxonomy in the action plan. (IIASA 2023)

2050 by helping direct investments towards low-carbon technologies (Schütze and Stede 2021).

The European Commission has established technical standards for tracking each environmental goal to provide an accurate list of ecologically friendly activities based on taxonomic legislation. The Delegated Act supplementing Article 8 of the Taxonomy Regulation was adopted by the European Commission in July 2021. This act brought greater precision to the content, methodology, and presentation of information that financial and non-financial undertakings were required to publish regarding environmentally sustainable activities in their business, lending, or investing (IIASA 2023). The Complementary Climate Delegated Act, which covers various nuclear and gas energy activities in EU taxonomy, was approved by the European Commission on February 2, 2022, adding an extra revision to the Taxonomy Regulation (European Commission 2022). Figure 3.16 points out the role of EU taxonomy in the action plan for necessary and complementary parts.

This figure demonstrates that the EU taxonomy has two parts as necessary and complementary. Necessary ones include the prudential rules and financial product standards. Complementary one includes private investment and public investment policy.

The taxonomy is focusing on greenhouse gas emissions. The current proposed taxonomy, which was created by a technical expert group (TEG) and enacted by EU delegated acts, is focused on actions that aid in mitigating and adapting to climate change (European Commission 2020).

One major challenge is the potential for the EU Taxonomy to create important administrative and compliance obligations. The extensive reporting requirements can be a problem, especially for smaller companies that may not have the resources to meet these demands. This could create entry barriers in the green finance sector. Highly skilled labour would be diverted to compliance tasks instead of focusing on productive innovation, hindering progress (Kooths 2022). Streamlining these processes and supporting smaller firms is essential to mitigate these effects and encourage widespread participation in sustainable finance initiatives.

Moreover, the taxonomy's strict performance thresholds must balance aim and practicality. While designed to ensure considerable contributions to sustainability, they must also be flexible enough to account for sectoral differences. Overly strict thresholds may discourage investments in transitional activities, while less strict thresholds risk greenwashing. Therefore, it is important to determine appropriate thresholds that incentivise sustainability improvements without imposing unrealistic standards is crucial (Schütze et al. 2020).

In addition, as scientific understanding and technological capabilities evolve, technical screening criteria must be periodically reviewed and updated to reflect the latest advancements. This requires a robust governance framework that involves a range of stakeholders, including scientific experts, industry representatives, and policymakers. Continuous updates ensure that the taxonomy remains relevant and effective, fostering innovation and adaptation in response to new information and emerging technologies (Schütze et al. 2020). Additionally, global coordination is crucial to prevent market fragmentation and facilitate cross-border investments, making the EU Taxonomy a potential global benchmark for sustainable finance (Schütze et al. 2020).

3.5.5 Global Reporting Initiative (GRI)

The GRI Standards are a comprehensive set of guidelines that allow organisations to report impacts and performance metrics related to sustainability. They establish a common language which supports transparency and develops an open dialogue about the impacts of organisations (de Villiers et al. 2022). The GRI Standards are designed to apply universally to organisations. They focus on sustainability and stakeholder engagement. Investors, policymakers, NGOs, employees, suppliers, customers, NGOs, and local communities use the GRI Standards (de Villiers et al. 2022).

The GRI framework identifies an organisation's economic, environmental, and social effects through material topics. Organisations determine these topics by evaluating their actual and potential impacts. Then, they evaluate the expectations of their stakeholders through a systematic process (GRI 2024).

The GRI Standards consist of three complementary series: Universal Standards, Sector Standards, and Topic Standards. Organisations need to follow all three Universal Standards. Then, use the Sector Standards based on their sector and select Topic Standards relevant to their material topics (GRI 2024). The GRI Standards offer Sector Standards tailored to specific industries, providing guidance on common material topics within those sectors. Once material topics are identified, organisations use Topic Standards to report detailed information on their impacts and management approaches related to each topic (GRI 2024).

Most organisations around the world use the GRI Standards voluntarily. However, as global emphasis on sustainability reporting grows and regulations like the EU Directive on non-financial reporting evolve, this leads to increased adoption of these standards among large, publicly traded companies (de Villiers et al. 2022).

According to the KPMG Survey of Sustainability Reporting, the GRI Standards are the most widely used global standards. The survey found that GRI Standards are the only reporting standards used by most companies in all regions globally. They used 75% in the Americas, 68% in Europe and Asia Pacific, and 62% in the Middle East and Africa (KPMG 2022). Furthermore, 78% of the world's 250 largest companies by revenue use the GRI to report on their sustainability, ESG impacts, and performance (KPMG 2022).

The GRI Standards offer organisations the opportunity to report their sustainability performance. They can choose to report according to the GRI Standards by preparing a standalone sustainability report. Alternatively, they can report referencing the GRI Standards using selected standards or their content in other public communications. In either case, organisations must publish a GRI Content Index. It acts as a road map for stakeholders by listing all required disclosures and their locations within the report or website (GRI 2024).

The GRI Standards offer a structured framework for disclosing a comprehensive range of sustainability-related information. Organisations should explain their methodologies to collect data so stakeholders can understand how the information was gathered and analysed (GRI 2024). This disclosure encompasses both qualitative and quantitative data, allowing stakeholders to understand the significant issues identified by the organisation and how they measure and manage their performance (de Villiers et al. 2022). This is especially important when reporting on complex issues like employee fluctuations or assessing regional water stress. Similarly, when reporting GHG emissions, organisations must classify and disclose their emissions according to Scope 1, Scope 2, and Scope 3 (GRI 2024).

On the other hand, small- and medium-sized enterprises (SMEs) face challenges when trying to use the GRI standards. These challenges include limited resources, making it difficult for SMEs to allocate the necessary staff and funds to meet GRI requirements. There is also a perceived lack of government support for sustainability reporting, making investing in GRI implementation less appealing for SMEs than for larger companies. Additionally, SMEs find it difficult to understand the concept of materiality, which is emphasised in the GRI framework. This lack of clarity can create challenges in determining which sustainability issues are material for their reporting (de Villiers et al. 2022).

3.5.6 International Sustainability Standards Board (ISSB)

The inclusion of sustainability considerations in investing decisions is growing in popularity. Feedback from numerous meetings with market participants indicates a growing demand for enterprises to offer high-quality, globally comparable information on sustainability-related risks and opportunities. Additionally, there is a great desire to address the dispersed set of voluntary sustainability-related norms and regulations that increase risk, expense, and complexity for businesses and investors (IFRS 2024).

Parallel to these growing demands and desires, The International Sustainability Standards Board (ISSB) was introduced by the International Financial Reporting Standards (IFRS) Foundation in November 2021 at the COP26 summit in Glasgow to satisfy the demand for ESG performance disclosures that are "high quality, transparent, reliable, and comparable". By developing a comprehensive worldwide baseline for ESG reporting that is "focused on the needs of investors and financial markets", the IFRS Foundation's ISSB initiative aims to standardise ESG standards (Invest Europe 2024). The guidelines are made to accommodate the requirements of all businesses, not just the most advanced. They give investors information comparable across borders and give businesses a clear notion of what they must publish to satisfy the demands of the world's capital markets (Vaessen 2023).

Negotiating the existing dispersed and voluntary landscape of sustainability-related criteria is challenging. ISSB standards build on recommendations from the Climate Disclosure Standards Board (CDSB), Task Force on Climate-Related Financial Disclosures (TCFD), and Sustainability Accounting Standards Board (SASB) to reduce the likelihood of double-reporting and prevent additional costs.

The Global Reporting Initiative (GRI) and the ISSB have also joined forces to guarantee complete interoperability between their standards. Combining both systems' advantages aims to provide thorough and cogent sustainability reporting. Both organisations aim to lessen the reporting requirements placed on businesses and enhance the general calibre and usefulness of the information provided by harmonising the investor-focused criteria of the ISSB with the multi-stakeholder viewpoint of the GRI. In order to advance global sustainability goals and promote more sustainable company practices, a consistent approach to sustainability reporting must be developed, as this collaborative effort demonstrates (Shaw 2024). This collaborative effort emphasises the importance of a unified approach to sustainability reporting, which is particularly relevant given the diverse landscape of existing frameworks and standards. Figure 3.17 presents an overview of the International Sustainability Standards Board (ISSB) and its role in relation to other prominent sustainability reporting frameworks.

Figure 3.17 illustrates the complex ecosystem of sustainability reporting frameworks and standards, highlighting the role of the International Sustainability Standards Board (ISSB). The ISSB is positioned at the top of the figure, emphasising its potential to provide a unified framework for companies to report on climate and other sustainability risks that may not be captured by traditional financial reporting. The lower portion of the figure lists several existing sustainability reporting frameworks and standards, including the Global Reporting Initiative, Climate Disclosure Standards Board, Task Force on Climate-Related Financial Disclosures, Value Reporting Foundation, and CDP (formerly the Carbon Disclosure Project). This visual representation highlights the current fragmentation in sustainability reporting and the need for a cohesive approach, which the ISSB aims to address by offering a comprehensive and standardised framework for sustainability disclosure.

The ISSB has four main goals:

1. To create guidelines for a worldwide minimum level of sustainability disclosures.
2. To satisfy investors' information needs.
3. To allow businesses to give the world's capital markets through sustainability information.

4. To promote interoperability with disclosures targeted at larger stakeholder groups or unique to a given jurisdiction (IFRS 2024).

To get input from stakeholders regarding its objectives for its upcoming two-year work plan, the ISSB released a Request for Information Consultation on Agenda objectives on May 4, 2023. The request covered four possible projects. Three studies have examined the opportunities and risks related to sustainability, including human rights, human capital, ecosystems, and biodiversity. Also, one study on reporting integration looks at ways to incorporate data into financial reports that go beyond IFRS S1 General Requirements for Disclosure of Sustainability-Related Financial Information and IFRS S2 Climate-Related Disclosures' connected information requirements (B Corporation 2024).

General standards for disclosure of opportunities and risks connected to sustainability that potentially impact a company are provided by IFRS S1. In addition to prescribing the format and substance of reports, it outlines the disclosure requirements. With an emphasis on climate-related risks and opportunities, IFRS S2 mandates that companies reveal how these factors may affect cash flow, capital availability, financial position, and performance over short-, medium-, and long-term periods. It offers information on what businesses must reveal to give capital markets and investors a similar understanding of these implications. Regulators will apply the IFRS S1 and S2 standards in different ways, depending on the particular requirements of respective countries, since they are voluntary. Recognising these expected variations, the ISSB will shortly publish an Adoption Guide outlining various routes (Bright 2024).

3.5.7 Science Based Targets Initiative (SBTi)

The Science Based Targets Initiative (SBTi) is a global collaboration between leading environmental and sustainability organisations, including the Carbon Disclosure Project (CDP), the World Resources Institute (WRI), the United Nations Global Compact (UNGC), and the World Wide Fund for Nature (WWF). Launched in 2015, the SBTi aims to mobilise the private sector to take ambitious climate action by providing a science-based framework that guides companies and financial institutions in setting greenhouse gas (GHG) reduction targets aligned with the goals of the Paris Agreement (Science Based Targets 2024).

The SBTi's primary goal is to enable companies to reduce their GHG emissions in line with the global effort to limit temperature rise to well below 2 °C, and ideally to 1.5 °C, compared to pre-industrial levels. The initiative provides clear guidance on how to set science-based targets (SBTs) that reflect the level of decarbonisation necessary to meet these objectives. The SBTi helps businesses navigate the increasingly complex regulatory landscape by providing a widely accepted standard that promotes competitiveness while ensuring alignment with global climate goals. It offers sector-specific guidance for industries with significant environmental impacts, such as financial institutions and businesses operating in land-intensive sectors like forestry, land, and agriculture (Giesekam et al. 2021). The SBTi provides a step-by-step process for companies to set these targets, including submitting a commitment letter, developing a target, submitting the target for validation, announcing the target, and disclosing their progress. In Fig. 3.18, a step-by-step process is shown. Finally, the SBTi empowers companies and financial institutions to voluntarily align their actions with the Paris Agreement, emphasising continuous improvement and transparent reporting on climate performance.

This process provides a standardised framework for organisations to contribute to global climate goals. It emphasises formal commitment, target development, external validation, and ongoing transparency in emissions reporting. The SBTi approach aims to ensure that organisational climate targets are in line with the latest climate science and global agreements on climate change mitigation.

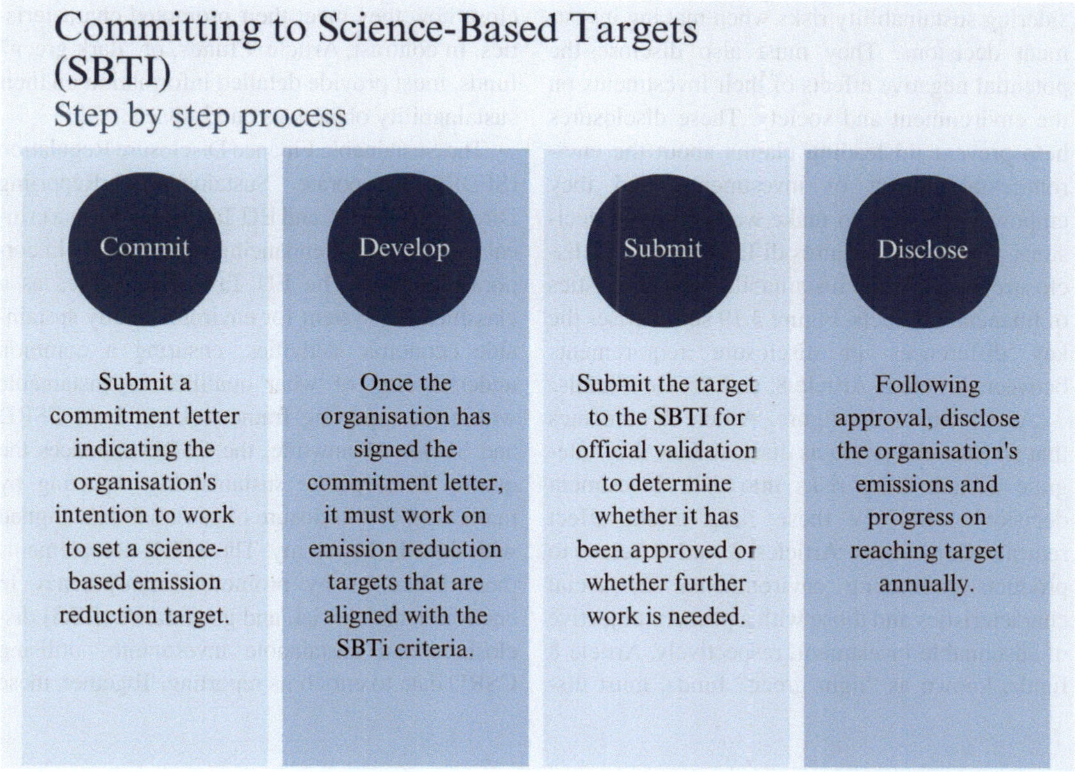

Fig. 3.18 Committing to science-based targets. (Science Based Targets 2024)

3.5.8 Sustainable Finance Disclosure Regulation (SFDR)

Under the EU Commission's action plan on sustainable finance, the taxonomy serves as the foundation for a number of related initiatives. By 2022, large businesses subject to the Non-Financial Reporting Directive (NFRD) will have to reveal the percentage of their operations that are aligned with the taxonomy (Schütze and Stede 2021).

Financial market participants that provide sustainably oriented financial products are required by the Sustainable Finance Disclosure Regulation (SFDR) to begin reporting their taxonomy-aligned investments or activities for various market segments by December 2021. Furthermore, EU Green Bond Standard and EU Eco-label (Hessenius et al. 2020) will both use the taxonomy as a common definition for green financial products (Schütze and Stede 2021).

The Sustainable Finance Disclosure Regulation (SFDR) is an important part of the European Union's sustainable finance framework. It aims to increase transparency in the financial sector's sustainability practices. SFDR provides a clear roadmap for investors by mandating detailed disclosures on how financial market participants (FMPs) and financial advisers integrate sustainability risks and consider the potential impacts of their investment decisions.

The SFDR does designate various categories into which financial goods must fall, ranging from activities with no sustainability considerations to initiatives with a particular and conscious sustainability goal (Cremasco and Boni 2022). The need to adopt SFDR arose to address these issues and highlights the need to formalise the way financial market participants shape and manage financial products by integrating sustainable elements defined by SFDR as "environmental, social and human resource issues, respect for human rights, fight against corruption and anti-bribery matters" (Cremasco and Boni 2022).

Financial market participants, including asset managers, insurance companies, pension funds, and financial advisers, are transparent about considering sustainability risks when making investment decisions. They must also disclose the potential negative effects of their investments on the environment and society. These disclosures help prevent misleading claims about the environmental impact of investments, and they empower investors to make well-informed decisions. The SFDR requires different levels of disclosure based on the sustainability characteristics of financial products. Figure 3.19 summarises the key differences in disclosure requirements between Article 6, Article 8, and Article 9 funds.

As shown in the figure, Article 6 mandates that all financial products disclose how they integrate sustainability risks into their investment decisions and how these risks might affect returns. Conversely, Articles 8 and 9 relate to products promoting environmental or social characteristics and those with a primary objective of sustainable investment, respectively. Article 8 funds, known as "light green" funds, must disclose how they meet their promoted characteristics. In contrast, Article 9 funds, or "dark green" funds, must provide detailed information on their sustainability objectives and impacts.

The Sustainable Finance Disclosure Regulation (SFDR), Corporate Sustainability Reporting Directive (CSRD), and EU Taxonomy form a critical framework for enhancing sustainability in corporate finance. The EU Taxonomy serves as a classification system for environmentally sustainable economic activities, ensuring a common understanding of what qualifies as sustainable within the reporting frameworks of both CSRD and SFDR. Meanwhile, the CSRD enhances the quality of corporate sustainability reporting by mandating the disclosure of essential data aligned with the EU Taxonomy. The SFDR complements these initiatives by promoting transparency in environmental, social, and governance (ESG) disclosures and sustainable investments, utilising CSRD data to enrich its reporting. Together, these

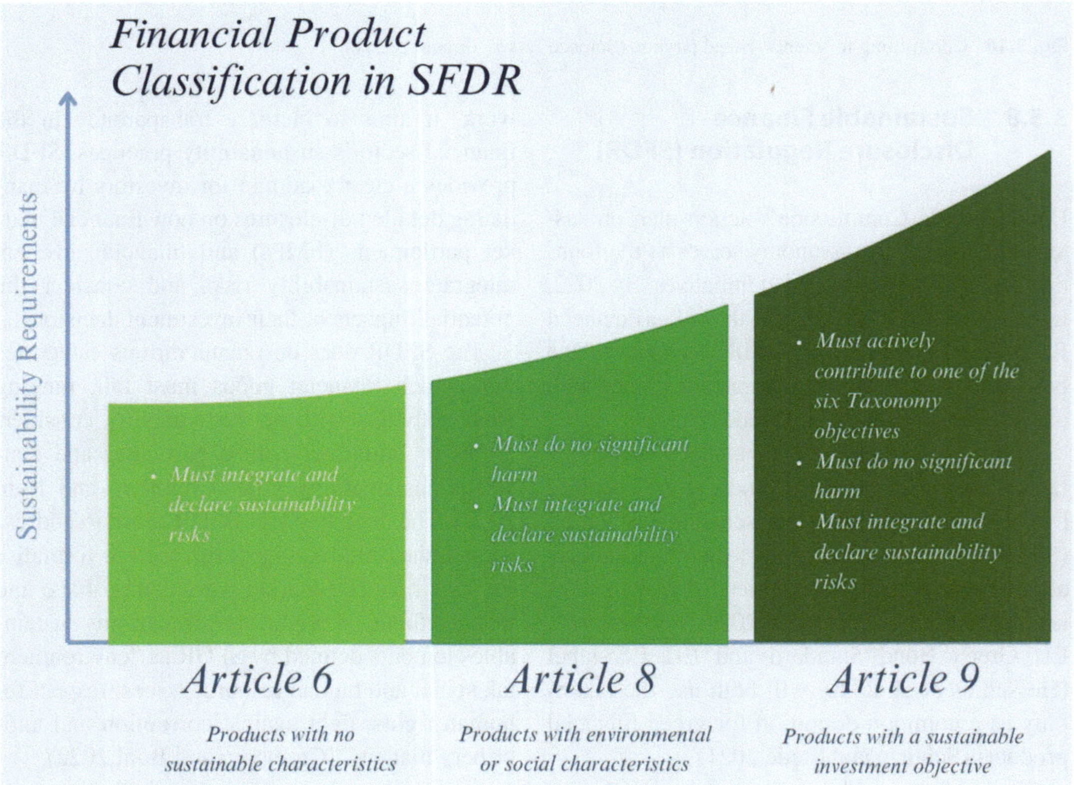

Fig. 3.19 Financial product classification in SFDR. (Carbometrix 2024)

	Purpose	Interrelations	Role
EU Taxonomy	Classifies sustainable economic activities within CSRD and SFDR.	Sets criteria for identifying sustainable activities for CSRD and SFDR reporting.	Aligns investments with Taxonomy-compliant activities.
CSRD	Enhances corporate sustainability reporting quality.	Supports SFDR with essential data.	Requires companies to report ESG metrics in line with EU Taxonomy for SFDR.
SFDR	Promotes transparency in ESG disclosures and sustainable investments.	Uses CSRD data to provide a broader perspective in its reporting.	Highlights ESG disclosures to encourage sustainable investments, integrating CSRD metrics into SFDR.

Fig. 3.20 EU sustainability reporting framework: CSRD, SFDR, and EU Taxonomy. (Barral 2023)

regulations help safeguard against misleading sustainability claims and provide investors with reliable information for informed decision-making. Figure 3.20 illustrates the interplay between these regulations, highlighting their respective purposes, interrelations, and roles.

The EU Taxonomy, CSRD, and SFDR represent interconnected yet distinct elements within the framework of sustainable finance, each playing a crucial role in advancing transparency and accountability. The EU Taxonomy serves as a classification system that defines what constitutes environmentally sustainable economic activities, providing the foundational criteria for sustainability reporting. The CSRD complements this by enhancing the quality of corporate sustainability disclosures, requiring companies to report essential ESG metrics that align with the Taxonomy's definitions. In contrast, the SFDR focuses specifically on promoting transparency in ESG disclosures and sustainable investments, utilising data from the CSRD to inform its reporting requirements. While each regulation has its unique purpose, classification, reporting quality enhancement, and transparency, they aim to prevent misleading sustainability claims and improve investor confidence. By aligning these frameworks, the SFDR ensures that financial products are accurately assessed against robust sustainability standards and empowers investors with reliable information, ultimately fostering a more sustainable and responsible financial ecosystem.

The SFDR improves transparency and investor confidence by delivering precise information on integrating sustainability risks and impacts into financial decisions. This aids in averting greenwashing, where financial products are deceptively marketed as environmentally friendly, by ensuring robust data back sustainability claims.

Conversely, the SFDR brings particular challenges, such as accessibility and data quality, with its benefits. Financial market participants may encounter difficulties in obtaining essential data from the companies they invest in, thereby complicating the reporting process. In addition, the SFDR's requirements entail legal uncertainties and complexities, potentially leading to increased operational costs and compliance complications. Moreover, the costs associated with meeting the SFDR's reporting requirements can be considerable, especially for smaller firms. Lastly, greenwashing is risky if the regulation's criteria are sufficiently stringent and transparent, which could result in the mislabelling of sustainable financial products.

3.5.9 Sustainability Accounting Standards Board (SASB)

The SASB Standards were taken up by the IFRS Foundation's International Sustainability Standards Board (ISSB) in August 2022. When the ISSB released the draft IFRS Sustainability Disclosure Standards, it showed how the industry-based SASB Standards would initially assist businesses in identifying the opportunities and risks associated with sustainability and providing metrics for use in their disclosures (SASB 2024).

The SASB Standards, a framework for environmental, social, and governance recommendations, identify sustainability challenges that may impact a company's financial performance and business value across 77 industries. These industry-specific guidelines include six disclosure issues and thirteen accounting metrics covering the five main facets of sustainability—environment, social capital, human capital, business model and innovation, and leadership and governance. In order to help firms identify, measure, manage, and report on the subset of ESG subjects that most directly impact long-term organisational value creation, sustainability disclosures like the SASB Standards help address this need (IBM 2024).

Investors need industry-specific and cross-industry disclosures because certain information, like climate risk governance, is more valuable for making decisions when relevant to an industry. In contrast, other information is necessary for comparison across all industries and independent of the business model. As demonstrated by their adoption by over 3100 companies in more than 80 countries worldwide, including 75% of the S&P Global 1200 Index, the 77 industry-based SASB Standards offer relevant, comparable information to investors and are economical for businesses (SASB 2024).

In 2022, 2231 businesses from 66 jurisdictions standardised their reporting to the SASB Standards, a 60% increase from the previous year. Even though Americans account for most reporters, the SASB Standards are widely and steadily used elsewhere. By 2022, almost sixty percent of SASB reporters were based outside of the USA, and a minimum of fifty percent more organisations worldwide were aligning their reporting to the SASB Standards in 2022 compared to 2021 (Bradley 2023). To identify sustainability-related risks, opportunities, and related metrics, organisations must refer to and consider the SASB Standards and the industry-specific disclosures mandated by the ISSB's Standards. Accordingly, businesses using the SASB Standards will have an advantage when implementing the first two ISSB Standards.

The SASB Standards are a strong foundation for businesses implementing the IFRS Sustainability Disclosure Standards. They are a significant component of the first two IFRS Sustainability Disclosure Standards, IFRS S1 General Requirements for Sustainability-Related Disclosures, and IFRS S2 Climate-Related Disclosures. Businesses using the TCFD Recommendations and SASB Standards will be well-equipped to use IFRS S1 and S2. As preparers and investors move towards adopting ISSB Standards, the ISSB will actively encourage them to do so by maintaining and improving the SASB Standards during their ongoing existence (SASB 2024).

3.5.10 Paris Agreement Capital Transition Assessment (PACTA)

The Paris Agreement Capital Transition Assessment (PACTA) is a free and open-source software that enables users to examine individual enterprises and evaluate how well financial portfolios match climate scenarios (PACTA 2025). PACTA assists financial institutions in comparing anticipated capital expenditures and company assets to what is required to mitigate global warming. By measuring alignment by sector or technology, the methodology guarantees that sector-specific factors are taken into account. For instance, it takes into account that some industries, like power generation, require reform, while others, like coal, require phase-out. Financial institutions can better understand their exposure to climate-related risks and spot possibilities for more sustainable financing with the help of PACTA's forward-looking insights. Power, coal mining, oil and gas upstream industries, auto manufacturing, cement, steel, and aviation are the sectors currently covered by PACTA that are related to climate change. Approximately 75% of the world's greenhouse gas emissions come from these sectors combined. More than 1500 financial institutions utilise PACTA globally, and central banks and supervisors use it to evaluate the businesses they regulate (pacta 2025). Working principles of PACTA can be summarised as follows (Fig. 3.21):

3.5.11 Harmonisation and Standards

Harmonisation in the context of sustainability metrics is the process of unifying definitions, methodologies, and procedures to achieve standardisation or comparability in calculation and

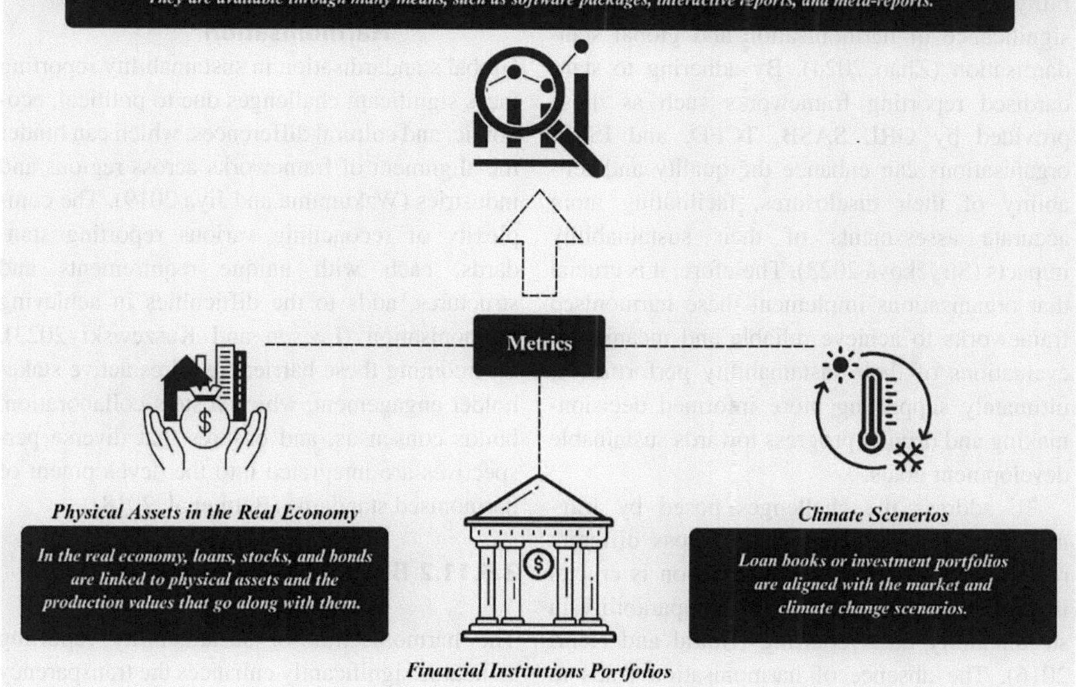

Fig. 3.21 Working mechanism of PACTA. (pacta 2025)

reporting of data related to sustainability. This standardisation is important for appropriate assessment and review in various organisations and regions for final decision-making towards sustainable development (Schumacher et al. 2020).

In this context, global standards play a vital role in sustainable finance as it allows consistency and comparability of the integration of ESG factors into financial decision-making. Key bodies like the Global Reporting Initiative (GRI), the Sustainability Accounting Standards Board (SASB), the Task Force on Climate-Related Financial Disclosures (TCFD), and the International Sustainability Standards Board (ISSB) establish these standards to support investors, companies, and other stakeholders in consistently assessing and comparing sustainability performance (Mujiani 2023; Strýčková 2023). This, in turn, ensures more consistency in the reporting and evaluation of ESG factors, which increases their reliability and transparency in the assessment of sustainability performance.

The necessity of addressing inconsistencies, ensuring transparency, and improving comparability in sustainability reporting highlights the significance of harmonisation and global standardisation (Zhao 2024). By adhering to standardised reporting frameworks such as those provided by GRI, SASB, TCFD, and ISSB, organisations can enhance the quality and reliability of their disclosures, facilitating more accurate assessments of their sustainability impacts (Strýčková 2023). Therefore, it is crucial that organisations implement these harmonised frameworks to achieve reliable and meaningful evaluations of their sustainability performance, ultimately supporting more informed decision-making and driving progress towards sustainable development goals.

To address the challenges posed by fragmented standards and metrics across different regions and industries, harmonisation is crucial in promoting consistency and comparability in sustainability data reporting (Boiral and Henri 2016). The absence of harmonisation leads to risks associated with the incomparability of sustainability data, hindering accurate assessments and effective decision-making processes. Moreover, harmonisation plays a vital role in reducing greenwashing, where companies exaggerate or misrepresent their sustainability efforts. By improving transparency and reliability in reporting, harmonisation enhances investor confidence in the accuracy of sustainability disclosures, ultimately supporting more credible and meaningful sustainability performance assessments (Gao 2024).

In the context of sustainable finance, harmonisation is essential for ensuring the validity and comparability of environmental, social, and governance (ESG) data, which is crucial for making well-informed investment decisions (Mujiani 2023). By aligning sustainability reporting standards, such as those set by the Global Reporting Initiative (GRI) and other standard-setting bodies, the risks of greenwashing are reduced, and investor confidence in the accuracy and integrity of ESG data is strengthened (Strýčková 2023). This convergence of standards enhances the overall reliability of sustainability disclosures, contributing to more transparent and credible financial decision-making.

3.5.11.1 Challenges to Achieving Harmonisation

Global standardisation in sustainability reporting faces significant challenges due to political, economic, and cultural differences, which can hinder the alignment of frameworks across regions and industries (Wakunuma and Jiya 2019). The complexity of reconciling various reporting standards, each with unique requirements and structures, adds to the difficulties in achieving harmonisation (Leeson and Kuszewski 2023). Overcoming these barriers requires active stakeholder engagement, which fosters collaboration, builds consensus, and ensures that diverse perspectives are integrated into the development of harmonised standards (Betti et al. 2018).

3.5.11.2 Benefits of Harmonised Standards

The harmonisation of sustainability reporting standards significantly enhances the transparency and comparability of ESG data, enabling stakeholders to conduct more robust evaluations of organisations' ESG performance. By standardising reporting frameworks and aligning diverse metrics, both corporations and institutional investors benefit from greater operational efficiency in sustainability assessments, leading to more informed capital allocation and refined risk management strategies (Rawhouser et al. 2017). Furthermore, harmonised standards are integral to advancing global sustainability objectives, including the UN Sustainable Development Goals (SDGs) and the Paris Agreement, by ensuring the consistency and verifiability of ESG disclosures, thus facilitating more effective monitoring of environmental and social impacts (de Villiers et al. 2022). The alignment of these frameworks ultimately enhances the credibility of sustainability reporting and strengthens global accountability mechanisms for sustainable development.

3.6 GHG and Carbon Accounting and Reporting

3.6.1 Introduction to GHG and Carbon Accounting

Companies face growing responsibilities to measure, report, and reduce their carbon emissions due to their increasing environmental impact and evolving regulatory requirements. This need has given rise to the practice of "carbon accounting", which involves measuring and reporting greenhouse gas (GHG) emissions caused by an organisation (Carbon Chain 2024). This crucial process relies on collecting data from various emission-producing areas, such as industrial processes and transportation, allowing businesses to effectively measure and manage their environmental impact (Buggie and Farbstein 2024).

To ensure the effectiveness and reliability of carbon accounting, a set of principles has been established that closely align with financial accounting best practices. These principles form the foundation for GHG emissions management and reporting, emphasising five core tenets: relevance, completeness, consistency, transparency, and accuracy (Brightest 2024). In practice, organisations are encouraged to maintain comprehensive digital records of all carbon-related assets, events, and transactions. This includes developing dashboards to track key sustainability performance indicators (KPIs) and conducting regular monitoring of carbon emissions on monthly, quarterly, and annual bases. Thorough documentation of assumptions, emissions factors, and data sources is also essential to maintain the integrity of the accounting process. By adhering to these principles and practices, companies can not only meet regulatory requirements but also gain valuable insights into their environmental performance, enabling them to make informed decisions and implement effective strategies for emissions reduction (Plana Earth 2024b).

Carbon accounting involves several key steps to accurately measure and manage an organisation's greenhouse gas emissions. Figure 3.22 illustrates these key steps:

As can be seen from the figure above, the process begins by defining emissions scopes, categorised as Scope 1 (direct emissions), Scope 2 (indirect emissions from purchased energy), and Scope 3 (other indirect emissions) (EPA 2024). Further details on the scopes are given in the GHG Protocol section. Next, relevant emissions data is collected and thoroughly reviewed for accuracy and completeness. The third step involves performing calculations to convert various types of emissions into their carbon dioxide equivalents (CO2e), providing a standardised measure of global warming potential. Finally, the resulting emissions inventory is analysed to identify opportunities for reduction, enabling organisations to develop targeted strategies for minimising their carbon footprint and improving overall environmental performance.

One critical aspect to consider in the process of carbon accounting for an organisation is the selection of carbon accounting methods. These methods vary in their approach, data requirements, and level of precision. The choice of method often depends on the availability of data, the resources at hand, and the specific emissions category being assessed. Some methods rely on direct measurements, while others use financial data or industry averages to estimate emissions. It's common for organisations to employ a com-

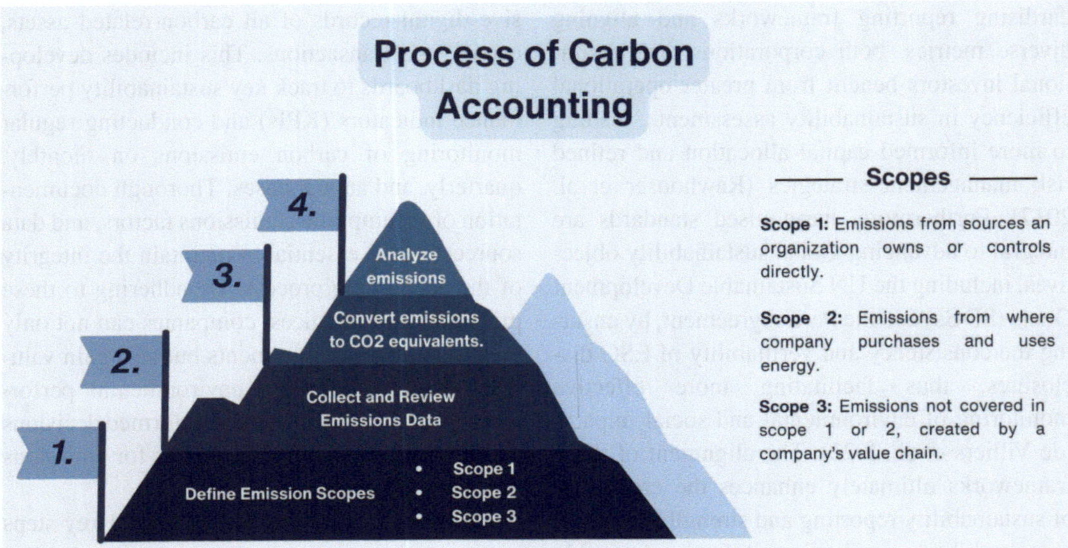

Fig. 3.22 Process of carbon accounting. (Brightest 2024; National Grid 2024)

Table 3.6 Carbon accounting methods (Plana Earth 2024b)

Method	Description
Physical-unit method	Describes activity data used to calculate GHG emissions based on physical units consumed (e.g., kWh and litres)
Spend-based method	Calculates GHG emissions by multiplying the monetary value of purchases by their respective emissions factors (e.g., euros spent on IT hardware × emissions per euro)
Supplier-specific method	Gathers product-specific cradle-to-gate emission data directly from suppliers. Most accurate but resource-intensive method
Average-data method	Calculates emissions by multiplying mass or relevant units by secondary (e.g., industry average) emission factors
Hybrid method	Combines supplier-specific data with secondary data to fill gaps; may mix physical- and spend-based methods

bination of these methods to achieve a comprehensive and accurate emissions inventory. Table 3.6 describes the main carbon accounting methods used in practice.

Table 3.6 outlines five distinct carbon accounting methods: physical-unit, spend-based, supplier-specific, average-data, and hybrid methods. The physical-unit method calculates emissions based on actual consumption of resources, such as kilowatt-hours of electricity or litres of fuel. The spend-based method, however, uses financial data to estimate emissions by applying emission factors to monetary values of purchases. The supplier-specific method, while resource-intensive, provides the most accurate results by gathering detailed emission data directly from suppliers. The average-data method relies on industry averages and secondary data sources, making it less precise but more widely applicable. Finally, the hybrid method combines elements of the other approaches, offering flexibility in situations where data availability varies. Each method has its strengths and limitations, and the choice often depends on the specific context and data availability of the organisation conducting the carbon accounting (Plana Earth 2024b).

3.6.2 GHG Reporting Standards and Frameworks

While understanding the fundamentals of greenhouse gas emissions and carbon accounting is crucial, it's equally important to recognise the standardised frameworks that guide how this information is reported and communicated. As the global focus on climate change intensifies, numerous organisations and regulatory bodies have developed comprehensive standards and protocols to ensure consistency, transparency, and comparability in GHG reporting. These frameworks provide organisations with structured approaches to measure, manage, and disclose their environmental impact. In this section, some of the most widely recognised and adopted GHG reporting standards and frameworks will be explored.

3.6.2.1 Carbon Disclosure Project (CDP)

CDP was established in 2000 as an international non-profit organisation known as the "Carbon Disclosure Project". Scope of the project was helping companies in disclosures concerning climate impact. In 2013, Carbon Disclosure Project rebranded only as CDP going forward to expand their focus further than climate change to obtain a wider environmental impact. Since then, CDP helps companies, cities, and governments in disclosing their environmental impact concerning climate change, water security, and deforestation (CDP 2023a; Carbon Trail 2023).

The CDP evaluates companies' environmental performance and reporting through a comprehensive scoring system. This system is based on four key categories, each building upon the previous one to provide a holistic assessment of a company's environmental efforts and disclosures.

The first category, "Disclosure", measures how successful the company is in reporting environmental data. The relevance and amount of data reported are crucial factors in this assessment. Moving to the second category, "Awareness" evaluates how well the company understands the link between environmental concerns and its own practices. This deeper level of understanding is essential for effective environmental management. The third category, "Management", assesses the concrete actions a company takes to improve its environmental impact. This category looks at the practical steps and initiatives implemented by the organisation to address environmental issues. The fourth and highest category is "Leadership", which recognises companies that go above and beyond in their environmental efforts and serve as examples for others in their industry (CDP 2023b).

It's important to note that each question in the CDP assessment evaluates the entity across a combination of these categories. To be evaluated at the higher-level categories, entities need to achieve a threshold score in the previous level, ensuring a comprehensive and progressive approach to environmental performance (Sustainserv 2023).

One of the key strengths of the CDP is its aim to align with other disclosure frameworks and standards. This alignment includes prominent frameworks such as the European Sustainability Reporting Standards (ESRS), the Task Force on Nature-Related Financial Disclosures (TNFD), the Task Force on Climate-Related Financial Disclosure (TCFD), and the International Financial Reporting Standards (IFRS) S2 Climate-Related Disclosures (Segal 2023). This compatibility significantly reduces the workload for organisations in terms of disclosure, as they can report through CDP and simultaneously fulfil many requirements of other frameworks.

3.6.2.2 Climate-Related Financial Disclosures and Risk Assessment

There are numerous frameworks and regulatory initiatives focusing on climate-related financial disclosures, risk assessment, and sustainability. The Securities and Exchange Commission (SEC) Climate Rules, Taskforce on Nature-Related Financial Disclosures (TNFD), and Central Bank Climate Stress Tests (e.g., the Network for Greening the Financial System (NGFS)) are worthy of mentioning by being widely adopted by companies, institutions, and regulatory bodies worldwide. The following figure provides a summary of the comparative assessment (Fig. 3.23).

3.6.2.3 Climate Registry General Reporting Protocol

The Climate Registry (TCR) designs and operates voluntary and compliance greenhouse gas (GHG) reporting programs globally and assists organisations in measuring, reporting, and verifying (MRV) the carbon in their operations in

Feature	*SEC Climate Rules*	*Taskforce on Nature-related Financial Disclosures (TNFD)*	*Central Bank Climate Stress Tests (e.g., NGFS)*
Primary Objective	*Investor disclosure: standardise material climate-related financial information for U.S. public companies.*	*Shift global financial flows away from nature-negative outcomes towards nature-positive outcomes.*	*Financial stability assessment: quantify the exposure of financial institutions to climate risks and support micro- and macroprudential measures.*
Primary Focus	*Climate-related risks (physical and transition) are impacting financial performance and strategy.*	*Nature-related dependencies, impacts, risks, and opportunities (physical, transition, systemic).*	*Macro-financial impacts of climate change (physical and transition risks) on the economy and financial system.*
Regulatory Status/Nature	*Mandatory regulatory framework for U.S. public companies (currently stayed).*	*Voluntary, market-led, science-based framework.*	*Supervisory guidance and reference scenarios; individual central banks may mandate stress tests for large financial institutions.*
Types of Risks Addressed	*Physical risks (acute/chronic); Transition risks.*	*Nature-related physical, transition, and systemic risks.*	*Physical risks (acute/chronic); Transition risks.*
Applicability/Target Entities	*U.S. public companies (Large Accelerated, Accelerated, Non-Accelerated Filers, Smaller Reporting, Emerging Growth Companies).*	*Organisations of all sizes, across all sectors and geographies, including financial institutions and companies.*	*Central banks and financial supervisors, as well as private financial institutions (such as banks and insurance firms), conduct internal exercises. Large financial institutions, including those with assets exceeding USD 100 billion (e.g., U.S. banks), are subject to mandatory tests.*
GHG Scope 1, 2, 3 Requirements	*Material Scope 1 & 2 required for LAF/AF; Scope 3 explicitly excluded.*	*Not directly specified, but the framework supports assessment of impacts that could include emissions.*	*N/A; provides scenarios that include emissions pathways as inputs for risk assessment.*
Relationship to TCFD/ISSB	*Good interoperability with ISSB standards.*	*Based on TCFD's four pillars, close alignment with ISSB standards to harmonise reporting.*	*Central banks and supervisors use NGFS scenarios to support TCFD-aligned disclosures and broader climate risk analysis.*

Fig. 3.23 Summary of the Securities and Exchange Commission (SEC) Climate Rules (sec 2024), Taskforce on Nature-Related Financial Disclosures (TNFD) (tnfd 2025), and Central Bank Climate Stress Tests (e.g., the Network for Greening the Financial System (NGFS)) (ngfs 2025)

order to manage and reduce it. TCR also consults with governments nationally and internationally on all aspects of GHG measurement, reporting, and verification (The Climate Registry 2023). TCR's Carbon Footprint Registry is aligned with international standards and provides a nexus between business, government, and nongovernmental organisations to share policy information and exchange best practices.

The General Reporting Protocol (GRP) is a comprehensive framework designed to ensure accurate and consistent reporting of greenhouse gas (GHG) emissions. At its core, the GRP is built upon five fundamental accounting and reporting principles that serve as the foundation for its requirements and guidance (The Climate Registry 2019). These principles are crucial in ensuring that GHG data provides a faithful, true, and fair representation of an organisation's emissions.

The first principle, Relevance, emphasises the importance of creating a GHG inventory that accurately reflects an organisation's emissions while meeting the decision-making needs of both internal and external stakeholders. This principle ensures that the reported data is not only accurate but also useful and applicable to its intended audience. Completeness, the second principle, requires organisations to account for and report all relevant GHG emissions and activities within their defined inventory boundary. This comprehensive approach prevents the omission of significant emission sources and provides a full picture of the organisation's environmental impact. The third principle, Consistency, is vital for enabling meaningful comparisons of emissions over time (Pocknell 2017). It calls for the use of consistent methodologies and clearly documented changes in data, inventory boundaries, or methods. This principle allows stakeholders to track progress and identify trends in an organisation's GHG emissions. Transparency, the fourth principle, emphasises the need for clear and factual reporting. Organisations are required to address all relevant issues coherently, disclose assumptions, and provide references to quantification methods and data sources. This principle ensures that intended users can make informed decisions with reasonable confidence in the reported information. The final principle, Accuracy, focuses on the quality of the quantification process. It requires organisations to ensure that their GHG emissions are neither systematically overstated nor understated, and that uncertainties are minimised to the extent practicable (The Climate Registry 2019). This principle is crucial for maintaining the integrity of the reported information and enabling users to make decisions with reasonable assurance.

By adhering to these five principles—Relevance, Completeness, Consistency, Transparency, and Accuracy—the GRP provides a robust framework for organisations to report their GHG emissions in a manner that is credible, comprehensive, and useful for decision-making. These principles work together to create a standardised approach to GHG reporting, facilitating better understanding and comparability of emissions data across different organisations and time periods.

The General Reporting Protocol (GRP) is structured into five distinct yet interconnected modules, each designed to provide essential components for TCR's Carbon Footprint Registry. These modules are intended to be used in conjunction with one another, creating a comprehensive framework for greenhouse gas (GHG) reporting (The Climate Registry 2019).

The first module, Introduction, serves as the foundation for understanding the GRP's structure and purpose. It provides an overview of the protocol and sets the stage for the more detailed modules that follow.

The second module, Inventory Boundaries, is crucial for defining the scope of an organisation's GHG inventory. This module provides guidelines for establishing organisational and reporting boundaries, enumerates the GHGs and types of GHG sources that may be included in the inventory, and explains the various levels of granularity at which emissions data can be made publicly available. By clearly delineating these boundaries, organisations can ensure that their GHG inventories are comprehensive and accurately represent their emissions profile.

GHG Emissions Quantification, the third module, forms the core of the GRP by offering methodologies for measuring emissions across various source categories. It covers direct emissions from combustion in fixed sources (such as boilers or furnaces), direct emissions from combustion in mobile sources (like vehicles), indirect emissions from the use of non-electric energy (e.g., purchased steam or heat), and direct fugitive emissions (such as leaks from refrigeration systems). This module provides organisations with the tools and techniques necessary to accurately calculate their emissions across these diverse sources.

The fourth module, Advanced Methods, builds upon the basic quantification techniques by offering more sophisticated approaches to emissions calculation. This module is designed for organisations with complex emissions profiles or those seeking to enhance the accuracy and detail of their GHG inventories.

The final module, Reporting an Inventory, guides organisations through the process of compiling and submitting a complete GHG inventory. It outlines additional required and optional data that should be included alongside the quantification of GHG emissions. This module ensures that the reported inventory is not only accurate but also provides context and additional information that enhances its usefulness for stakeholders.

Together, these five modules—Introduction, Inventory Boundaries, GHG Emissions Quantification, Advanced Methods, and Reporting an Inventory—form a comprehensive and structured approach to GHG reporting. By using these modules in tandem, organisations can develop robust, accurate, and transparent GHG inventories that meet the requirements of TCR's Carbon Footprint Registry and provide valuable insights into their environmental impact (The Climate Registry 2019).

3.6.2.4 GHG Protocol

The Greenhouse Gas Protocol (GHG) is a comprehensive global framework that helps businesses and governments assess and manage greenhouse gas emissions. The GHG Protocol, a fundamental component of GHG reporting standards and frameworks, provides defined methods and tools for accounting and reporting emissions, allowing enterprises to measure their environmental impact and execute effective climate plans (LoPucki 2022).

The GHG Protocol was created by a collaboration between the World Resources Institute (WRI) and the World Business Council for Sustainable Development. It was launched in 2001 and has since become the most popular international accounting tool for government and industry leaders to analyse, quantify, and manage greenhouse gas emissions (LoPucki 2022). The protocol has two main standards: the GHG Protocol Corporate Standard and the GHG Protocol Project Quantification Standard. The "Corporate Standard" provides a framework for businesses to create a GHG inventory that encompasses emissions from all operations, whereas the "Project Quantification Standard" provides methodology for quantifying greenhouse gas reductions from specific projects (WWF 2024).

Importance of the GHG Protocol

The GHG Protocol is an important part of the worldwide effort to prevent climate change. By establishing a standard methodology, it allows for more comparability and transparency in emissions reporting, which is critical for stakeholders like investors, consumers, and regulatory agencies. The Protocol's systematic approach assists organisations in identifying emission hotspots, establishing reduction targets, and increasing the credibility of their sustainability claims (WWF 2024).

Companies, for example, can track their emissions at three different levels (WWF 2024):

- Scope 1: Direct emissions.
- Scope 2: Indirect emissions from purchased electricity.
- Scope 3: Indirect emissions from the value chain.

This detailed tracking is critical since Scope 3 emissions frequently account for the majority of a company's overall emissions, making precise accounting essential for effective climate action (Singh et al. 2023).

Challenges and Limitations

Despite its widespread adoption, the GHG Protocol faces numerous hurdles. One key challenge is the heterogeneity in how firms understand and apply the protocol, notably for Scope 3 emissions, which frequently account for the bulk of a company's overall emissions (WWF 2024). The complexity of supply chains, combined with the lack of defined procedures for assessing Scope 3 emissions, can result in reporting errors and inaccuracies (Singh et al. 2023). Furthermore, while the GHG Protocol provides a strong framework, it does not require mandatory reporting, which might lead to varied levels of commitment across enterprises (LoPucki 2022).

Integration with Other Standards

The GHG Protocol is frequently combined with other reporting standards and frameworks, including the Sustainability Accounting Standards Board (SASB) and the Task Force on Climate-Related Financial Disclosures (TCFD). These connections broaden the scope of company reporting by tying GHG emissions data to financial performance and risk management initiatives. For example, the SEC's proposed regulation on obligatory GHG emissions reporting is consistent with the GHG Protocol, implying that adhering to this framework will become increasingly important for compliance and transparency in corporate reporting (LoPucki 2022).

At last, the GHG Protocol acts as a core element in the landscape of GHG reporting standards and guidelines. Its comprehensive approach to emissions accounting and reporting allows firms to better understand their environmental impact, set reduction objectives, and convey progress. However, the obstacles associated with its implementation underline the importance of continuous efforts to standardise procedures and improve the reliability of emissions reporting across businesses (Singh et al. 2023).

3.6.2.5 ISO 14064

ISO 14064 is a family of standards used in quantifying, monitoring, reporting, and validating GHG emissions and removals (ISO 2018). The standard aims to support sustainable development through a low-carbon economy and to benefit organisations. The standard covers all types of emissions from transportation to waste. For carbon accounting, it provides emission calculation methods. In terms of carbon reporting, it specifies the content and presentation of a GHG report. Thus, it creates a common ground for companies to track and monitor their carbon footprint (Duault 2024; LRQA 2024).

One of the primary benefits of these ISO standards is the enhancement of environmental integrity in GHG quantification. By providing standardised methodologies and best practices, ISO 14064 ensures that organisations are measuring their GHG emissions and removals in a manner that accurately reflects their environmental impact (Tunley Environmental 2023). This improved accuracy forms the foundation for all subsequent GHG-related activities and decision-making.

Furthermore, these standards significantly enhance the credibility, consistency, and transparency of GHG processes. They establish a common language and framework for quantification, monitoring, reporting, verification, and validation of GHG data. This standardisation not only improves the quality of reported information but also facilitates better communication and understanding among stakeholders, both within and outside the organisation.

The ISO 14064 standards also play a crucial role in facilitating the development and implementation of GHG management strategies and plans. By providing a structured approach to GHG accounting, these standards enable organisations to gain a clear understanding of their emissions profile, identify key areas for improvement, and develop targeted strategies for GHG management (Tunley Environmental 2023). This strategic approach is essential for organisations looking to reduce their environmental impact and contribute to global climate goals.

In addition to strategy development, these standards facilitate the implementation of concrete mitigation actions (ISO 2018). They provide guidance on how to quantify and report emission reductions or removal enhancements, enabling organisations to track the effectiveness of their mitigation efforts. This capability is particularly valuable for organisations engaged in carbon offset projects or those aiming to achieve carbon neutrality.

Lastly, the ISO 14064 standards greatly enhance an organisation's ability to track performance and progress in GHG management. By establishing consistent methodologies for quantification and reporting, these standards allow for meaningful comparisons of GHG data over time. This enables organisations to accurately measure their progress in reducing emissions or increasing removals, providing valuable insights for continuous improvement and demonstrating commitment to environmental goals (Tunley Environmental 2023).

These standards support various organisational needs, from identifying emission reduction opportunities and increasing profitability through reduced energy consumption to managing climate-related risks and opportunities. They facilitate participation in voluntary initiatives such as GHG registries and sustainability reporting, while also supporting activities in GHG markets and assisting with regulatory compliance in government GHG programmes.

The term "family of standards" reflects the interconnected nature of ISO 14064 (ISO 2023). While ISO 14064-1 provides the core framework for organisational GHG inventories, other components address specific aspects:

- ISO 14064-2: Focuses on GHG projects
- ISO 14064-3: Covers verification of GHG statements
- ISO 14065: Outlines requirements for validation and verification bodies
- ISO 14066: Addresses team competence for GHG validation and verification
- ISO 14067: Deals with product carbon footprints

3.6.2.6 Streamlined Energy and Carbon Reporting (SECR)

Streamlined Energy and Carbon Reporting (SECR) is a UK-based government policy that requires large companies to report on their annual energy use and greenhouse gas emissions, as well as any efficiency measures they have taken throughout the year. It was first introduced on April 1, 2019, and last updated in September 2023 (UK Government 2024; UL Solutions 2024).

The goal of SECR is to expand the benefits of carbon and energy reporting among UK companies (UL Solutions 2024). It also aims to improve alignment with other energy policies in the UK and provide greater transparency for stakeholders (Persefoni 2024; UK Government 2024).

The SECR guidance is for chief financial officers, accounting officers, and trustees in academy trusts (UK Government 2024). The key criteria for companies that need to comply with SECR are:

- 250 employees
- Annual turnover of more than £36 m.
- Annual balance sheet of over £18 m.

So SECR is for large businesses, both quoted and unquoted companies, as well as large limited liability partnerships (LLPs) in the UK (UL Solutions 2024).

According to SECR guidelines, an organisation will need to report:

- Energy use which includes gas, purchased electricity and transport fuel, along with GHG emissions.
- Information around methodology used to measure energy use and emissions.
- A narrative description of efforts taken to improve energy efficiency over the last 12 months.
- An intensity ratio which is used to compare emissions data with a business metric, to allow for comparison with similar businesses and previous years.

- Equivalent figures from previous years as a base for comparison on changes and improvements.

The policy investigates emissions on three scopes (UK Government 2024; PwC 2024).

- Scope 1: Direct GHG emissions.
- Scope 2: Energy indirect emissions.
- Scope 3: Other indirect emissions.

The following table illustrates how an academy trust might calculate figures (2022/23) (Table 3.7).

This table illustrates how an academy trust might calculate its carbon emissions for the 2022/23 academic year across three key areas: gas, electricity, and transport. For gas consumption, the trust used 171,584 kWh, falling under Scope 1 emissions. Using a conversion factor of 0.18256 kgCO2e per kWh for natural gas in 2023, this resulted in 31.32 tCO2e. Electricity usage of 56,169 kWh is categorised as Scope 2 emissions, with a UK-specific conversion factor of 0.207074 kgCO2e per kWh, leading to 11.63 tCO2e. Transport emissions, also Scope 1, come from two mini-buses travelling a total of 15,433 miles. Using a conversion factor of 0.27745 kgCO2e per mile for diesel vans, this contributed 4.28 tCO2e (UK Government 2024).

3.6.2.7 Task Force on Climate-Related Financial Disclosures (TCFD)

The Financial Stability Board (FSB) formed the Task Force on Climate-Related Financial Disclosures (TCFD) in December 2015 to improve the transparency of climate-related financial reporting. The TCFD intends to create a framework for businesses to consistently communicate climate-related risks and opportunities, thereby enhancing the quality of information available to investors, lenders, insurers, and other stakeholders (Nisanci 2021). The TCFD was established in response to the growing realisation that climate change poses a systemic danger to the global economy. Its principal goal is to create voluntary, uniform climate-related financial risk disclosures that corporations can utilise to provide information to stakeholders. The TCFD's guidelines are intended to be applicable to companies in all sectors and jurisdictions, with an emphasis on four important areas: governance, strategy, risk management, and metrics and targets (Zenghelis et al. 2016).

The TCFD framework encourages businesses to examine and disclose how climate-related risks and opportunities affect their operations and financial performance. Organisations that incorporate climate-related factors into their financial reporting can better inform investors and other stakeholders about their resilience to climate change and the possible financial consequences of climate-related risks (Eccles et al. 2018). Since its establishment, the TCFD has garnered substantial traction with businesses and investors. According to a 2020 report, 42% of corporations with a market capitalisation of more than USD 10 billion released information according to TCFD recommendations (Chua et al. 2022). Despite this development, many businesses continue to struggle with implementing the TCFD framework, particularly in terms of matching disclosures with recommended criteria. The extent of TCFD implementation varies greatly by industry, with certain industries having better compliance rates than others. For instance, the financial sector has made significant progress in incorporating climate-related risks into its governance and risk management frameworks, whereas industries

Table 3.7 Example emission calculations (UK Government 2024)

Energy source	Consumption	Scope	Emissions calculation
Gas	171,584 kWh	Scope 1	171,584 kWh * 0.18256 = 31.32 tCO2e
Electricity	56,169 kWh	Scope 2	56,169 kWh * 0.207074 = 11.63 tCO2e
Transport	17,076 kWh	Scope 1	15,433*0.27745 = 4.28 tCO2e

such as oil and gas confront unique challenges due to the nature of their operations (Aversa 2023). Organisations frequently struggle to assess climate-related risks and opportunities, limiting their capacity to make meaningful disclosures. The lack of established measurements and procedures for identifying climate hazards exacerbates the reporting process (David et al. 2021). As a result, many companies may give disclosures that lack openness and comparability, reducing their usefulness to investors and other stakeholders (Achenbach 2021).

The execution of TCFD guidelines is not without challenges. One key impediment is the lack of a globally acknowledged definition of climate-related risks, which can result in variations in reporting standards. Furthermore, many organisations may lack the skills and resources required to adequately review and report climate-related data (Hiung et al. 2023). Another difficulty is the changing regulatory landscape for climate-related disclosures. While the TCFD framework is optional, some countries are beginning to compel climate-related reporting, putting additional pressure on enterprises to meet both voluntary and legal standards (Lupu and Ivan 2024). This combined pressure can create misunderstanding and uncertainty about the expectations for climate-related disclosures. Furthermore, the constantly changing nature of climate science and policy makes it difficult for corporations to offer forward-looking disclosures. Companies must constantly alter their plans and disclosures to reflect new information and developments connected to climate change (Tyagi 2018).

Acknowledgment The author would like to acknowledge the help and contributions of K. Kağan Heper, Emrecan Ünal, İbrahim Ayrancı, Oğuz Eren Saklı, Alparslan Burak Başaran, Melis Bostancı, Selim Burak Cantürk, and Berkay Ispir in completing this chapter.

References

Abbasi MH et al (2023) A framework to identify and prioritise the key sustainability indicators: assessment of heating systems in the built environment. Sustain Cities Soc 95:104629. https://doi.org/10.1016/j.scs.2023.104629

Acemoglu D, Robinson JA (2013) Why nations fail: the origins of power, prosperity, and poverty. Crown Business, New York

Achenbach M (2021) Transparency of climate-related risks and opportunities: determinants influencing the disclosure in line with the Task Force on climate-related financial disclosures. Glocality 4(1). https://doi.org/10.5334/glo.32

Ajmal MM et al (2017) Conceptualizing and incorporating social sustainability in the business world. Int J Sustain Dev World Ecol 25(4):327–339. https://doi.org/10.1080/13504509.2017.1408714

Al-Thani MJ, Koç M (2024) In search of sustainable economy indicators: a comparative analysis between the sustainable development goals index and the green growth index: sustainability (2071-1050). Sustainability (2071-1050), [online] 16(4):1372. https://doi.org/10.3390/su16041372

Asif M et al (2013) An integrated management systems approach to corporate social responsibility. J Clean Prod 56:7–17. https://doi.org/10.1016/j.jclepro.2011.10.034

Atzori L, Iera A, Morabito G (2010) The internet of things: a survey. Comput Netw 54(15):2787–2805

Aversa D (2023) Climate change and climate-related financial disclosures in the banking sector. Risk Govern Control Fin Market Inst 13(1):70–94. https://doi.org/10.22495/rgcv13i1p6

B Corporation (2024) B lab's response to the International Sustainability Standards Board (ISSB) consultation. B Lab Europe. Available at: https://bcorporation.eu/news_article/b-lab-response-to-the-international-sustainability-standards-board-issb-consultation/. Accessed: 20 July 2024

Barral AC (2023) Understanding the relationship between CSRD, EU Taxonomy, and SFDR: a comprehensive guide, Greenomy. Available at: https://www.greenomy.io/blog/relationships-csrd-eu-taxonomy-sfdr. Accessed: 6 Oct 2024

Bell S, Morse S (2003) Measuring sustainability: learning from doing. Earthscan Publications, London

Betti G, Consolandi C, Eccles R (2018) The relationship between investor materiality and the sustainable development goals: a methodological framework. Sustainability 10(7):2248. https://doi.org/10.3390/su10072248

Bhat, M. A. (2024). Artificial Intelligence and Sustainable Development: Potentials and Risks. Springer Nature

Blesserholt J (2021) The sins of greenwashing. Available at: https://www.diva-portal.org/smash/get/diva2:1562569/FULLTEXT01.pdf. Accessed: 27 Apr 2024

Boiral O, Henri J (2016) Is sustainability performance comparable? A study of gri reports of mining organizations. Bus Soc 56(2):283–317. https://doi.org/10.1177/0007650315576134

Borgia, E. (2014). The Internet of Things vision: Key features, applications, and open issues. Computer Communications, 54, 1–31. https://doi.org/10.1016/j.comcom.2014.09.008

Borhan, N., & Elsadig, M. A. (2010). The role of information and communication technology for sustainable development: Evidence from Malaysia. International Journal of Economics and Finance Studies, 2(1), 1–12

Bradley L (2023) Five things you need to know about the SASB standards, KPMG. Available at: https://kpmg.com/xx/en/home/insights/2023/08/ifrs-blog-sasb-standards.html. Accessed: 23 July 2024

Branquart C (2023) CSRD insights: key challenges, common mistakes & tips, CSRD insights: key challenges, common mistakes & tips. Available at: https://www.greenomy.io/blog/csrd-reporting-expert-insights. Accessed: 28 June 2024

Bright F (2024) An introduction to ISSB's New International Sustainability Disclosure Standards (IFRS SDS). EcoAct. Available at: https://eco-act.com/blog/issb-disclosure-standards/. Accessed: 16 July 2024

Brightest (2024) Carbon accounting methods. Available at: https://www.brightest.io/carbon-accounting-methods. Accessed: 12 Sept 2024

Buggie, M. and Farbstein, E. (2024) Carbon accounting business benefits. Normative. Available at: https://normative.io/insight/carbon-accounting-business-benefits/#:~:text=Carbon%20accounting%20quantifies%20the%20climate,risk%2C%20and%20greenwashing%2Dfree. Accessed: 12 Sept 2024

Cahill J (2024) Compliance with CSRD: the challenges and strategic benefits for business. Savvi Recruitment. Available at: https://www.savvirecruitment.com/blog/compliance-csrd-challenges-strategic-benefits-for-business. Accessed: 30 June 2024

Cairns J, McCormick PV, Niederlehner BR (1993) A proposed framework for developing indicators of ecosystem health. Hydrobiologia 263(1):1–44. https://doi.org/10.1007/bf00006084

California Office of the Attorney General (2020) California Consumer Privacy Act (CCPA). Available at: https://oag.ca.gov/privacy/ccpa. Accessed: 4 Sept 2024

Carbometrix (2024) All about SFDR regulation and the challenges it represents for investment funds. [Online] Available at: https://carbometrix.com/insights/all-about-sfdr-regulation-and-the-challenges-it-represents-for-investment-funds/. Accessed: 5 Oct 2024

Carbon Chain (2024) Carbon reporting. Available at: https://www.carbonchain.com/carbon-reporting#:~:text=Carbon%20reporting%20is%20the%20process,GHG)%20emissions%20of%20an%20organization. Accessed: 12 Sept 2024

Carbon Trail (2023) What is the carbon disclosure project?. Available at: https://carbontrail.net. Accessed: 13 Sept 2024

Case Western Reserve University (2023) Data collection | Research Data. Available at: https://case.edu/research/resources/data-management/data-collection. Accessed 5 July 2024

CDP (2023a) About us. Available at: https://www.cdp.net/en/info/about-us. Accessed: 13 Sept 2024

CDP (2023b) CDP scores explained. Available at: https://www.cdp.net/en/scores. Accessed: 13 Sept 2024

Chen Y, Chang C (2012) Enhance green purchase intentions. Manag Decis 50(3):502–520. https://doi.org/10.1108/00251741211216250

Christensen HB, Hail L, Leuz C (2021) Mandatory CSR and sustainability reporting: economic analysis and literature review. Rev Acc Stud 26(3):1176–1248. https://doi.org/10.1007/s11142-021-09609-5

Chua WF, James R, King A, Lee E, Soderstrom N (2022) Task Force on climate-related financial disclosures (TCFD) implementation: an overview and insights from the Australian Accounting Standards Board dialogue series. Aust Account Rev 32(3):396–405. https://doi.org/10.1111/auar.12388

Clifft, B. (2023). Digital Governance and Ethical AI Implementation. MIT Press

ClimatePartner (2024) European Sustainability Reporting Standards (ESRS). ClimatePartner. Available at: https://www.climatepartner.com/en/knowledge/glossary/european-sustainability-reporting-standards-esrs. Accessed: 02 July 2024

Clune WH, Zehnder AJ (2018) The three pillars of sustainability framework: approaches for laws and governance. J Environ Prot 09(03):211–240. https://doi.org/10.4236/jep.2018.93015

Colantonio, Andrea (2009) Social sustainability: a review and critique of traditional versus emerging themes and assessment methods. In: Horner, M., Price, A., Bebbington, J. and Emmanuel, R., (eds.) Sue-mot conference 2009: second international conference on whole life urban sustainability and its assessment: conference Proce. Loughborough University, Loughborough, pp. 865–885. ISBN 9780947974

Coolset (2024) Understanding the European Sustainability Reporting Standards (ESRS)—updated June 2024. Available at: https://www.coolset.com/academy/european-sustainability-reporting-standards-esrs. Accessed: 05 July 2024

Cremasco C, Boni L (2022) Is the European Union (EU) sustainable finance disclosure regulation (SFDR) effective in shaping sustainability objectives? An analysis of investment funds' behaviour. J Sustain Financ Invest:1–19. https://doi.org/10.1080/20430795.2022.2124838

Dale VH et al (2015) A framework for selecting indicators of bioenergy sustainability. Biofuels Bioprod Biorefin 9(4):435–446. https://doi.org/10.1002/bbb.1562

David, F., Larcker, D. F., & Tayan, B. (2021). Corporate Governance Matters: A Closer Look at Organizational Choices and Their Consequences (4th ed.). Pearson.

de Freitas Netto SV et al (2020) Concepts and forms of greenwashing: a systematic review. Environ Sci Eur 32(1). https://doi.org/10.1186/s12302-020-0300-3

de Villiers C, La Torre M, Molinari M (2022) The global reporting initiative's (GRI) past, present and future: critical reflections and a research agenda on sustainability reporting (standard-setting). Pac Account Rev 34(5):728–747. https://doi.org/10.1108/par-02-2022-0034

Delmas MA, Burbano VC (2011) The drivers of greenwashing. Calif Manag Rev 54(1):64–87. https://doi.org/10.1525/cmr.2011.54.1.64

Dempsey N et al (2011) The social dimension of sustainable development: defining urban social sustainability. Sustain Dev 19(5):289–300. https://doi.org/10.1002/sd.417

Denuwara N et al (2022) Corporate economic performance and sustainability indices: a study based on the Dow Jones Sustainability Index. SN Bus Econ 2(7). https://doi.org/10.1007/s43546-022-00251-0

Dissanayake D, Tilt C, Xydias-Lobo M (2016) Sustainability reporting by publicly listed companies in Sri Lanka. J Clean Prod 129:169–182. https://doi.org/10.1016/J.JCLEPRO.2016.04.086. Accessed 5 July 2024

Dobility Inc. (2024). SurveyCTO Platform Documentation and User Guide. SurveyCTO

Dong, X., Chen, W., & Shi, Y. (2017). Big data analytics for supply chain management: A review. Journal of Business Research, 70, 264–276.

Doni F, Corvino A, Bianchi Martini S (2022) Corporate governance model, stakeholder engagement and social issues: evidence from the European oil and gas industry. Soc Responsib J 18(3):636–662. https://doi.org/10.1108/SRJ-08-2020-0336. [Accessed 5 July 2024]

Duarte F (2024) Amount of data created daily (2024). Explod Topics. Available at: https://explodingtopics.com/blog/data-generated-per-day. Accessed: 2 July 2024

Duault M (2024) ISO 14064: a reference in carbon accounting. Traace 2024. Available at: https://en.traace.co/post/iso-14064-a-reference-in-carbon-accounting. Accessed: 13 Sept 2024

Durand R, Paugam L, Stolowy H (2019) Do investors actually value sustainability indices? Replication, development, and new evidence on CSR visibility. Strateg Manag J 40(9):1471–1490. https://doi.org/10.1002/smj.3035

Duvernay G (2024) CSRD timeline: when should your company report?. [online]. Plana.earth.

Eccles, R. G., Ioannou, I., & Serafeim, G. (2018). The impact of corporate sustainability on organizational processes and performance. Management Science, 60(11), 2835–2857

EFRAG (2022) Draft ESRS 2 general disclosures. Available at: https://www.efrag.org/sites/default/files/sites/webpublishing/SiteAssets/07.%20Draft%20ESRS%202%20General%20disclsoures%20November%202022.pdf. Draft ESRS 2 General disclsoures November 2022.pdf. Accessed: 26 June 2024

Elkington J (1994) Towards the sustainable corporation: win-win-win business strategies for sustainable development. Calif Manag Rev 36(2):90–100

EPA (2024) Scope 1 and Scope 2 inventory guidance. Available at: https://www.epa.gov/climateleadership/scope-1-and-scope-2-inventory-guidance. Accessed: 12 Sept 2024

ESGpro. (2024). ESG Data Solutions and Methodology Guide. ESGpro Analytics

European Commission (2018) 2018 reform of EU data protection rules. Available at: https://ec.europa.eu/info/law/law-topic/data-protection/reform_en. Accessed: 4 Sept 2024

European Commission (2020) EU Technical Expert Group on sustainable finance Teg final report on the EU taxonomy. Available at: https://finance.ec.europa.eu/system/files/2020-03/200309-sustainable-finance-teg-final-report-taxonomy_en.pdf. Accessed: 08 May 2024

European Commission (2022) EU taxonomy: complementary climate delegated act to accelerate decarbonisation. Finance. Available at: https://finance.ec.europa.eu/publications/eu-taxonomy-complementary-climate-delegated-act-accelerate-decarbonisation_en. Accessed: 08 May 2024

European Commission (2023) The commission adopts the European Sustainability Reporting Standards. European Commission. Available at: https://finance.ec.europa.eu/news/commission-adopts-european-sustainability-reporting-standards-2023-07-31_en. Accessed: 02 July 2024

European Commission (2024a) Corporate sustainability reporting. Available at: https://finance.ec.europa.eu/capital-markets-union-and-financial-markets/company-reporting-and-auditing/company-reporting/corporate-sustainability-reporting_en. Accessed: 24 June 2024

European Commission (2024b) Corporate sustainability due diligence. Available at: https://commission.europa.eu/business-economy-euro/doing-business-eu/sustainability-due-diligence-responsible-business/corporate-sustainability-due-diligence_en#:~:text=On%2025%20July%202024%2C%20the,across%20their%20global%20value%20chains. Accessed 2 Sept 2024

Faber G, Sick V (2022) Identifying and mitigating greenwashing of carbon utilization products [preprint]. http://hdl.handle.net/2027.42/171930

Friede G, Busch T, Bassen A (2015) ESG and financial performance: aggregated evidence from more than 2000 empirical studies. J Sustain Financ Invest 5(4):210–233. https://doi.org/10.1080/20430795.2015.1118917

Gangadharan L, Valenzuela MR (2001) Interrelationships between income, health and the environment: extending the environmental kuznets curve hypothesis. Ecol Econ 36(3):513–531. https://doi.org/10.1016/s0921-8009(00)00250-0

Gao Y (2024) Watchdogs or enablers? Analyzing the role of analysts in esg greenwashing in China. Sustainability 16(11):4339. https://doi.org/10.3390/su16114339

Garcia-Blandon J, Argilés-Bosch JM, Ravenda D (2023) Leveraging stakeholder engagement for market value growth: empirical evidence on sustainable development leadership in Europe. Sustain Dev 32(1):336–347. https://doi.org/10.1002/sd.2662

Garfinkel S (2015) Database nation: the death of privacy in the 21st century. O'Reilly Media, Inc.

Generation Impact Global (2024) European Sustainability Reporting Standards, Generation Impact Global: we make it easier to do better. Available at: https://www.generationimpact.global/esrs#further. Accessed: 03 July 2024

GEO BON (2023) Biodiversity Habitat Index. [online]. Available at: https://geobon.org/ebvs/indicators/biodiversity-habitat-index/. Accessed 6 Sept 2024

Giesekam J et al (2021) Science-based targets: on target? Sustainability 13(4):1657. https://doi.org/10.3390/su13041657

Gond J-P, Kang N, Moon J (2011) The government of self-regulation: on the comparative dynamics of corporate social responsibility. Econ Soc 40(4):640–671. https://doi.org/10.1080/03085147.2011.607364

GRI (2024) Global reporting initiative (GRI) standards 2024: sustainability reporting. Global Reporting Initiative, Amsterdam

Grossman GM, Krueger AB (1995) Economic growth and the environment. Q J Econ 110(2):353–377. https://doi.org/10.2307/2118443

Gubbi, J., Buyya, R., Marusic, S., & Palaniswami, M. (2013). Internet of Things (IoT): A vision, architectural elements, and future directions. Future Generation Computer Systems, 29(7), 1645–1660. https://doi.org/10.1016/j.future.2013.01.010

Hák T, Moldan B, Dahl AL (2007) Sustainability indicators: a scientific assessment: a project of scope, the scientific committee on problems of the environment, of the International Council for Science. Island Press, Washington (D.C.)

Hák T, Moldan B, Dahl AL (2012) Sustainability indicators: a scientific assessment. Island Press, Washington

Hawn O, Chatterji AK, Mitchell W (2018) Do investors actually value sustainability? New evidence from investor reactions to the Dow Jones Sustainability Index (DJSI). Strateg Manag J 39(4):949–976. https://doi.org/10.1002/smj.2752

Hessenius, R., Vogel, P., & Metzler, G. (2020). Smart city data governance: Challenges and frameworks. Journal of Urban Technology, 27(4), 3–23

Hiung, L., Tan, J., & Wong, S. (2023). Artificial Intelligence Adoption and Sustainability Reporting in Emerging Markets. Journal of Cleaner Production, 423, 138912

Hopwood B, Mellor M, O'Brien G (2005) Sustainable development: mapping different approaches. Sustain Dev 13(1):38–52. https://doi.org/10.1002/sd.244

Hoque N, Biswas WK, Mazhar I, Howard I (2020) Life cycle sustainability assessment of alternative energy sources for the Western Australian Transport Sector [online]. Available at: https://www.researchgate.net/publication/342858606_Life_Cycle_Sustainability_Assessment_of_Alternative_Energy_Sources_for_the_Western_Australian_Transport_Sector. Accessed 6 Sept 2024

Hristov I, Searcy C (2024) Integrating sustainability with corporate governance: a framework to implement the corporate sustainability reporting directive through a balanced scorecard. Manag Decis [Preprint]. https://doi.org/10.1108/md-10-2023-1995

Hsien W, Yang C, Hwang M (2016) A survey of public auditing for secure data storage in cloud computing. Int J Netw Secur

Hsu T (2011) Skepticism grows over products touted as eco-friendly, Los Angeles Times. Available at: https://www.latimes.com/business/la-xpm-2011-may-21-la-fi-greenwash-20110521-story.html. Accessed: 26 Apr 2024

IBM (2024) What is the SASB? IBM. Available at: https://www.ibm.com/topics/sasb. Accessed: 23 Sept 2024

IEA (2014) Energy efficiency indicators: fundamentals on statistics. IEA, Paris

IFC (2023). ESG Guidebook. Available at: https://www.ifc.org/en/insights-reports/2021/ifc-esg-guidebook

IFRS (2024) International Sustainability Standards Board. IFRS. Available at: https://www.ifrs.org/groups/international-sustainability-standards-board/#about. Accessed: 14 July 2024

IIASA, (2023). International Institute for Applied Systems Analysis (IIASA) D5.2: assessment of policy frameworks for sustainable finance. [online]. Available at: https://pure.iiasa.ac.at/id/eprint/19281/1/deliverable-d5-2.pdf [Accessed 4 October 2024]

Inukollu V, Arsi S, Ravuri S (2014) Security issues associated with big data in cloud computing. Int J Netw Secur Appl 6:45–56. https://doi.org/10.5121/ijnsa.2014.6304

Invest Europe (2024) International Sustainability Standards Board (ISSB). [online]. Available at: https://www.investeurope.eu/invest-europe-esg-reporting-guidelines/who-is-who/international-sustainability-standards-board-issb/. Accessed 4 Oct 2024

Iozzelli and Sandoval Velasco (2023) Mandatory or voluntary? The hybrid nature of sustainability disclosure in the EU's Corporate Sustainability Reporting Directive (CSRD), European University Institute. Available at: https://cadmus.eui.eu/bitstream/handle/1814/75910/(13-09)%20RSC%20PP%202023.08.pdf?sequence=1&isAllowed=y. Accessed: 26 June 2024

ISO (2018) ISO 14064-1:2018 greenhouse gases — part 1: specification with guidance at the organization level for quantification and reporting of greenhouse gas emissions and removals. ISO, Geneva. Available at: https://www.iso.org/obp/ui/en/#iso:std:iso:14064:-1:ed-2:v1:en. Accessed: 13 Sept 2024

ISO (2023) ISO 14068-1:2023 greenhouse gas management and related activities—carbon neutrality—part 1: requirements and guidelines for organizations and products. ISO, Geneva

Jain M, Sharma GD, Srivastava M (2019) Can sustainable investment yield better financial returns: a comparative study of ESG indices and MSCI indices. Risks 7(1):15. https://doi.org/10.3390/risks7010015

Jones N et al (2019) Exploring the role of local community perceptions in sustainability measurements. Int J Sustain Dev World Ecol 26(6):471–483. https://doi.org/10.1080/13504509.2019.1638330

Kahn, R. L., & Cannell, C. F. (1983). Interviewing: A Practical Guide for Social Scientists. University of Michigan Press

Khan M, Serafeim G, Yoon A (2016) Corporate sustainability: first evidence on materiality. Account Rev 91(6):1697–1724

Kim Y, Park MS, Wier B (2012) Is earnings quality associated with corporate social responsibility? Account Rev 87(3):761–796

Kim J et al (2015) Measures of perceived sustainability. J Glob Scholars Market Sci 25(2):182–193. https://doi.org/10.1080/21639159.2015.1015473

Koons E (2023) How to avoid greenwashing, Energy Tracker Asia. Available at: https://energytracker.asia/how-to-avoid-greenwashing/. Accessed: 29 Apr 2024

Kooths S (2022) EU taxonomy: mission impossible. The Economists' Voice 19(2):243–249. https://doi.org/10.1515/ev-2022-0028

Kostakis I et al (2023) The investigation of EKC within CO2 emissions framework: empirical evidence from selected cross-correlated countries. Sustain Analyt Model 3:100015. https://doi.org/10.1016/j.samod.2023.100015

KPMG (2022) Survey of sustainability reporting 2022, KPMG. Available at: https://assets.kpmg.com/content/dam/kpmg/se/pdf/komm/2022/Global-Survey-of-Sustainability-Reporting-2022.pdf. Accessed: 04 Oct 2024

Küfeoğlu S (2021a) The home of the future: digitalization and resource management. Springer

Küfeoğlu S (2021b) The enabling technology: internet of things (IoT). In: Küfeoğlu S (ed) The home of the future: digitalization and resource management. Springer International Publishing, Cham, pp 109–140

Lehtonen M, Sébastien L, Bauler T (2016) The multiple roles of sustainability indicators in informational governance: between intended use and unanticipated influence. Curr Opin Environ Sustain 18:1–9. https://doi.org/10.1016/j.cosust.2015.05.009

Leeson, P., & Kuszewski, A. (2023). Behavioural insights for climate action. OECD Publishing

Lindwall C (2023) What is greenwashing?, Be a Force for the Future. Available at: https://www.nrdc.org/stories/what-greenwashing. Accessed: 26 Apr 2024

López MV, Garcia A, Rodriguez LA (2007) Sustainable development and corporate performance: a study based on the Dow Jones sustainability index. J Bus Ethics 75(3):285–300. https://doi.org/10.1007/s10551-006-9253-8

LoPucki LM (2022) Corporate greenhouse gas disclosures. https://scholarship.law.ufl.edu/facultypub/1211/

LRQA (2024) ISO 14064 greenhouse gas emissions verification. Available at: https://www.lrqa.com/en/iso-14064/. Accessed: 13 Sept 2024

Lupu A-C, Ivan OR (2024) Non-financial reporting frameworks in emerging economies. Ovidius University Annals Economic Sciences Series 23(2):745–750. https://doi.org/10.61801/ouaess.2023.2.91

Majláth M (2017) The effect of greenwashing information on AD evaluation. Eur J Sustain Dev 6(3). https://doi.org/10.14207/ejsd.2017.v6n3p92

Marr B (2024) How much data do we create every day? The mind-blowing stats everyone should read. Bernard Marr. Available at: https://bernardmarr.com/how-much-data-do-we-create-every-day-the-mind-blowing-stats-everyone-should-read/. Accessed: 2 July 2024

Mazzoni F, Bezerra M, Orr JJ, Soga K (2022) Sustainability in numbers by data analytics. Circular Econ Sustain

Mebratu D (1998) Sustainability and sustainable development. Environ Impact Assess Rev 18(6):493–520. https://doi.org/10.1016/s0195-9255(98)00019-5

Mlaba K (2021) Greenwashing: what is it and how to avoid it. Global Citizen. Available at: https://www.globalcitizen.org/en/content/greenwashing-what-is-it-and-how-to-avoid-it/. Accessed: 27 Apr 2024

Moldan, B., Janoušková, S., & Hák, T. (2007). Indicators of sustainable development: Past, present and future. Ecological Indicators, 17, 1–13. https://doi.org/10.1016/j.ecolind.2009.03.009

Monks RAG, Minow N (2011) Corporate governance. Wiley, Hoboken

Moran C, O'Neill C (2022) Here are the 7 sins of greenwashing, Environmental Research Institute | UCC ERI. Available at: https://www.ucc.ie/en/eri/news/here-are-the-7-sins-of-greenwashing.html. Accessed: 27 Apr 2024

Mujiani S (2023) Exploring the shift toward sustainable finance and financial reporting: an extensive analysis of emerging trends, research patterns, and implications for global sustainability. West Sci Account Fin 1(02):52–58. https://doi.org/10.58812/wsaf.v1i02.111

National Grid (2024) What are Scope 1, 2 and 3 carbon emissions? Available at: https://www.nationalgrid.com/stories/energy-explained/what-are-scope-1-2-3-carbon-emissions. Accessed: 12 Sept 2024

Ngfs (2025) The Network of Central Banks and Supervisors for Greening the Financial System (NGFS). Available at: https://www.ngfs.net/en

Nisanci DA (2021) FSB task force on climate-related financial disclosures. In: World Scientific Encyclopedia of Climate Change, pp 3–8. https://doi.org/10.1142/9789811213960_0001

Nondo C (2009) An empirical analysis of the interactions between environmental regulations and economic growth [Preprint]. doi:https://doi.org/10.33915/etd.4508

Nyilasy G, Gangadharbatla H, Paladino A (2013) Perceived greenwashing: the interactive effects of green advertising and corporate environmental performance on consumer reactions. J Bus Ethics 125(4):693–707. https://doi.org/10.1007/s10551-013-1944-3

Odobaša R, Marošević K (2023) Expected contributions of the European Corporate Sustainability Reporting

Directive (CSRD) to the sustainable development of the European Union. EU and comparative law issues and challenges series [preprint]. https://doi.org/10.25234/eclic/27463

OECD (2004) Measuring sustainable development: integrated economic, environmental and social frameworks. OECD, Paris

Osborne JW (2013) Best practices in data cleaning: a complete guide to everything you need to do before and after collecting your data. Sage

Pacta (2025) Paris Agreement Capital Transition Assessment. Available at: https://pacta.rmi.org/

Persefoni (2024) What is SECR? Streamlined Energy and Carbon Reporting Explained. Available at: https://www.persefoni.com/blog/what-is-secr-reporting. Accessed: 13 Sept 2024

Pertceva E, Zyulyaeva M (2019) Metrics for sustainable development. In: encyclopedia of sustainability in higher education, pp 1–7. https://doi.org/10.1007/978-3-319-63951-2_289-1

Plana Earth (2024a) Non-financial reporting directive (NFRD) Available at: https://plana.earth/policy/non-financial-reporting-directive-nfrd. Accessed: 27 June 2024

Plana Earth (2024b) Carbon Accounting. Available at: https://plana.earth/glossary/carbon-accounting#:~:text=Carbon%20accounting%20and%20reporting%20should,%2C%20consistency%2C%20transparency%20and%20accuracy. Accessed: 12 Sept 2024

Pocknell L (2017) The climate registry releases general reporting protocol and online reporting tool. GHG Protocol, 4 May, viewed 13 September 2024, https://ghgprotocol.org/blog/climate-registry-releases-general-reporting-protocol-and-online-reporting-tool

PwC (2024) PwC UK Streamlined Energy and Carbon Reporting (SECR). Available at: https://www.pwc.co.uk/services/audit/non-financial-assurance/streamlined-energy-and-carbon-reporting.html. Accessed: 13 Sept 2024

Queensland Government (2022) Water quality indicators. [online]. Available at: https://environment.des.qld.gov.au/management/water/health-indicators. Accessed 6 Sept 2024

QuestionPro (2023) Hard data vs soft data: What's the difference? Available at: https://www.questionpro.com/blog/hard-data-vs-soft-data/. Accessed: 4 Sept 2024

Rahman RA, Homayoun S (2020) Corporate economic, environmental, and social sustainability performance transformation through ESG disclosure. Sustainability 12(9):3910. https://doi.org/10.3390/su12093910. Accessed 5 July 2024

Ramesh G, Logeshwaran J, Aravindarajan V (2023) A secured database monitoring method to improve data backup and recovery operations in cloud computing. BOHR Int J Smart Comput Inform Technol 4:17

Ravat, F., Teste, O., Tournier, R., & Zurfluh, G. (2019). Data quality issues in data warehouse systems. Journal of Data and Information Quality, 11(3), 1–26

Rawhouser, H., Cummings, M., & Newbert, S. L. (2017). Social impact measurement: Current approaches and future directions. Academy of Management Perspectives, 31(2), 25–48

RIB Software SE. (2024). MTWO Construction Cloud: Integrated Digital Platform Overview. RIB Software

Rivery (2023) Big Data Statistics: how much data is there in the world? [online] Rivery. Available at: https://rivery.io/blog/big-data-statistics-how-much-data-is-there-in-the-world/ Accessed: 5 July 2024

Rödl and Partner, 2024. Interoperability of ESRS with Existing Sustainability Reporting Standards and Laws. [online] Available at: https://www.roedl.com/insights/esg-news/2024-3/interoperability-esrs-existing-sustainability-reporting-standards-laws [Accessed 4 October 2024]

Saheed O (2021) Hard data vs soft data: definitions, differences, and uses. Scraping Robot. Available at: https://scrapingrobot.com/blog/hard-data-vs-soft-data/. Accessed: 4 Sept 2024

SASB (2024) SASB standards overview, SASB. Available at: https://sasb.ifrs.org/standards/. Accessed: 23 July 2024

Schäfer, J. L., & Graham, J. W. (2002). Missing data: Our view of the state of the art. Psychological Methods, 7(2), 147–177. https://doi.org/10.1037/1082-989X.7.2.147

Schumacher K, Chenet H, Volz U (2020) Sustainable finance in Japan. J Sustain Financ Invest 10(2):213–246. https://doi.org/10.1080/20430795.2020.1735219

Schütze F, Stede J (2021) The EU sustainable finance taxonomy and its contribution to climate neutrality. J Sustain Finance Invest 14(1):128–160. https://doi.org/10.1080/20430795.2021.2006129

Schütze F et al (2020) The EU taxonomy and sustainability reporting: a critical analysis. [online]. Available at: https://www.econstor.eu/handle/10419/229644. Accessed 4 Oct 2024

Science Based Targets (2024) Set a Target. [online] Science Based Targets Initiative. Available at: https://sciencebasedtargets.org/set-a-target [Accessed 20 September 2024]

Sec (2024) Securities and Exchange Commission. Available at: https://www.sec.gov/files/rules/final/2024/33-11275.pdf

Segal M (2023) CDP to align reporting platform with European Sustainability Reporting Standards. ESG Today, 8 November. Available at: https://www.esgtoday.com/cdp-to-align-reporting-platform-with-european-sustainability-reporting-standards/. Accessed: 13 Sept 2024

Shaw C (2024) Guide to international sustainability standards board (ISSB): Anthesis Group. Anthesis. Available at: https://www.anthesisgroup.com/regulations-hub/international-sustainability-standards-board/. Accessed: 17 July 2024

Sikdar SK, Sengupta D, Mukherjee R (2017) Measuring Progress towards sustainability: a treatise for engineers. Springer

Simplilearn. (2024). Data Analytics Tutorial: Complete Guide. Simplilearn Solutions

Singh P, Mulay P, Patil Y (2023) Mapping climate change mitigation strategies adopted by industries: an overview from First Commitment of Kyoto Protocol (2009–2023). Environ Clim Technol 27:775–796

Sivarajah, U., Kamal, M. M., Irani, Z., & Weerakkody, V. (2017). Critical analysis of Big Data challenges and analytical methods. Journal of Business Research, 70, 263–286

Smeets, E. and Weterings, R. (1999) Environmental indicators: typology and overview

Social Investment Forum (SIF) (2012) Report on sustainable and responsible investing trends in the United States. Colombia, DC, USA, USSIF Foundation, The Forum for Sustainable and Responsible Investment

Solove DJ (2006) A taxonomy of privacy. Univ Pa Law Rev 154(3):477–560

Spangenberg JH (2002) Towards indicators for institutional sustainability: lessons from an analysis of agenda 21. Ecol Indic 2(1–2):61–77. https://doi.org/10.1016/s1470-160x(02)00050-x

Spangenberg JH (2004) Reconciling sustainability and growth: criteria, indicators, policies. Sustain Dev 12(2):74–86. https://doi.org/10.1002/sd.229

Statista (2022) Internet of Things (IoT) connected devices installed base worldwide from 2015 to 2025. Retrieved from https://www.statista.com/statistics/471264/iot-number-of-connected-devices-worldwide/

Statista (2024) Volume of data/information created, captured, copied, and consumed worldwide from 2010 to 2025. Available at: https://www.statista.com/statistics/871513/worldwide-data-created/. Accessed: 3 July 2024

Strähle J, Müller V (2016) Key aspects of sustainability in fashion retail. Springer Series in Fashion Business, pp 7–26. https://doi.org/10.1007/978-981-10-2440-5_2

Strýčková L (2023) Perception and identification of the general public with sustainable finance issues. https://doi.org/10.15240/tul/009/lef-2023-42

Sustainserv (2023) Navigating the CDP climate change questionnaire. Available at: https://sustainserv.com. Accessed: 13 Sept 2024

Sweep (2024) What are the European Sustainability Reporting Standards (ESRS)?, Sweep. Available at: https://www.sweep.net/insights/what-are-the-european-sustainability-reporting-standards-esrs. Accessed: 03 July 2024

The Climate Registry (2019) General reporting protocol V3. Available at: https://theclimateregistry.org/download/general-reporting-protocol-v3-05-2019/. Accessed: 13 Sept 2024

The Climate Registry (2023) Protocols. Available at: https://theclimateregistry.org/registries-resources/protocols/. Accessed: 13 Sept 2024

Thecsrdcompass (2023) CSRD: Benefits for businesses. The CSRD Compass. Available at: https://thecsrdcompass.com/csrd-benefits-for-businesses/. Accessed: 30 June 2024

Thecsrdcompass (2024) EU CSRD Implementation: Unravelling the Challenges. [online] Available at: https://thecsrdcompass.com/eu-csrd-implementation-unravelling-the-challenges/. Accessed 4 Oct 2024

Thurmond T (2023) Classifying data: what you need to know. KirkpatrickPrice. Available at: https://kirkpatrickprice.com/. Accessed 5 July 2024

Tnfd (2025) The Taskforce on nature-related financial disclosures. Available at: https://tnfd.global/metrics/#overview

Townsend I, Kennedy S (2004) Poverty: measures and targets, research paper 04/23. Economic Policy and Statistics Section, House Of Commons Library, London

Tunley Environmental (2023) BS ISO 14068:2023—The new standard for the transition to Net Zero. Available at: https://www.tunley-environmental.com/en/insights/bs-iso-140682023-the-new-standard-for-the-transition-to-net-zero. Accessed: 13 Sept 2024

Tyagi S (2018) What comes next: responding to recommendations from the Task Force on climate-related financial disclosures (TCFD). APPEA J 58(2):633. https://doi.org/10.1071/aj17179

UK Government (2024) Streamlined Energy and Carbon Reporting (SECR) for academy trusts. Available at: https://www.gov.uk/government/publications/academy-trust-financial-management-good-practice-guides/streamlined-energy-and-carbon-reporting. Accessed: 13 Sept 2024

UL Solutions (2024) Understanding SECR: streamlined energy and carbon reporting. Available at: https://www.ul.com/resources/understanding-secr-streamlined-energy-and-carbon-reporting. Accessed: 13 Sept 2024

UNCED (1992) Rio declaration on environment and development. Report of the United Nations conference on environment and development. Distr. General, August 12, 1992, A/CONF.151/26 (Vol. I)

UNCTAD (2022) The core SDG indicators. Available at: https://unctad.org/system/files/official-document/tdsbisd5_en.pdf

UNEP FI (2024) European Sustainability Reporting Standards (ESRS), UN Environment Programme Finance Initiative. Available at: https://www.unepfi.org/impact/interoperability/european-sustainability-reporting-standards-esrs/. Accessed: 01 July 2024

United Nations (2024) Greenwashing—the deceptive tactics behind environmental claims, United Nations. Available at: https://www.un.org/en/climatechange/science/climate-issues/greenwashing. Accessed: 27 April 2024

United Nations Development Group (2021) Data privacy, ethics and protection guidance. United Nations Development Group. Available at: https://unsdg.un.org/sites/default/files/UNDG_BigData_final_web.pdf

Vaessen M (2023) ISSB standards are now live! KPMG. Available at: https://kpmg.com/xx/en/home/insights/2023/06/issb-ifrss1-ifrss2-talkbook.html. Accessed: 16 July 2024

Voigt P, Von dem Bussche A (2017) The EU General Data Protection Regulation (GDPR): a practical guide. Springer. https://doi.org/10.1007/978-3-319-57959-7. Accessed: 4 Sept 2024

Vollero A et al (2016) Avoiding the greenwashing trap: between CSR communication and stakeholder engagement. Int J Innov Sustain Dev 10(2):120. https://doi.org/10.1504/ijisd.2016.075542

Wakunuma, K., & Jiya, T. (2019). Digital ethics in developing economies: Challenges and opportunities. Information Systems Frontiers, 21(4), 917–932. https://doi.org/10.1007/s10796-019-09952-y

Walsh P (2024) CSRD reporting for UK companies: compliance & benefits. Benchmark Gensuite. Available at: https://benchmarkgensuite.com/ehs-blog/uk-csrd-reporting-requirements/. Accessed: 30 June 2024

Wickham, H., François, R., Henry, L., & Müller, K. (2017). dplyr: A Grammar of Data Manipulation. R package version 0.7.0.

Widyawati L (2019) A systematic literature review of socially responsible investment and environmental social governance metrics. Bus Strateg Environ 29(2):619–637. https://doi.org/10.1002/bse.2393

Williams R (2023) How to measure & communicate your company's sustainability effort. Chandler Macleod. Available at: https://www.chandlermacleod.com/how-to-measure-and-communicate-your-company-s-sustainability-effort. Accessed: 12 May 2024

World Economic Forum. (2017). Digital Transformation Initiative: Unlocking $100 Trillion for Business and Society. WEF, Geneva

WWF (2024) Greenhouse gas accounting efforts undermined by disparate tools and frameworks. The Markets Institute At WWF I Business Case

Zenghelis, D., Agarwala, M., & Coyle, D. (2016). Final report of the Wealth Economy Project. Bennett Institute for Public Policy, University of Cambridge

Zhang L et al (2018) The influence of greenwashing perception on green purchasing intentions: the mediating role of green word-of-mouth and moderating role of green concern. J Clean Prod 187:740–750. https://doi.org/10.1016/j.jclepro.2018.03.201

Zhao, L. (2024). Machine Learning for Environmental Risk Assessment. Elsevier

Zumente I, Lāce N (2021) ESG rating—necessity for the investor or the company? Sustainability 13(16):8940. https://doi.org/10.3390/su13168940. Accessed 5 July 2024

Sustainable Finance Instruments and Mechanisms 4

Abstract

This chapter focuses on different tools and methods that enable sustainable finance in practice. It starts by explaining key sustainability concepts in finance, including green, blue, social, and sustainability-linked. The roles of these concepts in shaping financial products and strategies are clarified. Key terms such as use of proceeds, transition financing, KPI-linked finance, and green pure players are defined to create a clear framework. This chapter then looks at various sustainable finance instruments, such as green and social bonds, sustainability-linked loans, funds, green deposits, and derivatives. It also discusses contemporary topics like carbon pricing and trading. Additionally, it emphasises the importance of sustainability indices, venture capital, private equity, and Islamic finance structures in mobilising capital for sustainable development. This chapter reviews allocation and impact reporting practices, along with risk management and insurance mechanisms. This overview shows how financial innovation and organised methods are essential for aligning capital markets with global sustainability goals.

The author would like to acknowledge the help and contributions of Özgür Bulut Akanay, Sinem Özyiğit, Buğra Ateş, İsmail Zavrak, Atakan Atmaca, Berkay Yavuz, and M. Burcu Hıdımoğlu in completing this chapter.

4.1 Sustainability Concepts in Finance

As sustainability becomes a major pillar in the financial sector, a comprehensive grasping of its core concepts is essential. This section explores the sustainability concepts in finance by integrating environmental, social, and governance (ESG) factors into financial decision-making. By addressing projects such as renewable energy, energy efficiency, pollution prevention, and climate change adaptation, sustainable finance aims to promote and deliver substantial environmental and social benefits in financial mechanisms. Sustainable finance comprises four main concepts: green principles, social impact, sustainability, and sustainability-linked initiatives. These pillars are critical in resource allocation and ensuring financial activities align with broader sustainability objectives.

4.1.1 Green and Blue

At the 2013 Green Bond Symposium, investors pointed out that while shared characteristics define what qualifies as "green" in general, a prescriptive standard set of green definitions would not meet each investor's needs (World Bank 2018). This situation could prevent new and existing issuers from financing projects focused on climate action and the environmental benefits they are willing to support. Schoenmaker and

S. Küfeoğlu, *Financing the Green Transition*, Sustainable Development Goals Series,
https://doi.org/10.1007/978-3-032-10440-3_4

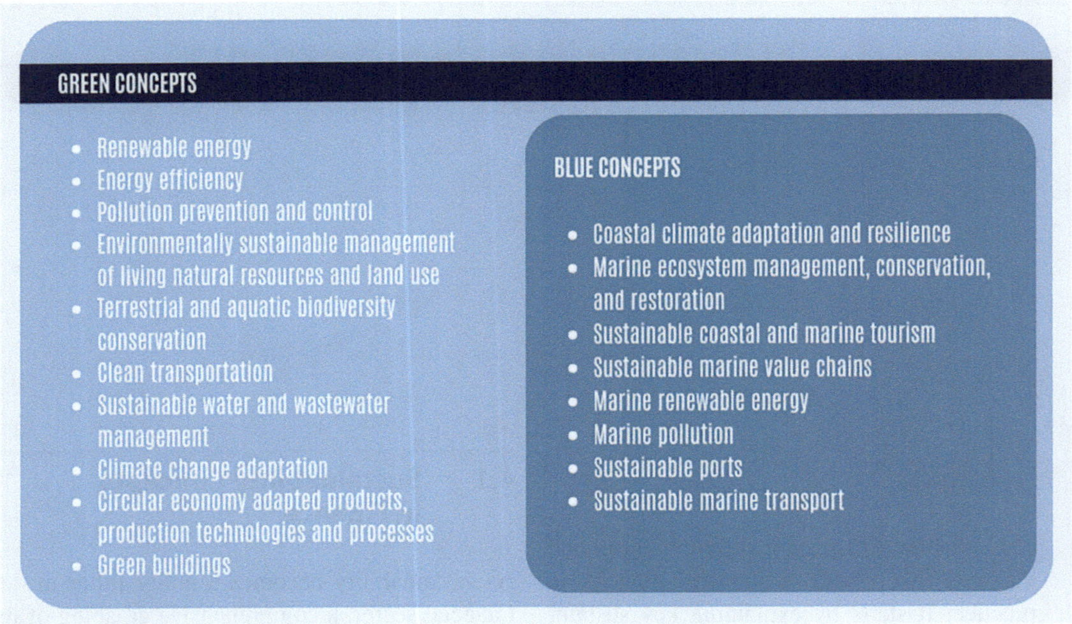

Fig. 4.1 Green concepts and blue concepts in finance ICMA (2022)

Schramade (2021) state that the dilemma about the definition of green has not changed over the years since the components that make up a green bond are open to debate. The green concept in finance has numerous standards and multiple ways to interpret it. Nevertheless, some standards and principles have been established and widely acknowledged in recent years. In Green Bond Principles (Green 2021), the "green" label denotes projects that minimise negative environmental impacts. The study by Green Loan Principles (Green 2021) discloses that all specified Green Projects must demonstrate evident environmental advantages, be subject to evaluation, and be suitably quantified by the borrower. Also, the Sustainability Bond Guidelines (Green 2021) affirm that just as some green projects have benefits to society, social projects can also have green benefits. Accordingly, issuers should determine whether a use of proceeds bond is categorised as a social bond, green bond, or sustainability bond, considering the principal aims of the underlying projects.

Within the scope of the green economy, the blue label specifically represents projects that reduce or eliminate negative impacts on coastal and marine resources (ICMA 2023). The motivation for creating the subcategory of blue is to encourage the protection of coastal and marine resources and, hence, to increase the share of investments in them. Although the first blue bond was issued in 2018 by the government of Seychelles (World Bank 2019a), blue bond principles were published later in 2023 (IFC 2022a; ICMA 2023). The International Capital Market Association (ICMA) introduced a non-comprehensive list of ten green and eight blue categories, demonstrated in Fig. 4.1 (ICMA 2022).

4.1.2 Social

Corporate strategies are critical in solving the challenges facing businesses and the economy. For instance, Shalhoob and Hussainey (2022) pointed out that companies that embraced ESG during COVID-19 enjoyed strategic, financial, reputational, and competitive leverage. The considerably sizeable private sector implies they significantly influence sustainable development. For instance, the 2022 UNDP report showed that

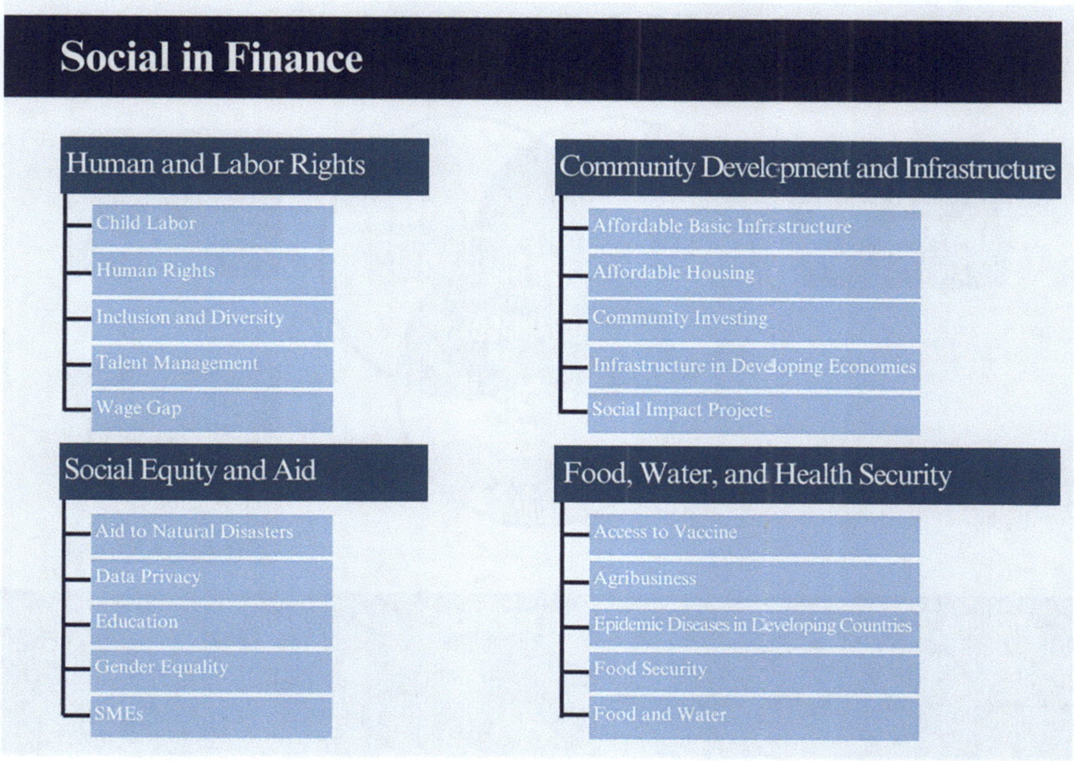

Fig. 4.2 Social concepts in finance

micro, small, and medium enterprises (MSMEs) constitute 90% of the private sector and over 50% of jobs globally (UNDP2023). United Nations Department of Economic and Social Affairs (UNDESA) further reported that these MSMEs serve as a source of livelihood for four billion people in the base pyramid, whose annual earnings are USD 3000 annually (UNDESA 2024). This claim implies they have more potential to contribute to SDGs. Even so, MSMEs still need to be addressed despite their need to attract and retain employees, investors, and customers through ESG strategies and reporting (Fenwick et al. 2022). There is an increasing need to address social values beyond the financial return, which has been the traditional focus of businesses in the capital market. The infographic presented in Fig. 4.2, entitled Social Concepts in Finance, visually summarises the principal themes within the social pillar of sustainable finance.

In finance, social factors focus on the welfare and development of society. These factors primarily address human and labour rights, community welfare, social equity, as well as the assurance of food, water, and health security which is illustrated in Fig. 4.2.

4.1.3 Sustainability

Sustainability in finance refers to supporting sustainable projects with the necessary financial instruments which cover ESG topics (European Commission 2024a). Its initiatives tackle interrelated issues holistically by incorporating social and environmental aspects. It is vital to finance sustainable projects because doing so will enable the effective implementation of solutions to these problems. Figure 4.3 shows how sustainable finance covers ESG topics (Küfeoğlu 2024b).

As described in Fig. 4.3, sustainability has three main pillars: environmental, social, and governance, also commonly called ESG, which incorporate many subtopics, one and all. Besides,

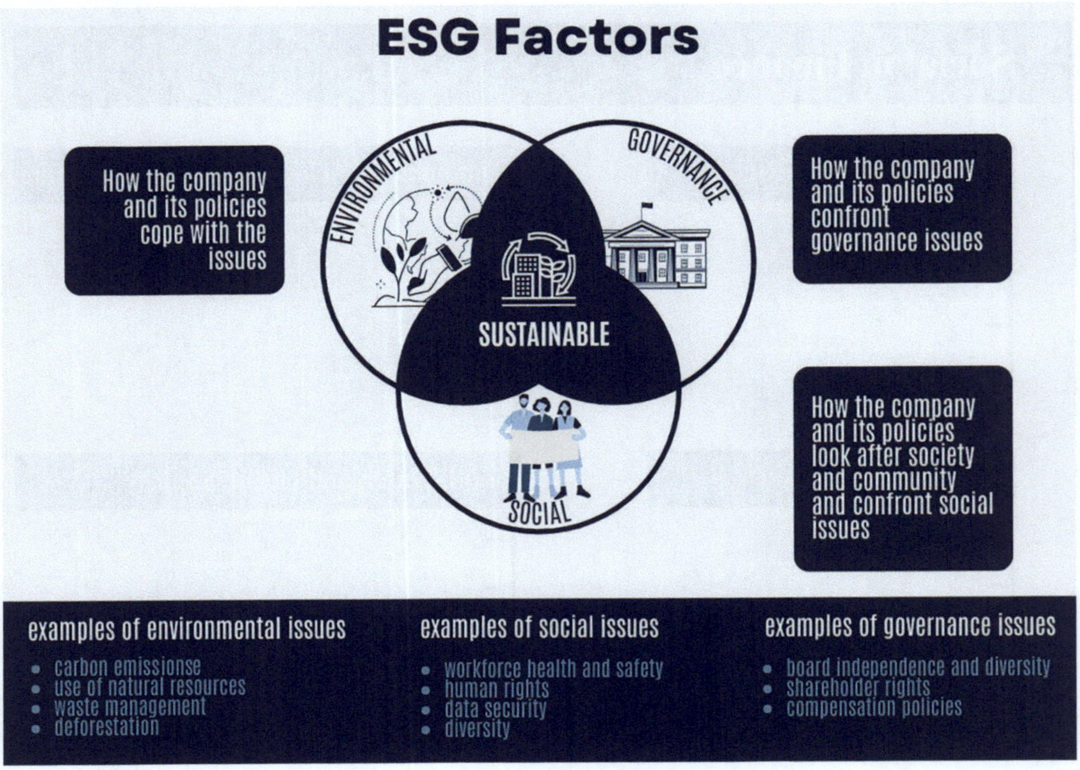

Fig. 4.3 Sustainability concepts in finance from an ESG perspective (Küfeoğlu 2024a)

in sustainability projects, ESG are often inseparable and feed each other. Although ESG factors are expected to be met simultaneously for a project to be sustainable, projects that meet them individually can also be called sustainable projects. That means ESG criteria can be met intersectionally or discretely. Allocating resources to sustainable initiatives, such as the construction of affordable housing, facilitates the effective implementation of solutions to societal and environmental problems.

Risks related to ESG are the most significant danger to corporations nowadays. Assets are expected to exceed USD 50 trillion by 2025; therefore, money is readily available to invest in this market (World Economic Forum 2023a). However, most programs have yet to produce the expected return. Businesses should assign a tangible value to their ESG impacts, with data directly attributable to the sustainable nature of the company. Markets must pay attention to these proof points and provide signals to guide producers' future ESG strategies. In large sectors of the global economy, the payoff for such practices is substantial for those focused on long-term value creation (Küfeoğlu 2024b).

In conclusion, sustainability in finance emphasises the importance of providing financial support to projects covering ESG aspects. This approach pursues to address societal and environmental challenges comprehensively. Investments in sustainable projects are crucial in facilitating effective solutions to these issues. While projects are generally expected to meet all ESG factors simultaneously to be deemed sustainable, those that fulfil these factors individually can also be considered sustainable. Ultimately, the significance of ESG factors for companies focused on long-term value creation must be balanced. The impact of these practices in primary sectors of the global economy is considerable.

4.1.4 Sustainability Linked

Sustainability-linked finance fundamentally encourages companies to achieve ESG goals (Orden and Calonje 2022). The market for financing tied to sustainability has expanded as businesses come under more pressure to lower their carbon footprint and perform better in ESG issues. By presenting chances to reduce capital expenses, financing boosts businesses to meet ESG objectives (MSCI 2022). Moreover, sustainability-linked financing supports improved ESG performance by linking financial products such as bonds and loans to sustainability criteria. These products adjust pricing through interest rates based on achieving set ESG targets, encouraging companies to meet sustainability goals. The firm's corporate sustainability strategy should align with these targets and be aspirational. Incentive structures vary, such as lowering interest rates for meeting objectives or increasing them for failing to do so (Orden and Calonje 2022). However, success in accessing the market hinges on the accurate use of ESG data by both borrowing companies and lending banks. If ESG data is inaccurate or a reliable sustainability rating is unavailable, there is a risk of misinterpreting an organisation's progress in environmental and social aspects. This result could lead to misguided lending decisions based on incorrect perceptions of the company's achievements (MSCI 2022).

Sustainability-linked finance comprises Sustainability Performance Targets (SPTs) and predefined Key Performance Indicators (KPIs). These components are instrumental in guiding companies towards more responsible actions and evaluating their achievements (ICMA 2020b). SPTs are tailored goals set by businesses to address specific ESG issues, such as diversifying staff for a tech company or reducing water usage for a manufacturing firm. These targets provide clear directions for sustainable development and are often measurable and time-bound to track progress effectively. They require collaboration among management, staff, investors, suppliers, and clients to ensure support, accountability, and progress towards sustainable development goals. Through performance goals measured by predefined performance indicators, companies can demonstrate their commitment to sustainable development and positive change and create long-term benefits for all stakeholders (Papulová et al. 2021). Figure 4.4 summarises numerous examples of indicative KPIs for sustainability-linked financing.

This infographic highlights example KPIs crucial for assessing sustainability efforts. Energy saving reflects the reduction in energy consumption due to efficiency improvements and conservation practices, demonstrating the effectiveness of energy-saving initiatives. Renewable energy generation shows the amount of energy produced from renewable sources, such as solar or wind, emphasising the shift towards cleaner energy and reduced fossil fuel dependence. Total installed capacity tracks the cumulative capacity of installed renewable energy systems, indicating the scale of sustainable energy infrastructure. Estimated annual reductions and/or avoided emissions quantify the greenhouse gas emissions reduced or avoided through sustainability measures, underlining the impact on climate change mitigation. Qualitative improvements in freshwater supply and/or wastewater treatment evaluate enhancements in water quality and treatment processes, highlighting advancements in managing water resources. Finally, the amount of treated wastewater and/or supplied freshwater measures the volume of wastewater treated and supplied, providing insight into the efficiency of water management practices and resource availability.

Conversely, bonds and loans connected to sustainability are priced based on how soon the issuer meets its ESG goals. They might operate as a catalyst for businesses to adopt sustainable, low-carbon procedures. Financial institutions support, encourage, and reward their borrowers to make positive changes using this strategy (KPMG 2022). Financial institutions encourage companies to prioritise a mission tour by offering performance-related financial benefits, such as low interest rates or good credit terms. This support allows companies to invest in sustainable technologies, reduce their carbon footprint, and adopt environmentally friendly practices (OECD

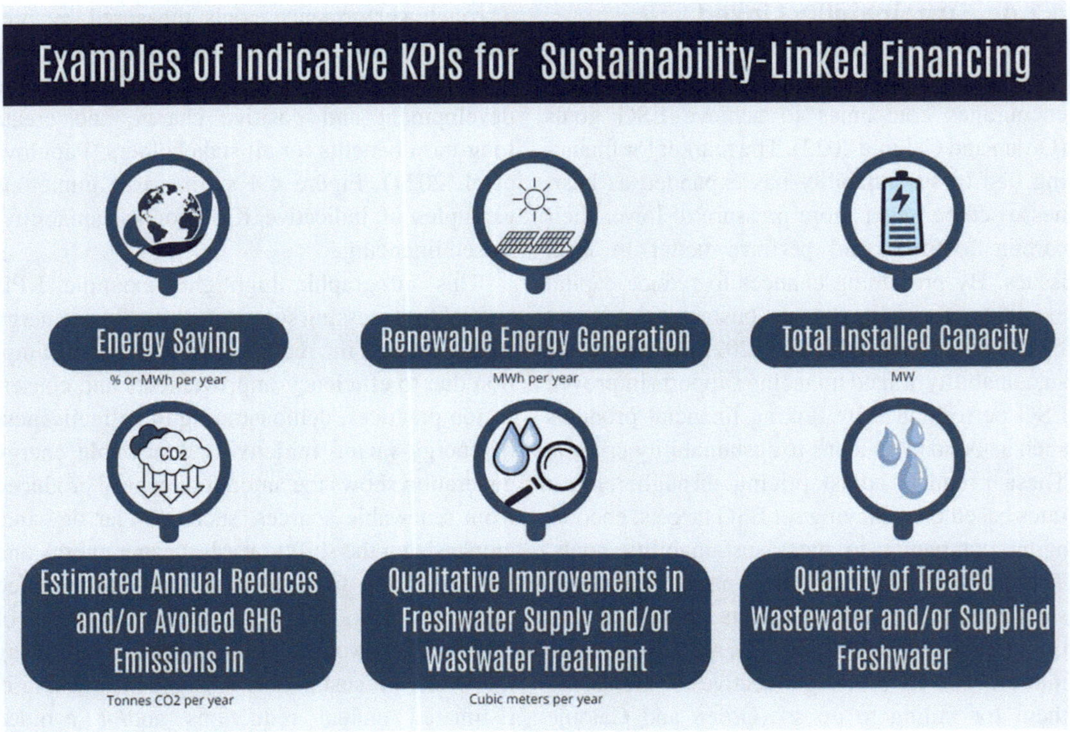

Fig. 4.4 Examples of indicative KPIs for sustainability-linked financing (IPCC 2021)

2021). Moreover, sustainable-linked must contribute to the needs and reality of the company. The company can use this as a loan. People struggling with debt-related security issues should be encouraged to do so. The goal is not necessarily security. If the borrower meets the agreed-upon sustainability targets, they may benefit from more favourable financial terms; if they do not achieve the targets, they may face penalties or higher costs (Ul Haq and Doumbia 2023).

One of the critical aspects of sustainability-linked finance is the structure of incentives and penalties tied to the performance against these KPIs. This mechanism ensures that companies are motivated to achieve their sustainability targets, enhancing accountability and transparency. The specific KPIs chosen can significantly influence the financial terms, such as interest rates, depending on the performance outcomes (Johnson and Toledano 2023). In the context of penalties, it is essential to understand the various penalties that may be imposed if the agreed-upon KPIs still need to be met. These penalties are designed to hold companies accountable and to incentivise continuous improvement in their sustainability practices. The type of penalties can be seen in Table 4.1.

Financial penalties for failing to meet sustainability targets can include several measures: a coupon step-up increases future interest payments by a predetermined amount if targets are missed, while a coupon step-down reduces payments if targets are met. A redemption premium adds an extra payment to the redemption price if sustainability goals are not achieved. Additionally, a portion of the bond's issuance may be donated to a preferred charity if targets are unmet. The bond might also be redeemed early at a predetermined price if the issuer fails to meet the targets. Lastly, the issuer may be required to purchase carbon emission credits equivalent to a specified percentage of the bond's principal amount if sustainability targets are not met (Ul Haq and Doumbia 2022).

When sustainability is linked to finance, it emerges as a viable option for integrating ESG into financial decision-making. Using SPTs and KPIs allows for setting the necessary goals, mea-

Table 4.1 Types of penalties (Ul Haq and Doumbia 2022)

Types of financial penalties	Definition
Coupon step-up	Coupon step-up penalties entail an increase in interest payments in the future by a predetermined amount (sometimes expressed in basis points) if the issuer fails to reach its predetermined sustainability targets by the target observation date
Coupon step-down	A coupon step-down occurs when interest payments are reduced by a predefined amount (sometimes expressed in basis points) if the issuer satisfies its predefined sustainability criteria by the target observation date
Redemption premium	A redemption premium is an additional amount that the issuer must pay, in addition to the redemption price, if it fails to meet its pre-established sustainability targets by the target observation date
Donation	Donation means that a portion of the bond's issuance, up to a predetermined amount, will be given to the issuer's preferred charity or organisation if the issuer cannot reach its pre-established sustainability targets by the target observation date
Early redemption	At a predetermined redemption price, the bond will be redeemed early if the issuer cannot satisfy its pre-established sustainability targets by the target observation date
Purchase carbon emission credits	Suppose the issuer has not met its pre-established sustainability targets by the target observation date. In that case, the issuer buys carbon offsets for a sum equal to a pre-specified percentage of the bond's principal amount

suring progress, and enhancing transparency and accountability (Boffo and Patalano 2020). In addition, sustainability-linked finance leads to the formation of a reduced carbon expenditure and the development of a socially responsible business. In the coming years, sustainability-linked financing will keep expanding and changing. By implementing innovative ideas, investing in sustainable developments, and following the best practices available, companies can generate added value for their contributors and community (IFC 2022b). In summary, sustainable finance addresses global environmental and social challenges. Embedding ESG factors into financial practices enables developing and supporting projects, contributing to a more sustainable and equitable future.

This section highlights the essential concepts within sustainable finance, emphasising the importance of green principles, social impact, overall sustainability, and sustainability-linked initiatives. As sustainable finance continues to grow and mature, its influence on promoting long-term sustainability goals will become increasingly significant, guiding financial activities towards meaningful and positive outcomes for society and the environment.

4.2 Fundamental Terms in Sustainable Finance

As the worldwide economy pushes more and more towards sustainability, sustainable finance becomes increasingly vital for the continuity of capital. This growing field incorporates ESG issues into financial decision-making to promote projects and activities that positively impact society and the environment. Nevertheless, one must have a solid grasp of its foundational terminologies to navigate this complicated terrain successfully. Use of proceeds (UoP), transition financing, key performance indicator (KPI) linked finance, and green pure players are crucial terms in sustainable finance to assist the understanding of the frameworks and structures of financial mechanisms. These terms are the cornerstone of sustainable finance and are all essential in directing resource allocation and guaranteeing that financial operations align with more general sustainability objectives.

4.2.1 Use of Proceeds (UoP)

Businesses must describe their plan for using sustainable finance instruments, including the investment categories. This procedure, called "Use of Proceed (UoP)", simply specifies how and where the financial source that is received will be used. A UoP statement is a brief overview of what monetary or non-monetary sources are supposed to be used by a company (Zettelmeyer et al. 2022). In other words, UoP gives an overview of the business areas where the corporation plans to invest its resources. According to the allocation of proceeds example from green bonds in 2019, Fig. 4.5 shows that 33% of the green bonds were used for energy projects, indicating a notable emphasis on initiatives to promote sustainable energy solutions. In addition, 30% is allocated to building projects, and 22% went towards transportation projects (Fatin 2019).

A company plan on the UoP determines certainty among investors regarding the company's survival (McGuinness 2019). However, research shows that most companies need more details on the intended UoP, resulting in no useful information to investors as per the requirements of the law (Hammer et al. 2024). Information on the UoP is necessary to eliminate information asymmetries and protect investors, especially during the initial public offering (IPO) process (Hammer et al. 2024).

Alyasa-Gan and Che-Yahya (2022) revealed that a company can use proceeds primarily to exploit growth opportunities or other activities such as debt repayment. A company's UoP outlines how it intends to allocate funds obtained from a financing round or investment. For example, a pharmaceutical startup securing USD ten million in venture capital may designate USD five million for research and development to advance drug discovery and testing. Another USD three million might be earmarked for manufacturing scale-up to facilitate production and distribution. The remaining USD two million could be allocated to expanding the sales and marketing team to promote newly developed products.

Similar to the context of the country for IPOs, seeking finance from international institutions, a well-thought-out UoP is vital in streamlining

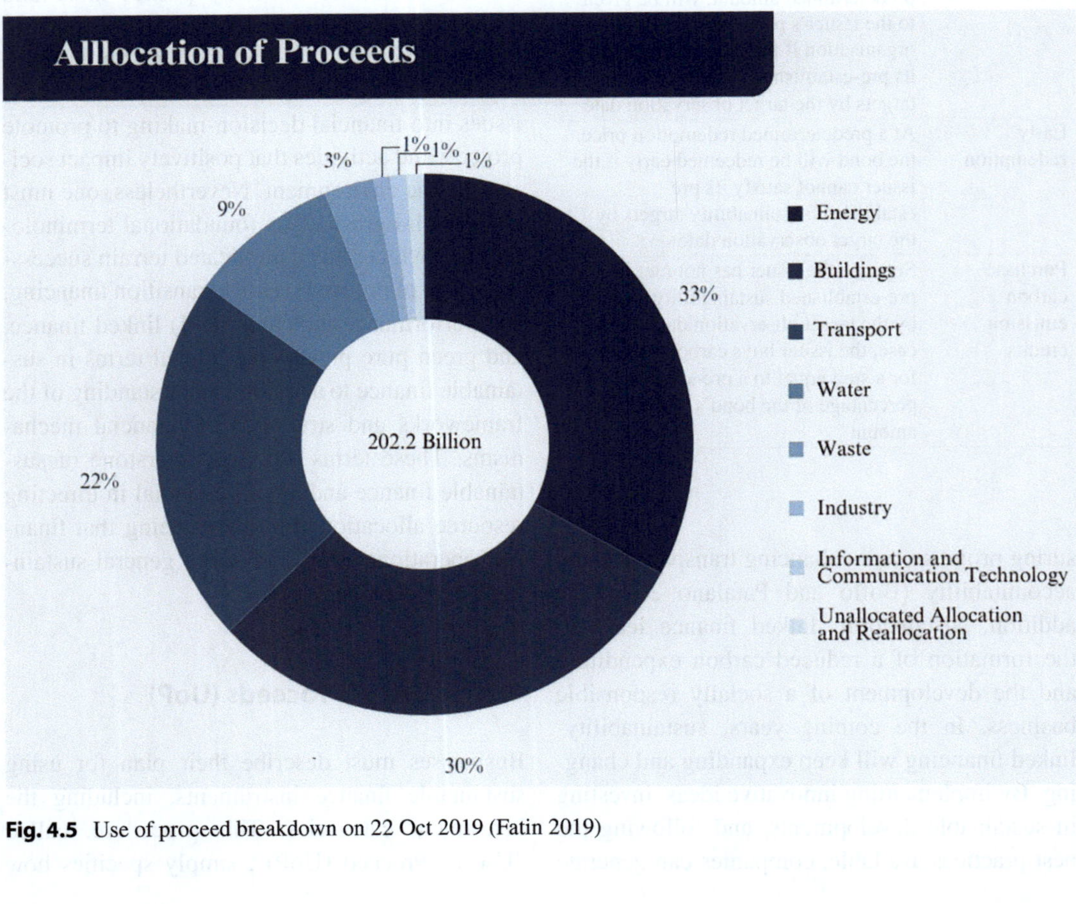

Fig. 4.5 Use of proceed breakdown on 22 Oct 2019 (Fatin 2019)

operations while focusing resources on key growth areas. Therefore, a company must adhere to a strategic UoP to maximise investment impact while improving investors' confidence. In the research conducted by Alyasa-Gan and Che-Yahya (2022), strategic use of proceeds is essential in creating value and reducing financing risks that potentially harm business survival.

To explain this in an example, the African Development Bank issues green bonds to countries focusing on inclusive growth and the transition to green growth. ICMA highlights the frameworks applied in capital markets, showing the need for information on UoP to determine the issuance of finance to various sectors of the economy (ICMA 2020a). Moreover, the 2022 green bond allocation by the African Development Bank revealed that the energy sector had the most prominent use of proceeds at 47%, followed by building and water at 42% and 5%, respectively, as shown in Fig. 4.6 (KMRC 2022).

Statements regarding using proceeds can also highlight areas where the corporation does not intend to allocate funds. For example, consider a company that issues a green bond to support the transition to a net-zero global economy. In this way, the corporation may particularly avoid investments in sectors, such as gambling, nuclear power generation, weapons, and alcohol, which are considered incompatible with sustainable development goals in their exclusion lists (HSBC 2023). Thus, companies can ensure transparency and reliability to their investors by specifying the products they will invest in and the products they will avoid investing in.

As pointed out, UoP is a principle that guides raising funds from financial instruments such as green bonds or sustainable-linked loans. It implies that UoP is crucial in directing financial investments towards environment and socially responsible projects. Since UoP bonds target specific projects, but repayment remains tied to the issuer, they attract investors who focus on their sustainable transition. In explaining the UoP, we have linked it to the context of a company and international financial institutions issuing green

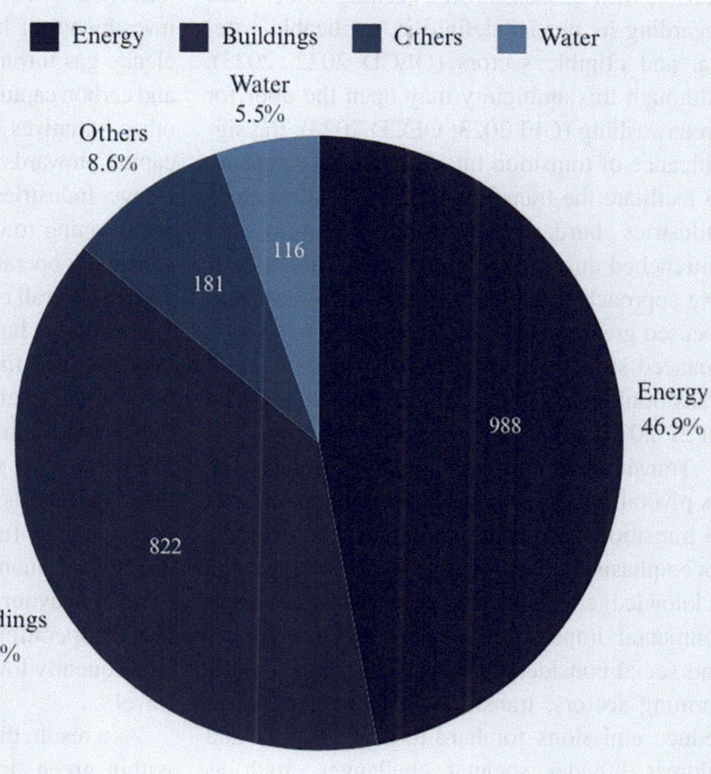

Fig. 4.6 Composition of green bond issuance by sector (KMRC 2022)

bonds to countries. The resulting trust and accountability from the UoP principal drive capital towards activities with positive environmental and social impacts, which align with sustainable finance's objective.

4.2.2 Transition Financing

Unlike conventional green finance, which predominantly targets projects directly aligned with environmental objectives, transition finance addresses sectors facing formidable challenges in decarbonisation, such as cement, steel, aviation, and shipping (OECD 2022). Its nomenclature reflects its fundamental purpose: to guide carbon-intensive industries through a gradual yet definitive transition towards sustainability (OECD 2023). This distinction emphasises the importance of navigating the intricacies inherent in shifting entire sectors away from reliance on fossil fuels and other environmentally detrimental practices.

Transition finance remains a topic of ongoing debate, with consensus still needing to be made regarding its precise definition, applicable criteria, and eligible sectors (OECD 2022, 2023). Although this ambiguity may open the door for greenwashing (GFI 2023; OECD 2023), the significance of transition finance lies in its capacity to facilitate the transformation of disadvantaged industries burdened by high emissions and entrenched unsustainable practices. This distinctive approach, diverging from the more narrowly focused green finance, emphasises the need for a nuanced strategy to address the complexities of transitioning key sectors towards sustainability (IPSF 2023).

Transition finance's designation stems from its pivotal role in managing the complex process of transitioning industries towards sustainability. Its emphasis on the transition to green reflects an acknowledgement of the need to balance environmental imperatives with economic stability and social considerations. By investing in transforming sectors, transition finance purposes to reduce emissions for hard-to-abate sectors and address broader societal challenges, such as unemployment and revenue loss for local governments (McCrossan 2024). It recognises that achieving sustainability requires a holistic approach that integrates environmental, economic, and social dimensions.

In this sense, green bonds, sustainability-linked bonds, transition bonds, loans with use of proceeds, and loans with KPIs have been frequently employed across various industries in transition finance (OECD 2022, 2023). For example, Nippon Yusen Kabushiki Kaisha, a Japanese shipping company, concluded a USD 300 million loan in 2024 to fund its vessel projects, minimising greenhouse gas emissions (NYK 2024). Also, Shell obtained a USD 10 billion revolving loan facility with a KPI in December 2019 to reduce its energy products' net carbon footprint (Shell 2019).

Other than obtaining loans, several companies have issued transition bonds in the context of transition finance. For instance, Mitsubishi Heavy Industries' issuance of its inaugural transition bond in September 2022 reflects a strategic commitment to funding projects aimed at emission reduction. The USD 71 million raised will support investments in hydrogen gas turbines, high-efficiency gas turbines fuelled by liquid natural gas, and carbon capture and storage technology, among other initiatives (McCrossan 2024). By directing capital towards these endeavours, Mitsubishi Heavy Industries demonstrates its dedication to transitioning towards cleaner energy sources and enhancing operational efficiency, thereby contributing to overall emissions reduction efforts.

Similarly, Japan Airlines' pioneering venture into transition finance further emphasises the sector's commitment to emission reduction. The airline's issuance of transition bonds in March 2022, followed by a subsequent issuance 15 months later, highlights its dedication to upgrading its fleet to more fuel-efficient aircraft (McCrossan 2024). By channelling bond proceeds towards these endeavours, Japan Airlines aims to modernise its operations, reduce fuel consumption, and consequently lower emissions associated with air travel.

As a result, the emergence of transition finance within green finance relies on decarbonisation, which is not easily conceivable in some specific sectors, creating an individual pathway to back up their transition towards more green practices in hard-to-abate sectors. Green finance should support the decarbonisation of the carbon-intensive sectors; however, a consensus regarding transition finance's precise definition, applicable criteria, and eligible sectors is required to prevent greenwashing as well as support transition finance.

4.2.3 Key Performance Indicator (KPI) Linked Finance

Key performance indicator (KPI) linked finance is a financial instrument that motivates companies to lower their total debt costs while enhancing their ESG performance. This innovative financial tool promotes ethical company conduct and environmental stewardship by matching cash rewards with performance targets. By combining financial incentives with ESG goals, KPI-linked finance encourages companies to adopt more moral and sustainable business practices, reaching broader environmental and social goals (ICMA 2020c).

KPIs should be critical to the issuer's core business plan and sustainability. They should fall under the purview of management's influence and directly address relevant ESG issues within the industry sector. This guarantees that the selected KPIs align with the issuer's operational environment and more general sustainability goals, promoting responsibility and openness in pursuing sustainable practices (ICMA 2020a). Carefully choosing KPIs is essential to the legitimacy of the sustainability-linked bond (SLB) market. It is imperative to refrain from incorporating KPIs that lack credibility to guarantee the success of this financial instrument. Robust and significant KPIs representing real sustainability improvements rather than flimsy measures are necessary for SLBs to be effective (Bryan 2023).

KPIs can be generated from internal measures that track company-specific goals and external standards like industry benchmarks and worldwide sustainability frameworks. Well-selected KPIs can boost investor confidence and guarantee that the SLB market contributes to social and environmental goals (ICMA 2020c). Borrowers must select KPIs that align with ESG objectives and key sustainability issues for effective KPI-linked financial structures. These KPIs should be measurable and data-driven for accurate tracking, enhancing accountability, and transparency. Lenders focus on KPIs that improve stakeholder engagement, operational efficiency, and risk reduction (Eccles and Serafeim 2013). Likewise, borrowers evaluate KPIs for scalability and long-term impact to promote sustainability across their value chain. Matching KPIs to stakeholder expectations ensures responsiveness to community, employee, investor, and consumer concerns. Additionally, KPIs are designed to encourage continuous sustainable performance, often linked to financial incentives (Hristov and Chirico 2019). Open to various entities, including corporations and governments, this financing requires collaboration between borrowers and lenders to establish relevant KPIs. Effective monitoring and communication are crucial, ideally with external oversight, to ensure transparency and accountability throughout the financing period (Dorleans 2018).

Finally, following industry norms and legal frameworks, borrowers consider the verification and reporting obligations linked to specific KPIs, striving for transparent and unambiguous reporting. By doing this, it is possible to effectively inform stakeholders on the advancement of sustainability targets, hence boosting their confidence in the KPI-linked financial structures. Borrowers can efficiently choose and adjust KPIs that maximise financial results and promote significant progress towards sustainability by following these specific guidelines.

The following requirements should be satisfied by the KPIs (ICMA 2020c):

- Relevance and Materiality: They should have a direct bearing on the primary business operations of the issuer and be of great strategic significance for present and upcoming undertakings. Critical elements of ESG performance that are essential to the issuer's business operations and industrial sector must be included in these KPIs.

- Measurability and Consistency: A consistent methodological approach must make the KPIs measurable or quantifiable. This guarantees that performance tracking and data collecting remain consistent across reporting periods, allowing for long-term trend analysis and relevant comparisons.
- External Verifiability: The KPIs must be verified by independent third parties for their measurement and reporting to be considered externally verifiable. Building confidence among stakeholders, external verification increases the sustainability performance disclosures' dependability and trustworthiness.
- Benchmarking Capability: The KPIs should be able to be benchmarked, preferably against peers in the industry or external standards and benchmarks. This position makes it possible to evaluate the issuer's sustainability performance compared to accepted standards or best practices, making it easier to gauge the issuer's ambition and advancement towards sustainability objectives.

It is recommended that issuers choose a KPI or several KPIs they have previously reported on in sustainability reports, annual reports, or other non-financial reporting disclosures, so that investors may assess the KPIs' past performance. Issuers should, to the greatest extent feasible, supply historical, externally validated KPI values spanning at least the preceding three years in cases where the KPIs have not been previously revealed (ICMA 2020a). To encourage sustainable corporate practices, KPI-linked finance is a revolutionary strategy in the finance sector that combines financial incentives with ESG objectives. With economic incentives for achieving predetermined KPIs, this cutting-edge financial instrument pushes businesses to match their long-term success with general societal and environmental goals. Businesses are incentivised to strengthen their sustainability efforts by explicitly connecting financing terms to ESG performance. This situation can result in lower lending costs and increased investor trust (ICMA 2023).

In summary, KPI-linked finance is more transparent and accountable when its KPIs are carefully chosen and managed. Businesses can successfully inform stakeholders about their sustainability performance by following set principles and standards. In addition to fostering trust, this openness guarantees that the financial arrangements support long-term goals. Ultimately, KPI-linked finance provides a strong foundation for combining financial and ESG objectives, encouraging a continuous improvement culture in sustainable practices, and favourably impacting international environmental and social goals (World Economic Forum 2020).

4.2.4 Green Pure Players

Pure play companies are companies whose business activities focus exclusively on the green economy (ICMA 2021a). They concentrate on producing a specific range of products. In addition, there are different definitions of "pure play". According to IFC (2023), related to the revenue composition of the firm, that firm is a "Pure Play" if more than 90% of its income comes from climate-related activities.

Pure play investments can attract investors interested in companies that are easier to research and may yield higher returns. However, according to a Jefferies LLC survey, ESG investors are less interested in pure-play green companies and more focused on businesses switching to renewable energy (Nasdaq 2022). Pure play companies, like other companies, create resources in various ways to finance their business. For this source of financing to be green, there are multiple criteria that they must meet, such as being a renewable energy infrastructure producer and having recyclable and environmentally friendly products (ICMA 2022). At the same time, only certain loans they use, such as business loans, are green. The net revenues from green bond offerings can also be used to finance loans for general corporate purposes for "pure play" companies (Danske Bank 2021). In the case of pure-play loans, banks seek to document the financing of a particular project or asset. Various industry infrastructures and renewable energy companies can be analysed in this context. The following examples allow a better explanation of this subject.

The green pure play concept covers specific businesses on sustainable practices and renewable energy. Their reliance on green financing mechanisms such as green loans emphasises a concerted effort towards responsible capital allocation. In addition, the developing environment of green finance and the increasing adoption of renewable technologies emphasise the significance of pure companies in advancing the energy transition, as well as the growing interest of investors in sustainable investment instruments. As demand for sustainable solutions continues to grow, pure companies gain importance in shaping the future of the green economy.

To summarise this part, the main terms covered under this title, which are use of proceeds, transition financing, green pure players, and KPI-linked finance, showcase the diverse approach needed to promote sustainable finance. Every notion has a distinct function in directing funds towards initiatives that promote social responsibility and environmental sustainability for the habitable earth. The terms in use sustainable financial mechanisms promote confidence among investors and stakeholders by guaranteeing accountability, transparency, and alignment with ecological, social, and governance objectives. The mechanisms of sustainable finance explained in detail in the next part will be crucial in directing the shift towards a more robust and sustainable global economy as the world continues to face urgent ecological and social challenges.

4.3 Sustainable Finance Instruments and Mechanisms

4.3.1 Incentives

Incentives foster investment into social and environmental causes and are thus integral to sustainable finance. These mechanisms facilitate aligning economic activity with global sustainability goals by reducing barriers, lowering costs, and enhancing the attractiveness of sustainable endeavours. When applying these incentives with appropriate care, policymakers, financiers, and international organisations can significantly change the direction of investment to environmentally friendly and socially focused projects (European Commission 2024a).

4.3.1.1 Tax Incentives

Regarding investment in sustainable strategies, tax incentives work as one of the most effective policies. These are all tax-related benefits such as tax reliefs, pension-orientated financing, tax deductions, or early cancelled depreciation, which all impact bringing down the costs associated with sustainable projects. In the USA, the Investment Tax Credit enables investors in solar systems to reduce their federal tax liability by 26% of the cost incurred for installation, making the initial costs of solar energy quite low (U.S. Department of Energy 2023). Such a value has been critical in fuelling demand in the solar industry, from smaller domestic operating systems to large-scale commercial installations.

Production tax credit (PTC) also enables the generation of electricity per kilowatt hour, which is generated from certified renewable energy sources like wind and biomass, making it another key consideration. This encouraged the rapid expansion of wind resources development across the USA, which has today gained the capacity to contribute a significant portion of the nation's renewable electric power energy supplies (U.S. Energy Information Administration 2023).

4.3.1.2 Subsidies and Grants

Governments, international organisations, or development banks offer subsidies and grants as a form of direct monetary benefit. This incentive is more suited to new technology or projects with high starting costs or associated risks. The total budget of the Horizon Europe program of the European Union for the years 2021–2027 is EUR 93.5 billion, with 35% of its budget to be directed to climate projects aimed at facilitating the transition towards clean energy through research and innovation (European Commission 2024a).

Subsidies are most frequently aimed at solving urgent problems of access to renewable energy or sustainable agriculture in developing countries. For example, the Global Environment Facility (GEF) has provided funding to assist in in situ biodiversity, land, and water management to promote improvement (Global Environment Facility 2022). As a result of GEF funding, solar microgrid systems have been set up in isolated areas, reducing the use of diesel generators (Global Environment Facility 2022).

4.3.1.3 Loan Guarantees and Concessional Financing

Loan guarantees and other special loans are also important in limiting the financial exposure that can arise with sustainable investments. Although risky projects exist, these loan guarantees help cover the risk posed to lenders from potential default by borrowers, hence facilitating the provision of credit to such ventures. The United States Department of Energy Loan Programs Office (LPO) provides loan guarantees to clean energy projects that cannot receive ordinary finance due to being too new or too complicated to understand (US. DOE Loan Programs Office 2022).

Concessional financing provides loans at lower interest rates or more favourable terms than the market. This is often achieved through multilateral development banks, including Asian Development Banks (ADB), which are important in enabling such financing to support climate mitigation and adaptation projects. The efforts of the ADB have been instrumental in the development of renewable energy across Pacific Islands and made it possible to implement projects aimed at decreasing diesel power dependency through an increase in solar and other renewable energy systems (Asian Development Bank 2023).

4.3.1.4 Performance-Based Incentives

Performance-based incentives are strategies that reward specific sustainability efforts with targeted finance. For example, sustainability-linked loans (SLLs) and sustainability-linked bonds (SLBs) provide cheaper finishes or interest based on the ability of the borrower to fulfil a given set of key performance indicators (KPIs). For example, Schneider Electric has included achieving stringent environmental objectives, including carbon cuts and energy intensity improvements, as one of the targets in its funding costs for the EUR 2 billion sustainability-linked loan (Schneider Electric 2021). In Rabobank's case, its "green loans" allow farmers who practice organic farming or water conservation to enjoy lower interest rates than those who do not practice such eco-friendly measures (Rabobank 2022).

4.3.1.5 Blended Finance

Blended finance refers to a mix of private and public funding, which may or may not include subsidies (Klein 2016). This instrument has also been major players in furthering efforts towards an international fight against biodiversity loss and forest degradation. In June 2022, donor countries committed over USD 5.33 billion to the Global Environment Facility for supporting big projects to address environmental issues in the next four years (Global Environment Facility 2022).

In renewable energy, blended finance mechanisms have played a significant role in scaling up solar power solutions in sub-Saharan Africa by combining public funds with private investments, as highlighted by the IFC's Scaling Solar program (IFC 2023).

4.3.1.6 Regulatory and Voluntary Market Incentives

Regulatory aspects such as mandates and standards are necessary to create a paradigm shift across industries. Article 2 of the European Union's Sustainable Finance Disclosure Regulation (SFDR) further requires asset managers to explain how they consider sustainability risks in their investment decisions, appealing to better accountability and transparency (European Commission 2022).

Kicking certification programs like Leadership in Energy and Environmental Design (LEED) as well as the Forest Stewardship Council (FSC) certification have non-regulatory advantages as well as voluntary market incentives, which help to improve the reputation and extend the markets of firms employing certification in sustainability that are accepted worldwide (U.S. Green Building Council 2024; Forest Stewardship Council 2024). In Fig. 4.7, other than financial incentives, administrative and reputational incentives that support sustainable finance are illustrated.

Incentives are key players in advancing sustainable finance. These mechanisms provide multiple financial and regulatory means to reduce costs, increase margins, and promote compliance with worldwide sustainable development goals in conjunction with or on the part of business enterprises. Thus, by using such incentives, governments, financial organisations, and international bodies can direct funds to those projects aimed at strengthening and enhancing the equity of the global economy.

4.3.2 Bonds

A bond is a financial instrument that allows its holder to lend money and receive periodic coupon payments for a defined period. At the end of the period, the borrower must repay the principal amount to the bondholder. Other terms can describe bonds, such as fixed-income securities, as these do not create much volatility when interest is paid out at regular and expected intervals (Fabozzi 2013a, b). As Mishkin and Eakins (2018) argue, governments sometimes sell bonds to support their spending that wouldn't be covered by the taxation segment of the economy and by corporations to obtain funds required to develop further or widen their activity. When a bond is bought, the corporation or the government, which claims to be the bond investor, is a creditor of the issuer.

Bonds are often considered safe havens since they present a lower risk than equities, especially low-risk investment-grade bonds (Brigham and Ehrhardt 2016). This trait makes them appealing to investors who want to protect their capital while obtaining a consistent revenue stream, which is particularly advantageous in retirement. Furthermore, such bonds minimise the risk of a portfolio because they offer reasonable protection against riskier securities such as stocks (Elton et al. 2019). The returns of bonds, particularly those of government and top-quality corporate bonds, are stable and contribute to financial security and diversification, complementing the needs of an investor seeking risk and return balancing. Key information about bonds is given in Fig. 4.8.

The financial sector has developed instruments that create socio-economic returns aligned with development goals as societies are increas-

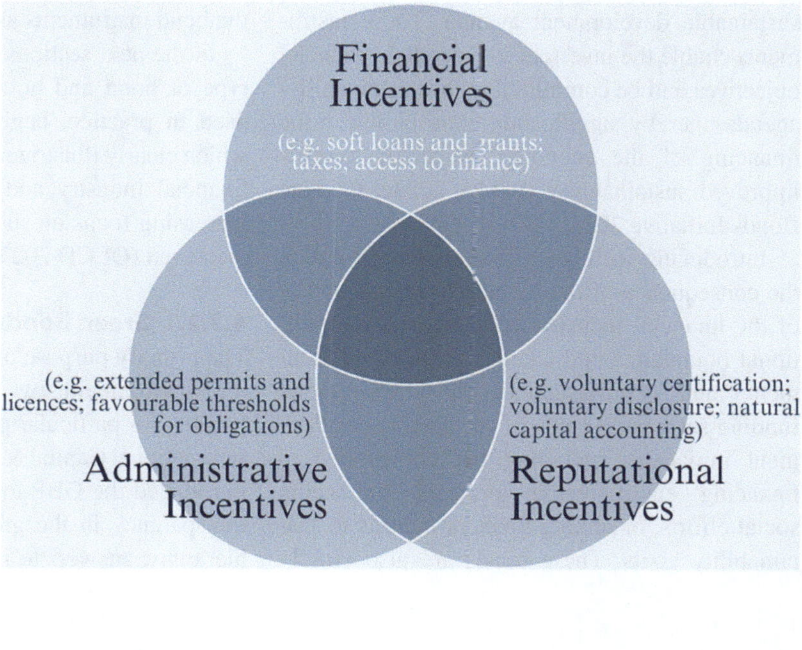

Fig. 4.7 Incentive mechanisms for supporting sustainable finance (Kristjanson 2020)

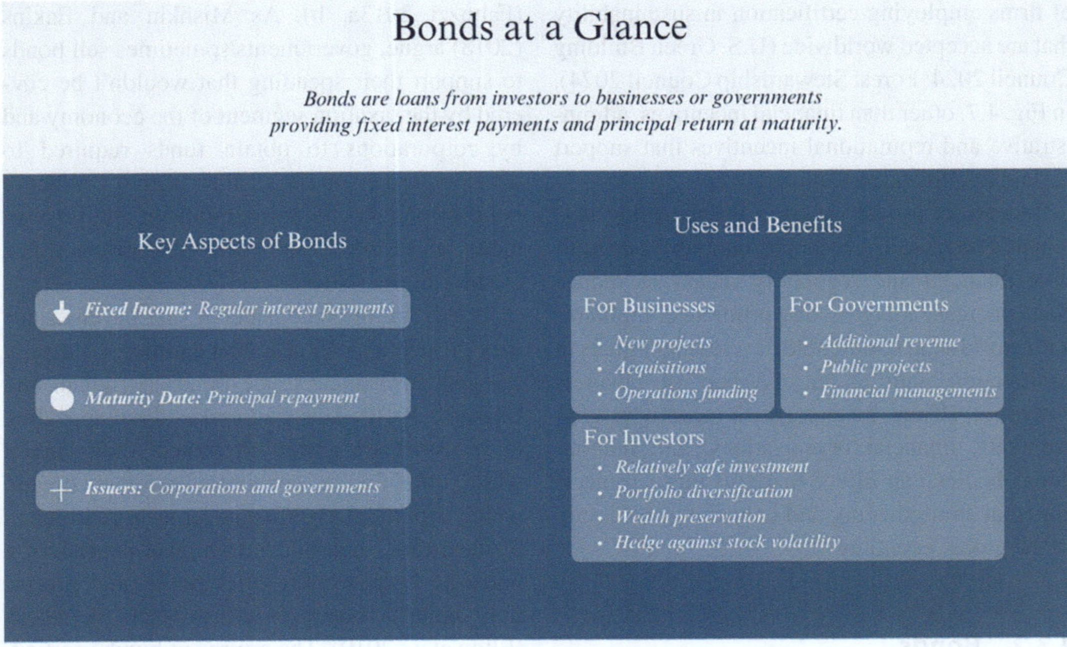

Fig. 4.8 Overview of bonds: key features, uses, and benefits (Climate Bond Initiative 2022)

ingly aware of environmental and social issues. Although in the past, bonds were regarded as instruments for generating constant capital returns on investments, their use has now shifted to being the most significant enabler in achieving environmental and social goals. Thus, new bond types have emerged, including green, sustainability-linked, transition, and social bonds, which address different components of the global sustainable development agenda. These instruments enable the investors to meet their financial objectives and be committed to the sustainability agenda, thereby significantly contributing to the financing of the economy transition towards improved sustainability (ICMA 2024c; Climate Bonds Initiative 2023).

Introducing such types of specialised bonds is the consequence of the increasing responsibility of the financial industry to go beyond its traditional boundaries and address pressing environmental and social issues. Global bonds provide funding for a wide range of sustainable development initiatives, including but not limited to financing environmental projects, enhancing social efforts, or linking financial returns to sustainability goals. These bonds are also closely related to the UN SDGs, overcoming international developmental challenges and integrating the concept of sustainable development into the financial industry (UNDP 2023). In the words of the Climate Bonds Initiative, these financial products direct a great amount of capital to useful purposes while improving the economy's efficiency, transparency, and accountability (Climate Bonds Initiative 2023). Figure 4.9 summarises the bond instruments as follows.

In the next sections, different aspects of each type of bond and how they are intended to be used in practice, beginning with green bonds, which clearly illustrate the changing nature of the financial industry and its tools because of the increasing focus on sustainable development are discussed (OECD 2023).

4.3.2.1 Green Bonds

The primary purpose of green bonds is to benefit the environment by providing the necessary funding for particular projects. In this regard, the International Capital Market Association (ICMA) introduced the GBP to guarantee adherence and transparency in the green bond market. In this hierarchy, answers to issuers' problems are sim-

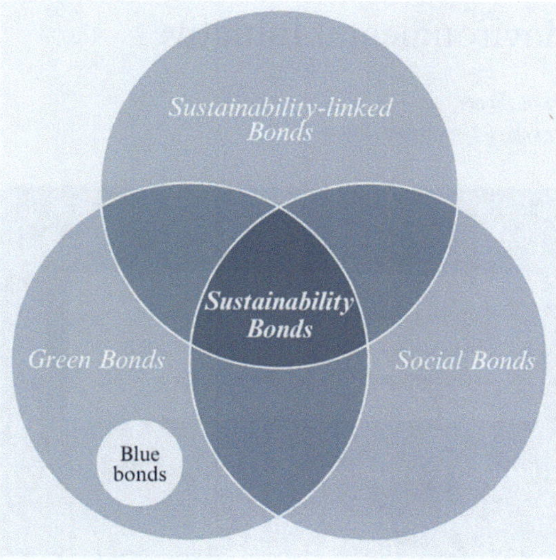

Fig. 4.9 Sustainable bond instruments

plified by outlining what constitutes a green bond and what projects can be financed by its proceeds (ICMA 2023). Figure 4.10 summarises the green bond mechanism.

Another area of focus concerning green bonds is the use of proceeds. Money raised from the bond issue should be used in green-eligible projects such as renewable energy, energy efficiency, sustainable waste management, clean transport, and conservation. This type of revenue generation for green projects is crucial since green bonds underpin worldwide actions on climate change, carbon emission reduction, and sustainability (Climate Bonds Initiative 2023).

Several countries joined the green bond market in 2023. Among these were Albania, Cyprus, Macedonia, and Montenegro, which issued their

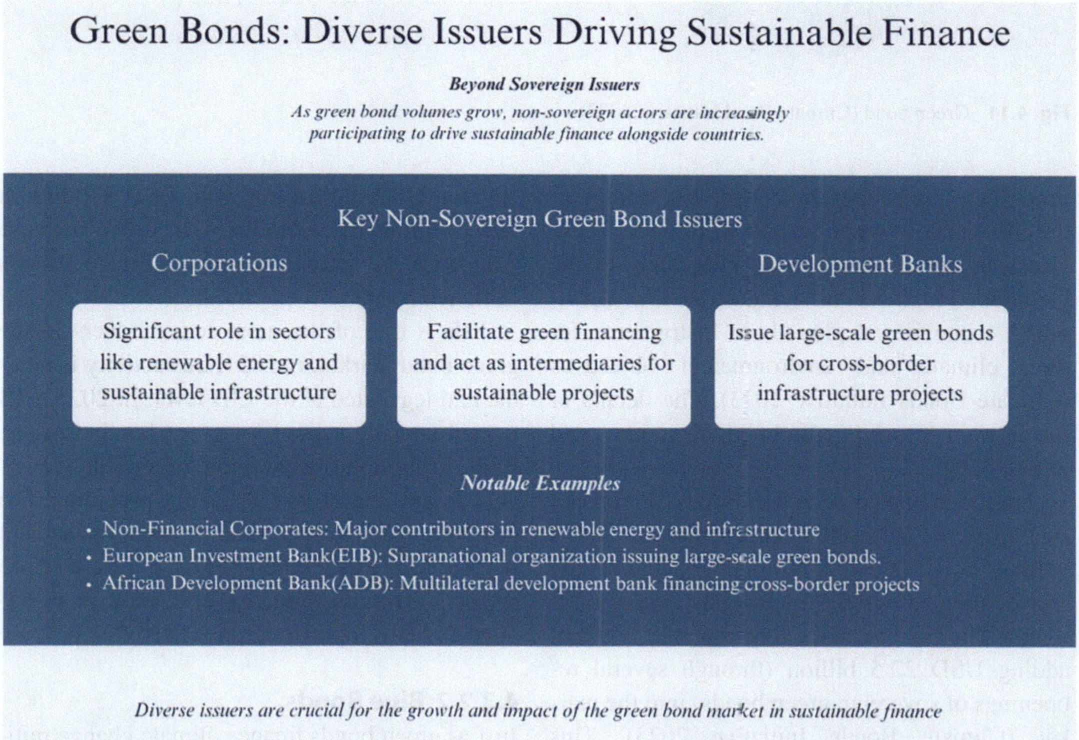

Fig. 4.10 Green bond mechanism (Climate Bond Initiative 2022)

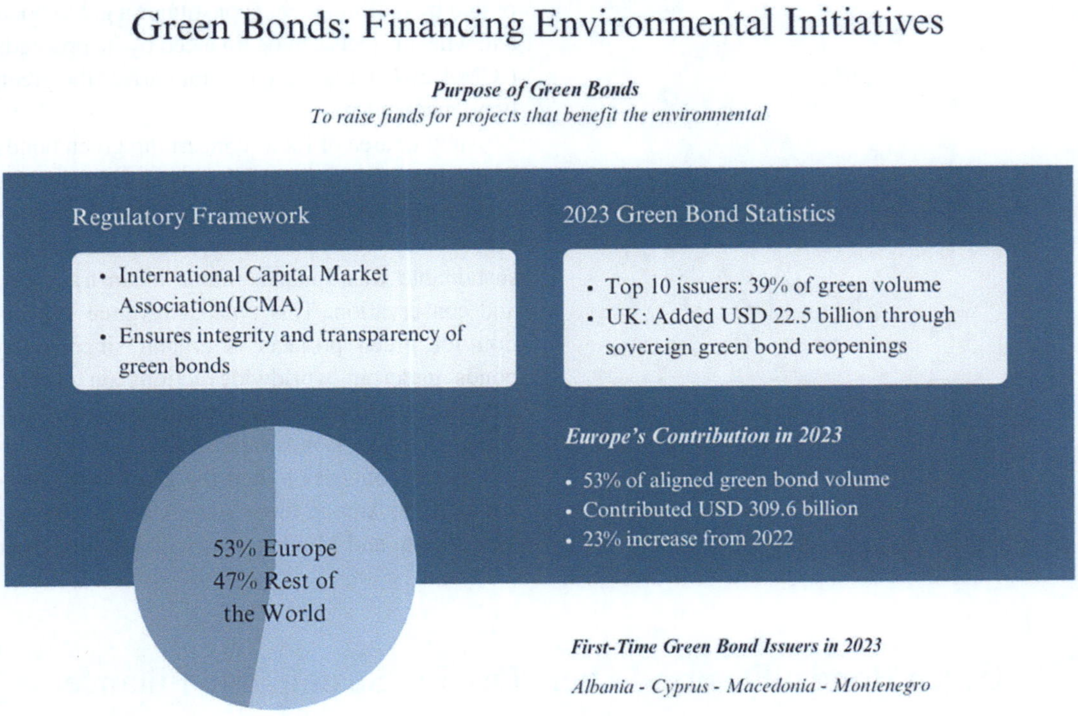

Fig. 4.11 Green bond (Climate Bond Initiative 2022)

first green bonds, thereby collectively adding to the emergence of green finance worldwide. These issuances depict the growing geography of the green bond markets and demonstrate how developing economies deploy these instruments for their climate and environmental objectives (Climate Bonds Initiative 2023). The details of the abovementioned green bond are summarised in Fig. 4.11.

The green bond market seems to be dominated by Europe. In 2023, the market responded with a volume of USD 309.6 billion, with the region representing 53% of the total aligned green bonds issued. The UK was at the forefront of this effort, adding USD 22.5 billion (through several reopenings of sovereign green bonds) into the market (Climate Bonds Initiative 2023). This remarkable increase illustrates the essential energy transition within Europe as regulations steer the increased funding, more vigorous climate initiatives, and rising levels of interest from the investment community. As of December 2023, the value of global issues of GSS+ bonds stood at USD 4.4 trillion, with USD 870 billion raised within the calendar year, a 3% increase compared to previous year (Climate Bonds Initiative 2024).

EIB is one of the most active issuers in the green bond market; in 2023, this activity is notable and legislated at the EIB level. In 2023, EIB tapped into the market for green bonds to issue EUR 2.6 billion, focusing on large-scale renewable energy and energy efficiency programs. The proceeds from this issuance financed solar and wind power and emission-reducing infrastructure projects promoting energy efficiency in developed and developing countries (EIB 2023).

4.3.2.2 Blue Bonds

Just as green bonds finance climate change mitigation projects, blue bonds are now a subclass of them and are issued for projects related to the conservation and management of maritime and coastal ecosystems. These bonds have become a vital instrument for confronting issues such as overfishing, pollution of the ocean and other water bodies, and coastal erosion, all of which endanger biodiversity and the socio-economic livelihood of communities that depend on fisheries resources. Blue bonds, however, are different from green bonds in that to these categories of bonds, blue bonds relate capital to ocean and water-focused undertakings.

So far, the International Finance Corporation (IFC) has been a leading entity in developing blue bonds, seeking to guide the management structure for blue finance. One such early blue bond was the Seychelles Sovereign Blue Bond, launched in 2018 and raised USD 15 million from international investors through the World Bank. The proceeds were used to enhance sustainable fisheries, the marine environment, and the state of the Seychelles blue economy (World Bank 2018). This bond provides a track for other upcoming blue bonds intended for the conservation of the oceans along with other marine resources.

Recently, IFC has done some admirable work in advancing blue finance. For example, in 2022, the IFC offered USD 50 million in funding for the first blue bond issued by a Thai commercial bank. This bond raised money to be used on initiatives such as promoting sustainable shipping, wastewater treatment, and conservation practices in tourism (IFC 2022c). Also, to further the goals of blue finance, IFC's guidelines focused on the blue economy towards the development of the market, assisting the clarity that investments are on important objectives of sustainable development goal 14 on life below water, which seeks to protect and sustainably manage oceans, seas, and marine resources.

Blue bonds represent an important step towards achieving a more sustainable ocean economy by directing investment to vital marine conservation projects. These instruments support environmental sustainability, foster economic growth, and improve livelihoods in coastal and island nations dependent on aquatic resources.

4.3.2.3 Sustainability Bonds

Sustainability bonds are similar to green bonds but with slight variations. It includes projects of environmental and social importance and impacts specific projects with an environmental focus. Sustainability bonds allow the issuers to target aims encompassing more than one area of sustainability as identified plainly by the categories of the United Nations Sustainable Development Goals (SDGs). The general projects for which sustainability bonds are used include social housing, basic services, renewable energy, and energy-saving measures (Climate Bonds Initiative 2023).

Latin America stands out as a key region in the sustainability bond market, with 47% of its debt market composed of aligned Green, Social, and Sustainability (GSS+) bonds by the end of 2023. It has been the only region steadily increasing its sustainability bond volume over the last 4 years (Climate Bonds Initiative 2024). This growth is particularly notable in countries like Brazil, where sustainability bonds have been used to support sustainable agriculture and renewable energy projects, helping the country reduce its reliance on hydroelectric power (World Bank 2024c).

4.3.2.4 Transition Bonds

Transition bonds are a unique financial instrument that supports the decarbonisation of high-carbon industries. Unlike green bonds, which focus exclusively on financing already low-carbon projects, transition bonds enable industries to transform their operations incrementally, moving from high-emission practices towards more sustainable business models. These bonds are crucial for the energy, steel, and cement sectors, where complete decarbonisation will take time.

The ICMA Climate Transition Finance Handbook (ICMA 2023a) outlines key guidelines for transition bonds. One important distinction is the flexibility offered in the use of proceeds (UoP) model; instead of solely funding green projects, transition bonds finance projects that facilitate a shift to more sustainable practices, even in traditionally carbon-intensive sectors. In these cases, issuers commit to using the bond proceeds for projects aligned with the goals of the Paris Agreement (ICMA 2023).

For example, Japan issued the world's first sovereign Climate Transition Bond in 2023. This bond supports the country's effort to reduce greenhouse gas emissions by 46% by FY 2030, compared to 2013 levels, and to achieve carbon neutrality by 2050. The bond emphasises UoP, where proceeds are directed towards decarbonising key industrial sectors such as steel and manufacturing (Amari 2024). Japan's Climate Transition Bond Framework explains how funds are earmarked for projects supporting the gradual shift from fossil fuels to renewable energy sources. This example shows how transition bonds, with their flexibility in UoP, allow incremental progress towards sustainability. They help high-emission industries fund projects that align with their transition strategies while maintaining operational continuity.

4.3.2.5 Social Bonds

The proceeds from social bonds would go exclusively to finance or refinance new and existing social projects that meet the requirements outlined by the Social Bond Principles (SBP) provided by the International Capital Market Association (ICMA). The countries of the region that are eligible for such projects provide affordable housing, access to basic amenities, or employment generation for the vulnerable segments of the population.

For instance, Equitable Bank raised EUR 500 million through regulatory-covered bonds for May 2027 in the first issuance. One of the objectives of the issuance was to finance social housing projects, which were intended to broaden people's access to equitable housing. This was the first social bond issued under the sustainable bond framework of the bank (Equitable Bank 2024). Also, the Republic of Iceland issued the first sovereign bonds in the international debt securities market with a gender angle worth EUR 50 million. This bond was used to fund projects to improve women's living standards, increase income equality, and promote gender equality through parental leave reforms (BNP Paribas 2024).

In Bolivia, BancoSol launched the country's first gender bond, Bonos Sociales Avanza Mujer BancoSol 1, raising USD 30 million to support women-led micro and small enterprises (MSEs). This bond funds at least 4500 MSEs, supporting BancoSol's Avanza Mujer strategy. The initiative seems consistent with the UN's SDGs, particularly the themes of gender equality and women's economic empowerment (IFU 2024). Also, in sub-Saharan Africa, the IFC supported NMB Bank in Tanzania's first gender bond issuance. The bond offer of USD 32 million was oversubscribed by 197%, which justifies the interest in socially responsible investments (IFU 2024). This allows them to be part of the solution to social problems, be protected, and obtain returns similar to those of a normal bond.

4.3.2.6 Sustainability-Linked Bonds (SLBs)

Sustainability-linked bonds (SLBs) are such bonds where specific lending terms like the interest rate are linked to accomplishing defined sustainability targets. Unlike the use of proceeds (UoP) bonds, which allocate funds to specific environmental or social projects, SLBs provide issuers with flexibility by linking financial incentives to KPIs related to ESG objectives. This structure helps align the issuer's sustainability strategy with its financial commitments (ICMA 2024c).

The ICMA's Sustainability-Linked Bond Principles (2024) further emphasise setting measurable KPIs that reflect the issuer's core sustainability challenges. These goals should be ambitious and aligned with sectoral or global standards, ensuring that companies meet financial targets and contribute to meaningful progress on environmental or social fronts (ICMA 2024c).

A good example of this trend is Snam's EUR 1.5 billion two-tranche offering in 2024, which comprised EUR 500 million in green bonds and EUR 1 billion in SLB. The bond's coupon rate is directly linked to Snam's achievement of reducing its Scope 1, 2, and 3 emissions within the three defined time frames (2030 (S1&2 target date), 2032 (S3 target date), and 2034 (SLB Maturity)). Should this emission reduction scheme not be achieved, the company will face a financial penalty at maturity. This issuance demonstrates how increasingly financial mechanisms are deployed to further the cause of sustainability and the wider decarbonisation objectives sought globally (Snam 2024). The transaction complies with the revised ICMA 2024 SLB framework and is a good example of how SLBs can be used for corporate decarbonisation. Figure 4.12 summarises the distribution of the bonds in the world in 2023.

In the bond distribution in the world in 2023, more than half is seen to be green bonds. This is followed by social, sustainable, and sustainable-related bonds. Transition bonds remain under 1% of the total.

4.3.3 Allocation and Impact Reporting

Allocation and impact reporting are crucial in ensuring the transparency and credibility of sustainable bonds, such as green, social, and sustainability-linked bonds. Allocation reports focus on how bond proceeds are used, providing details about how funds are distributed across eligible projects. This level of transparency helps investors ensure that their capital is directed to the intended environmental or social initiatives. Regular updates on allocations, often provided annually, are essential to building trust and preventing any form of greenwashing, a practice where the sustainability of projects is overstated (ICMA 2023).

Conversely, impact reporting quantifies the actual results of the projects these bonds fund. It evaluates whether the objectives—reducing carbon emissions or increasing access to social infrastructure—are being met. This requires quantifiable metrics, such as CO_2 reductions, or social benefits, such as the creation of jobs, in line with the bond's goals. Impact reporting helps track whether the bond's promises translate into tangible environmental or social benefits. Often,

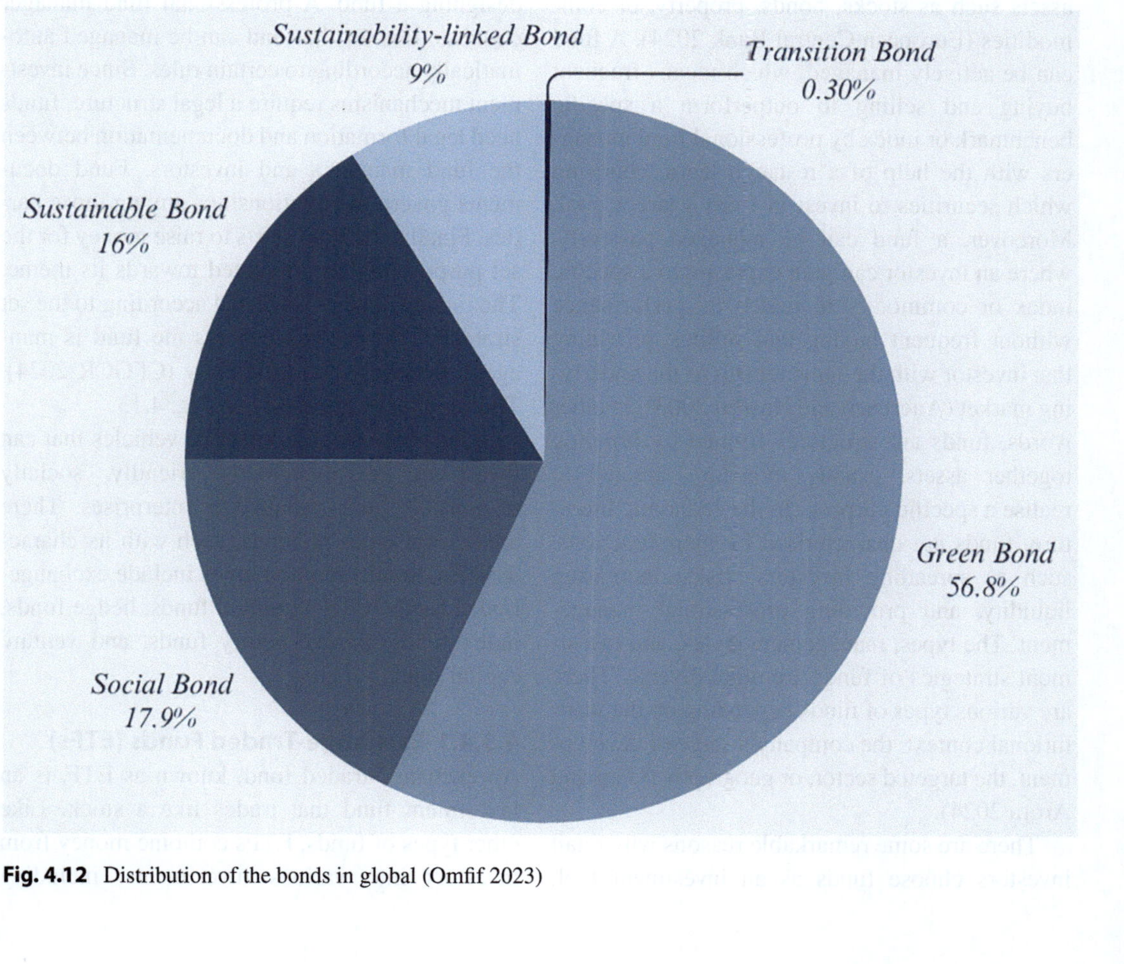

Fig. 4.12 Distribution of the bonds in global (Omfif 2023)

these reports are subjected to third-party verification to ensure accuracy and accountability, a practice supported by institutions like the World Bank and the International Finance Corporation (IFC 2024).

The combination of allocation and impact reporting is critical for maintaining the integrity of the sustainable bond market. It ensures that the funds support environmentally and socially beneficial projects and generate measurable and verifiable outcomes contributing to global sustainability goals.

4.3.4 Funds

In the world of finance, funds are one of the most significant investment mechanisms, with its EUR 17.969 billion investment fund assets outstanding in the Euro area in 2024 Q1 that enable different investors to invest in a diversified portfolio of assets such as stocks, bonds, property, or commodities (European Central Bank 2024). A fund can be actively managed, which means frequent buying and selling to outperform a specific benchmark or index by professional fund managers with the help of a research team, choosing which securities to invest in from a larger pool. Moreover, a fund can be managed passively, where an investor can gain exposure to a specific index or commodity to match its performance without frequent buying and selling, providing that investor with the same returns as the underlying market (Auerbach and Hassett 2003). In other words, funds are structures formed by bringing together assets, usually monetary, created to realise a specific purpose. In the economic literature, funds are characterised by their functions, such as spreading investors' risks, increasing liquidity, and providing professional management. The types, management styles, and investment strategies of funds are quite diverse. There are various types of funds depending on the institutional context, the company's stage of development, the targeted sector, or geography (Garg and Arora 2024).

There are some remarkable reasons why retail investors choose funds as an investment tool. Investors protect their investments from potential risks by dividing their portfolios into financial products through funds. In addition, funds provide access to professional management, allowing experienced fund managers to make informed investment decisions based on rigorous analysis. Funds suit investors needing more time or expertise to construct and monitor portfolios. They require a smaller investment than picking individual stocks or bonds, allowing the investor to access a greater number and variety of securities (Schröder 2004).

First, it is necessary to determine the fund's themes. These themes, such as education or pension funds, can be different. The purpose is also instrumental in creating the theme and naming of the fund. Next, finding investors or institutions that will contribute to the fund is necessary. These investors usually invest money with the expectation of a certain return. Moreover, deciding how the fund will be managed is a critical step when designing a fund. A professional fund manager could do this, or the fund can be managed automatically according to certain rules. Since investment mechanisms require a legal structure, funds need legal formation and documentation between the fund managers and investors. Fund documents govern the relationships among those parties. Finally, the fund starts to raise money for the set purpose and is orientated towards its theme. The money raised is invested according to the set strategy. This process ensures the fund is managed effectively and efficiently (CFGCR 2024). The fund types are shown in Fig. 4.13.

Funds are versatile financial vehicles that can invest in environmentally friendly, socially responsible, and sustainable enterprises. There are several types of funds, each with its characteristics. Some common types include exchange-traded funds (ETFs), mutual funds, hedge funds, index funds, private equity funds, and venture capital funds.

4.3.4.1 Exchange-Traded Funds (ETFs)

An exchange-traded fund, known as ETF, is an investment fund that trades like a stock. Like other types of funds, ETFs combine money from investors into various investments, including

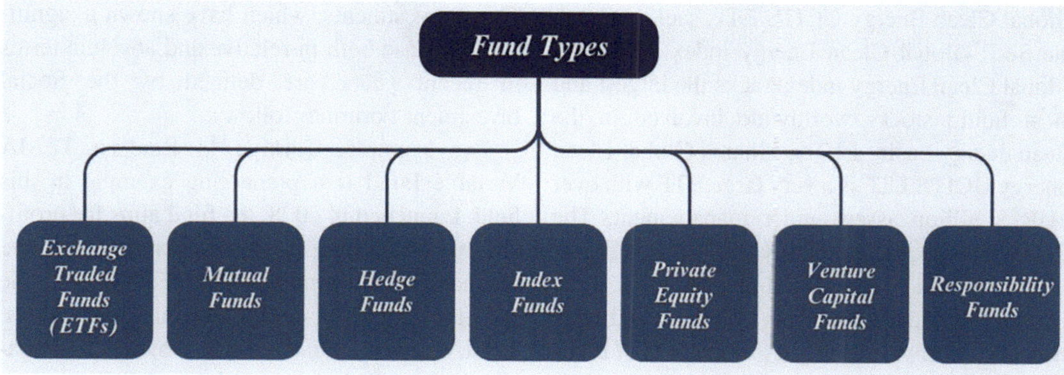

Fig. 4.13 Fund types

Fig. 4.14 An example of an exchange-traded fund portfolio

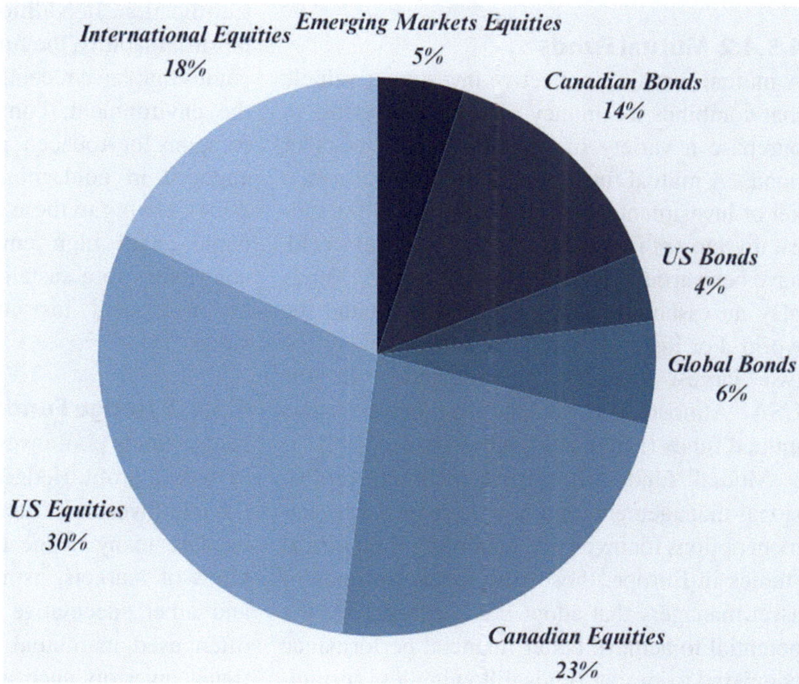

stocks, bonds, and other securities. By spreading the fund's cash across different securities, ETFs can generally provide investors with diversification, which can help manage risk. Like stocks, ETFs can be traded on the stock market and are usually affordable investment options. They may use passive or active management strategies. Because ETF shares are traded on a stock exchange, they're bought and sold like stocks throughout the day and may incur commissions. Additionally, ETFs commonly feature reduced costs compared to other fund varieties. Within ETFs, the "expense ratio" represents a portion of the yearly fee imposed on investors to fund the management and operational costs of the fund (Ben-David et al. 2017). Example of ETF portfolio distribution is presented in Fig. 4.14.

ETFs can focus on sustainability, investing in projects prioritising environmental and social impact. For example, clean energy ETFs are exchange-traded funds that invest in stocks in the alternative energy sector, including solar energy, wind, hydroelectric, and geothermal companies. One of the green energy ETFs, The iShares

Global Clean Energy UCITS ETF, seeks to track the S&P Global Clean Energy index. The S&P Global Clean Energy index tracks the largest and most liquid stocks worldwide involved in the clean energy business. The iShares Global Clean Energy UCITS ETF is a very large ETF with over EUR 2 billion assets under management. The ETF was launched on 6 July 2007 and is domiciled in Ireland (ETF Database 2024). A large amount of financing is provided to sustainable projects worldwide owing to these kinds of ETFs that invest in companies working in design, manufacturing, and deploying renewable energy generation technology.

4.3.4.2 Mutual Funds

A mutual fund is a collective investment vehicle that combines the money of several investors to purchase a variety of securities, like stocks or bonds. A mutual fund can be thought of as a basket of investments. Mutual funds, one of the oldest investment instruments in the financial world, have been around for more than 200 years. Funds play an essential role in economies around the world. For instance, mutual funds are one of the two largest financial intermediaries in the USA. Almost 50% of American families own mutual funds (Elton and Gruber 2013).

Mutual funds offer diversification, professional management, and a wide range of investment options for investors. According to empirical studies in Europe, those who invest in funds of asset managers that adopt ESG criteria have the potential to achieve better financial performance compared to similar funds following less sustainable investment policies. ESG criteria offer a competitive advantage as they shape investment decisions by considering non-financial data. Moreover, funds with high ESG scores have lower total costs despite more extensive analysis and additional costs (Abate et al. 2021).

Like stock, bond, and index funds, there are different investment instruments in the field of sustainability specifically designed to invest in companies and projects that prioritise ESG criteria. Social responsibility investments (SRI) are among the most prominent instruments. SRI has become a rapidly growing trend worldwide. These investments, which have shown a significant increase both in relative and absolute terms in recent years, are defined by the Social Investment Forum as follows:

For example, Türkiye İş Bankası TEMA Variable Fund is a pioneering example in this field. Launched in 2008, the fund aims for profitability by investing in environmentally sensitive companies and supports TEMA Foundation, one of Turkey's leading environmental organisations. TEMA Variable Fund does not invest in companies in the ISE Industry Index and ISE Services Index unless they have TS EN ISO 14001 Environmental Management System Certification. In addition, to ensure environmental sustainability, the fund does not invest in companies that have recently been fined for polluting the environment, companies that manufacture weapons or tobacco products, and companies engaged in gold mining (Türkiye İş Bankası 2024). Owing to these criteria, investing in companies with high environmental performance contributes to a sustainable future and becomes an investment instrument prioritising ethical value.

4.3.4.3 Hedge Funds

Hedge funds pool investors' money and invest it to make a profit. Hedge funds generally use more flexible investment strategies than mutual funds because many hedge funds aim to profit in all kinds of markets, using leverage, short selling, and other speculative investment practices not often used in mutual funds. Typically, institutional investors such as pension funds and rich individuals are the only ones allowed access to hedge funds. They are less regulated than mutual funds and have access to sophisticated tactics. Hedge fund managers are shielded against fraud but do not necessarily have to register with or submit reports to the SEC (U.S. Securities and Exchange Commission 2024). Additionally, hedge funds are growing in importance and prevalence in the financial world. In 1990, hedge funds managed approximately USD 39 Billion in assets, and current estimates put hedge fund assets at USD 2.5 Trillion (Getmansky et al.

2015). Hedge funds and mutual funds differences are illustrated in Fig. 4.15.

Although hedge funds are attractive due to their high return potential, their high risks cause significant controversy in the financial system. To invest in these funds, one must have a certain net worth or income or be an investment expert. Despite being labelled as hedges, hedge funds involve high risks and may adversely affect the economic policies of developing countries. Conversely, they support price-setting mechanisms by providing liquidity to financial markets and benefit investors by creating diversification in global portfolios (U.S. Securities and Exchange Commission 2024). Therefore, investors who can afford to take high risks are offered significant earning opportunities regardless of economic conditions. The hedge fund working principle is detailed in Fig. 4.16.

Inclusive Capital Partners was founded as a company focused on social investment. Based in San Francisco, Inclusive Capital Partners aims to invest in companies that respond to societal needs by considering environmental and social impacts. Inclusive Capital holds a stake in Nikola Corporation, one of the most important companies in Inclusive Capital's portfolio, valued at over USD 142 million, equivalent to 10.2% of the company's total shares (Reuters 2023). Nikola stands out as an important company in the portfolio of many hedge funds focussing on ESG and impact investing. However, it was announced in November 2023 that the fund would be closed.

4.3.4.4 Other Funds

An index fund is a form of investment that includes mutual funds and conceptually exchange-traded funds designed to emulate the performance of a general market index. Generally, an index fund is often appraised against the trend of the specific index it is supposed to track. These include examples such as the technology index, the health index, and even the sustainability index. The goal of investment in index funds is to track the performance of some market indices. For example, the S&P 500 includes the best 500 stocks on the U.S. market, while the Nasdaq comprises 100 technology stocks with huge capitalisations in the U.S. market. This kind of investment automatically results in the diversification of the portfolio, which then tracks the chosen index (SEC 2018).

Their other reason for the reduced expense ratio is the infrequent buying and selling of investments as opposed to most actively managed funds. They could also outperform most actively managed

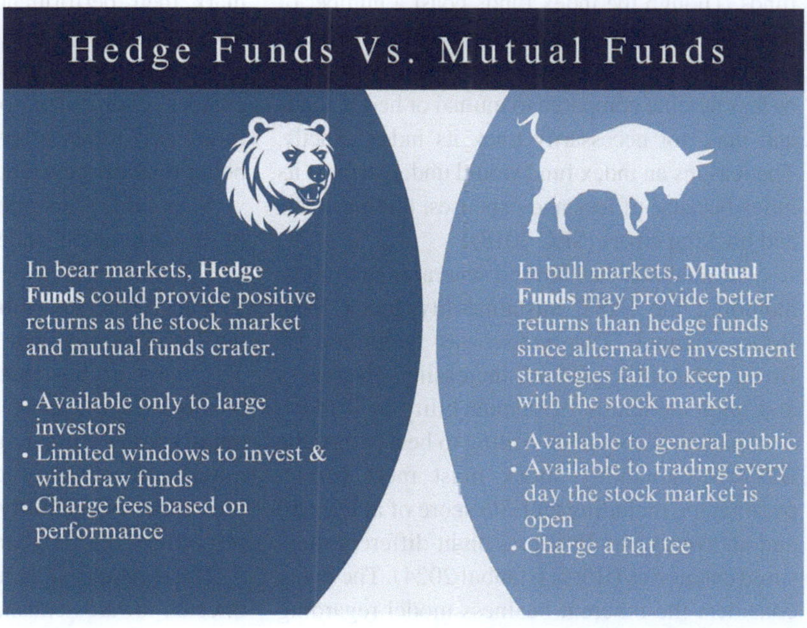

Fig. 4.15 Hedge funds and mutual funds differences (The Motley Fool 2024)

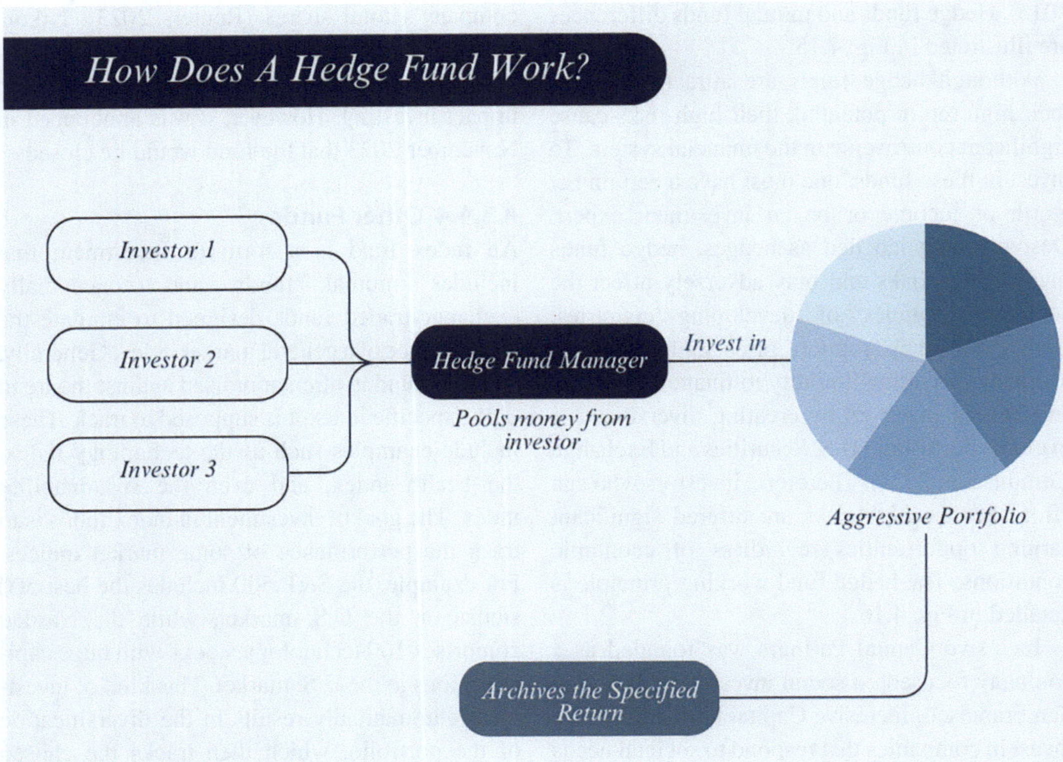

Fig. 4.16 How hedge funds work

funds in the long run. Due to these characteristics, index funds are considered less volatile or riskier than other fund types like hedge funds and mutual funds. Though the index funds boast a number of benefits, they also pose the same risks characterised by the tracked indexes. An index fund would be less flexible compared to mutual or hedge funds and may not necessarily track its index exactly. The reasons an index fund would underperform its index because of fees and expenses, trading costs, and tracking errors (SEC 2018).

Sustainability has birthed several index funds, including the BIST Sustainability Index from Borsa Istanbul. This index was facilitated by Borsa Istanbul's goal of increasing awareness and implementation of sustainability by Turkish Companies. Companies willing to be a part of the BIST Sustainability Index must meet certain parameters, including an ESG score of at least 50 and an average of 26 across eight different measured categories (Borsa Istanbul 2024). The index considers the inherent business model regarding global climate change and other environmental concerns, as well as health and social considerations, without restricting the business model's vision and judgement. Companies can benchmark their performance in sustainability practices within their own countries and across the globe. By introducing the index, Borsa Istanbul enables companies to set targets to improve future performance based on aiming to perform better than they currently do.

Vanguard offers another outstanding example of a sustainable index fund, Vanguard ESG U.S. Stock ETF, which tracks the MSCI USA Extended ESG Focus Index and puts investments into companies which meet the ESG guidelines. The Vanguard Sustainability Index Fund focuses on three specific areas: the environment, the community, and corporate governance. This incorporated approach seeks to enhance investment that promotes sustainability but also adheres to social responsibility (Vanguard 2024).

Private equity (PE) funds are an investment tool directed towards enhancing the worth of latent or underperforming companies by investing in them and then selling their stock to make a profit. In general, a private equity fund is a collection of funds from various investors aimed at investing in established private companies. The goal of private equity firms is to generate a return by enhancing the operational performance of the business and then selling it (Kaplan and Sensoy 2015). As a result of such developments, PE funds are professionally managed and specialised, becoming shareholders of companies with the potential to increase their value. Investment ends after alterations in management and company strategy are made, and there has been growth in the company's value.

As one of the most prestigious private equity funds, Parnassus also pays attention to sustainability criteria when making investments. When investing in companies, the analysts of Parnassus assess the dimensions of profit, coping structures, relations with employees and clients, and ecology (Parnassus Investments 2024). With such an understanding, the goal is to create bases that will help companies perform better in the future while still stressing social and environmental concerns. Figure 4.17 summarises the basics of private equity funds as follows.

Venture capital (VC) funds are private investment vehicles that support the growth of startup companies by investing in companies with high growth potential. These funds, which generally focus on innovative startups in fast-growing sectors such as technology, healthcare, and biotechnology, offer high return potential to their investors while also making significant contributions to the economy. Although VC is a fund that contributes to the development of startups, it carries high risk. Because the companies' future success is uncertain, investors generally prefer such investments with a long-term perspective and the capacity to take risks (Gompers and Lerner 2001).

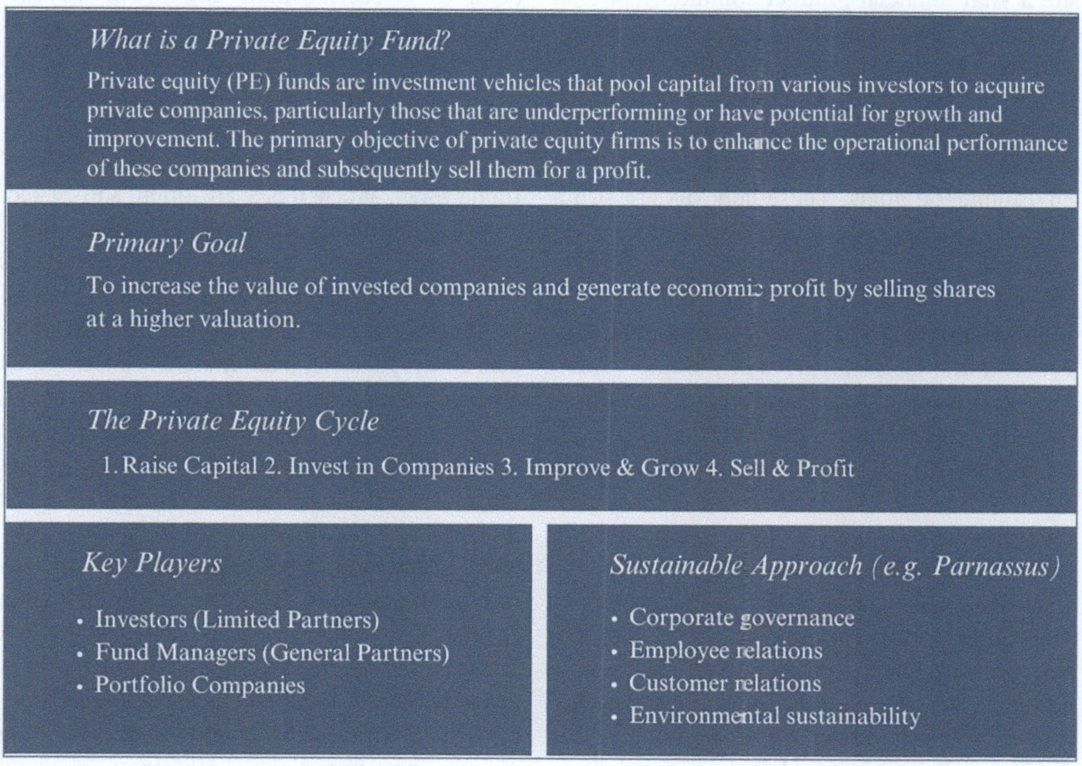

Fig. 4.17 Summary of private equity funds (Küfeoğlu 2024a)

For instance, Regeneration.VC is a global venture capital firm investing in Consumer ClimateTech innovation. Their investment strategy targets are returning biological resources to living ecosystems to support new biological growth, returning technical materials to productive next use, and eliminating the concept of waste. They are investing in Nature Coatings, which aims to convert timber industry by-products into carbon-negative, biobased black pigment (Regeneration.VC 2024). Figure 4.18 explains the differences between private equity and venture capital.

4.3.5 Green Deposits and Repo Markets

4.3.5.1 Deposits

A deposit refers to money placed in a financial institution expecting to earn interest or be used as collateral for specific transactions. Deposits are a vital financial product commonly offered to individuals and businesses, enabling banks to extend loans and investments that fuel economic growth. Green deposits represent an innovative evolution, targeting individuals and institutions seeking to align their capital with environmental goals. These deposits are structured as use of proceeds (UoP) products, ensuring that funds are channelled into projects defined within a Green Asset Pool, such as renewable energy, sustainable infrastructure, or conservation efforts (ICICI Bank 2024; Deutsche Bank 2023). Compared to traditional deposits, green deposits guarantee that the capital will exclusively finance sustainable projects, offering benefits like eco-friendly impact, ethical satisfaction, and portfolio diversification (FirstRand Bank 2022).

Green deposits also reflect growing investor demand for sustainable finance solutions, where transparency and alignment with UN SDGs play a crucial role. As the industry evolves, the UoP model remains central in guaranteeing that funds are allocated towards pre-approved green initiatives, highlighting the importance of trust and accountability in sustainable banking frameworks.

4.3.5.2 Repo Markets: Green Repos and Sustainability Impacts

A repurchase agreement (repo) is a short-term contract involving the sale of securities with a commitment to repurchase them at a higher price at a future date. Financial institutions and central banks use this tool to manage liquidity efficiently in money markets. Repos also allow financial institutions to raise short-term cash while maintaining exposure to the underlying securities (IMF 2024).

Green repos offer a specialised adaptation of this market, enabling banks to use the proceeds from repo transactions exclusively for green or sustainable investments. Deutsche Bank, for instance, executed one of the first green repo transactions in Europe by channelling proceeds into renewable energy projects, such as wind and solar farms and energy-efficient commercial buildings (Deutsche Bank 2021). This transaction demonstrated how repos can support sustainable finance by ensuring that capital flows remain consistent with green financing frameworks while enhancing market liquidity.

Unlike green deposits, which focus on UoP frameworks, green repos are often KPI-linked, ensuring that funds are tied to specific environmental performance metrics, such as reductions in carbon emissions or energy efficiency improvements (IMF 2024). The flexibility of repo markets allows financial institutions to convert green assets into cash without disrupting their sustainability objectives, facilitating efficient capital management aligned with environmental goals (BlackRock 2024). The development of green repos reflects the broader shift towards sustainable finance innovations, offering financial institutions reputation benefits, access to new market segments, and first-mover advantages in ESG markets. However, challenges remain, such as ensuring that the collateral and transaction structures meet strict sustainability standards while maintaining operational efficiency (Federal Reserve Bank of New York 2024).

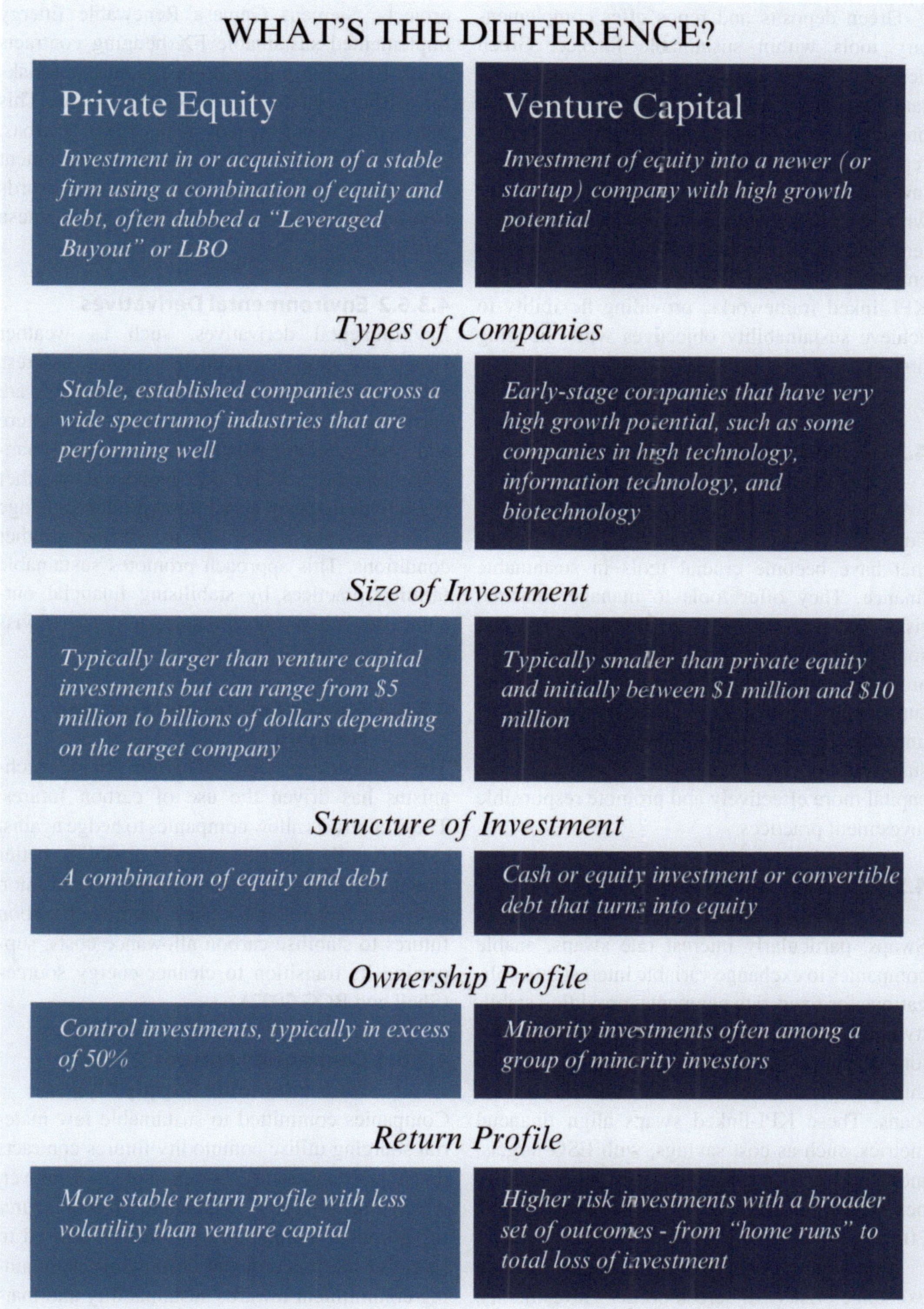

Fig. 4.18 Private equity vs. venture capital (Küfeoğlu 2024a)

Green deposits and repos offer complementary tools within sustainable finance. Green deposits cater primarily to individual and corporate customers, ensuring funds support green projects over a defined tenor. In contrast, green repos provide a dynamic solution for institutional investors, allowing them to maintain liquidity while ensuring that capital flows remain consistent with ESG principles. While green deposits emphasise UoP models, green repos focus on KPI-linked frameworks, providing flexibility to achieve sustainability objectives while hedging financial risks.

4.3.6 Derivatives in Sustainable Finance

Futures and swaps are examples of derivatives that have become crucial tools in sustainable finance. They offer tools to manage financial risks associated with environmentally and socially responsible projects. These instruments are critical in stabilising financial outcomes and supporting sustainability initiatives by mitigating uncertainties. By linking financial instruments to sustainability targets and KPIs, firms can access capital more effectively and promote responsible investment practices.

4.3.6.1 Swaps and Currency Management

Swaps, particularly interest rate swaps, enable companies to exchange variable interest rate obligations for fixed-rate payments, providing stability in cash flows. This is especially important for long-term sustainability projects financed through green bonds or sustainability-linked loans. These KPI-linked swaps align financial metrics, such as cost savings, with ESG targets, incentivising companies to meet sustainability benchmarks throughout the loan period (ISDA 2022).

Similarly, currency swaps mitigate foreign exchange risks in cross-border investments, ensuring the financial viability of international projects. Siemens Gamesa Renewable Energy implemented sustainable FX hedging contracts worth EUR 174 million to hedge currency risks for offshore wind turbines sold in Taiwan. This transaction, coordinated with BNP Paribas, aligned with the UN Sustainable Development Goals and directed premium payments towards forestry projects in Spain (Siemens Gamesa 2019).

4.3.6.2 Environmental Derivatives

Environmental derivatives, such as weather swaps and emission reduction swaps, address environmental risks directly. These products are particularly relevant in sectors like agriculture and energy, where climatic factors affect financial outcomes. Swiss Re pioneered weather derivatives that help agricultural producers hedge against revenue losses due to adverse weather conditions. This approach promotes sustainable farming practices by stabilising financial outcomes and mitigating climate-related risks (Swiss Re Institute 2024b).

4.3.6.3 Carbon Futures and Emission Hedging

The increasing relevance of carbon pricing mechanisms has driven the use of carbon futures. These contracts allow companies to hedge against carbon credit price fluctuations, enabling better strategic planning and investments in emission reduction technologies. Shell employed carbon futures to stabilise carbon allowance costs, supporting its transition to cleaner energy sources (Shell and BCG 2022).

4.3.6.4 Commodity Futures for Sustainable Supply Chains

Companies committed to sustainable raw material sourcing utilise commodity futures contracts due to price instability. For example, at Unilever, which aims for 100% sustainable palm oil sourcing purposes, commodity futures are utilised to level out the procurement costs while maintaining commitment towards sustainability and market risks (Unilever 2024).

4.3.6.5 The Role of ESG-Linked and KPI-Driven Derivatives

KPI-linked derivatives have been reported within sustainable finance. These prices related to the derivatives are favourable for price-environmentally linked debt instruments, category of instruments that submersion sustainability targets achievement into the financial structure, for instance, in ESG-linked debt instruments, where price competitiveness is achieved based on some predetermined criteria (ISDA 2022). These instruments enable companies such as marinas to try to meet their targets and increase their efforts while protecting themselves from the risk management tools provided.

To conclude, the role of derivatives in sustainable finance is diverse, be it in terms of risk management, liquidity of the market, or facilitating transition to energies. Firms can reap many benefits from these, including managing uncertainties and being efficient and sustainable in accessing capital markets. The transition to a sustainable economy is one of the tags that come with integrating the different kinds of derivatives into businesses, particularly those linked to KPIs.

4.3.6.6 Transactions in Sustainable Finance Instruments and Mechanisms

In sustainable finance, transactions are critical in facilitating financial operations aligned with ESG principles. This section examines key non-cash financial instruments, including trade loans, invoice financing, guarantees, letters of credit, and supply chain finance (SCF) mechanisms. These instruments ensure access to liquidity and manage financial risks while encouraging participants to pursue sustainable practices. Structure of non-cash transactions and supply chain finance is demonstrated in Fig. 4.19.

4.3.6.7 Trade Loans in Sustainable Finance

Trade loans provide suppliers liquidity and buyers with working capital, stabilising financial operations in ESG-compliant activities. These loans distinguish themselves by focusing on sustainable trade transactions and channelling capital towards projects that benefit the environment or society. For instance, trade loans reached USD 182.2 billion in Q2 2024, up from USD 79.5 billion in 2023, with forecasts predicting a 4.5% annual growth rate through 2032 (KPMG 2024). By offering preferential terms, such as extended repayment periods, financial institutions encourage companies to pursue sustainable products and projects.

4.3.6.8 Invoice Financing and Sustainable Cash Flow Management

Invoice financing addresses cash flow challenges by using unpaid invoices as collateral to access immediate funds. This mechanism is vital in supporting sustainable businesses offering liquidity while aligning with ESG goals. The global invoice factoring market, valued at USD 2.32 billion in 2023, is expected to reach USD 4.55 billion by 2030, growing at a compound annual rate of 10.11% (MMR 2024). By facilitating sustainable cash flow management, this financing mechanism enables businesses to remain financially viable while contributing to broader sustainability objectives.

4.3.6.9 Guarantees and Letters of Credit

Guarantees provide lenders with protection by assuring repayment even if the borrower defaults. In 2024, the World Bank Group issued guarantees worth USD 10.3 billion, with plans to increase annual guarantees to USD 20 billion by 2030 (WBG 2024b). Green letters of credit, or sustainable versions of LCs, support environmentally friendly transactions while securing payment obligations in international trade. These LCs finance projects such as renewable energy installations and sustainable agriculture initiatives. The LC confirmation market is projected to reach USD 4.4 billion by 2024, reflecting the increasing emphasis on sustainable trade finance (The Business Research Company 2024).

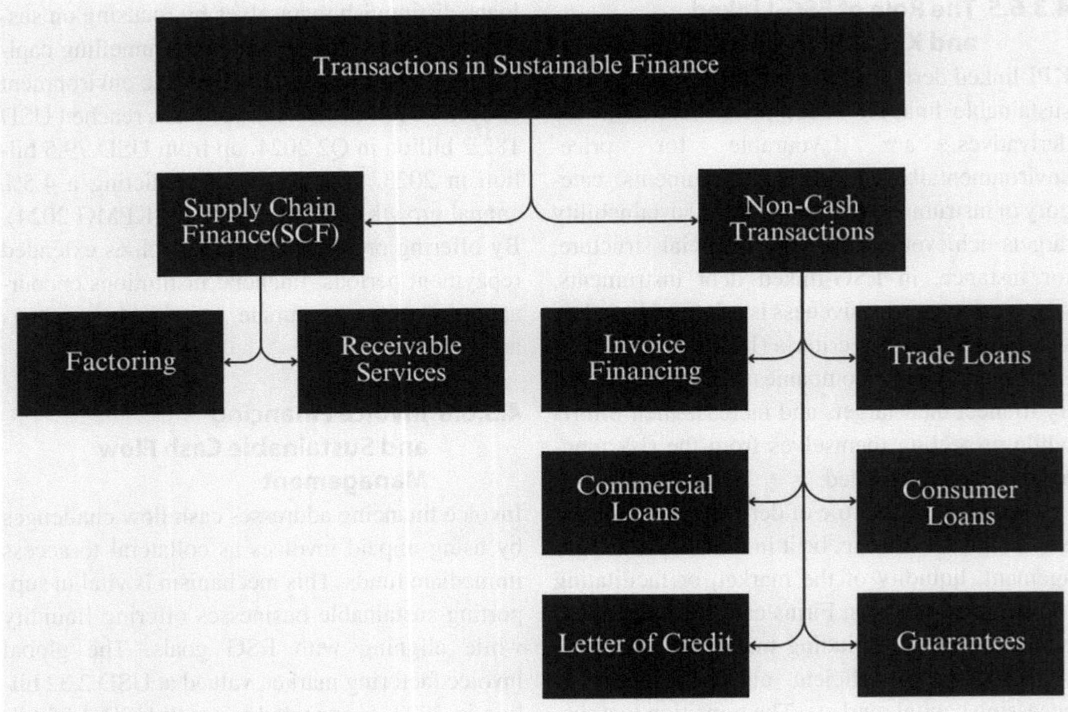

Fig. 4.19 Structure of non-cash transactions and supply chain finance in sustainable finance (Küfeoğlu 2024a)

4.3.6.10 Supply Chain Finance and ESG Integration

Supply chain finance (SCF) uses receivable discounting and factoring to maximise liquidity across ESG-aligned value chains. Factoring allows suppliers to sell receivables at a discount, enhancing cash flow while ensuring supply chain stability. The global factoring market, valued at USD 3876.8 billion in 2024, is expected to grow to USD 5238.82 billion by 2028, with a 7.8% CAGR (Research and Markets 2024). These instruments enable open-account trade relationships, reducing financial risks and fostering sustainability across the supply chain.

The integration of KPI-linked instruments within SCF further enhances sustainable practices. These financial products tie financial terms to ESG performance metrics, incentivising companies to meet sustainability goals. For example, BNP Paribas has structured loans and SCF products with performance-based terms, encouraging firms to align financial outcomes with ESG objectives (Loan Market Association 2023a, b). This trend reflects the growing emphasis on measurable sustainability outcomes in financial transactions.

Technology is increasingly important in SCF, enhancing transparency and monitoring ESG performance across supply chains. Blockchain-based platforms, for instance, enable real-time transaction tracking and ensure compliance with ESG criteria (Medina et al. 2023). Moreover, public institutions and development banks collaborate with financial institutions to provide guarantees, addressing SMEs' challenges in accessing SCF. The mechanism details of supply chain are shown in Fig. 4.20.

Transactions in sustainable finance encompass various instruments, from trade loans and invoice financing to guarantees and SCF mechanisms. These tools are essential for maintaining liquidity, managing financial risks, and promoting sustainable industry practices. The use of KPI-linked structures further amplifies the impact of financial instruments by directly linking financial performance to sustainability outcomes.

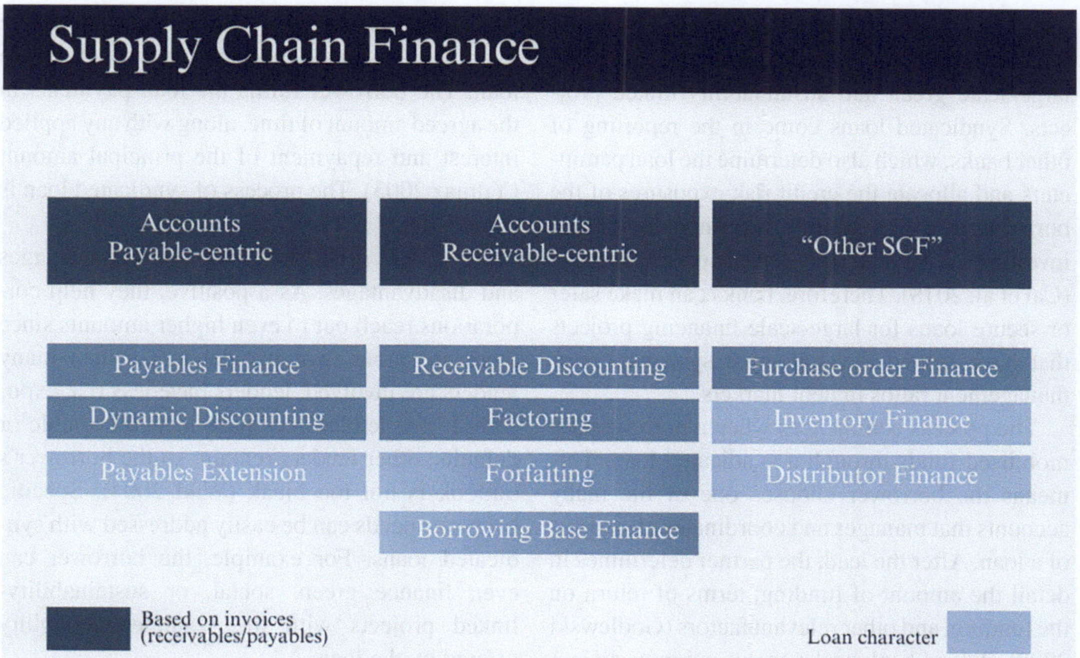

Fig. 4.20 Supply chain finance mechanism (Küfeoğlu 2024a)

Through these mechanisms, financial institutions and businesses contribute meaningfully to achieving sustainable development goals.

4.3.6.11 Loans (Retail and Commercial)

Loans are one of the most important financial instruments for providing project financing. These financial instruments are crucial for bridging the funding gap and enabling the completion of large-scale projects that would not otherwise be possible. A loan is a financial agreement whereby a bank usually gives money to a borrower, who then promises to repay the borrowed money plus interest over a predetermined period (Ivashina et al. 2020). Due to the loans supplying the funds required to launch large-scale endeavours and mitigate financial risk by matching loan repayment with project cash flow, loans are essential for funding sustainable projects. According to academics, bank loans are the primary funding source for long-term, sustainable investment projects—these loans fund sustainability-focused projects, including carbon reduction and renewable energy programs (Lalon 2015). Bank loans are essential to the success of these ecologically beneficial projects since they provide financial support in advantageous conditions, supporting global sustainability efforts. Bank loans are not only used by individuals for small expenses but also serve as a significant source of financing for corporations looking to develop their businesses through various purposes, such as sustainable projects. Large-scale loans are available in this situation to fund large-volume projects. For instance, syndicated and multilateral loans are frequently employed techniques in funding large-scale projects.

4.3.7 Methods of Banks Finding Resources from International Banks

4.3.7.1 Syndicated Loans

Banks often find themselves unable to finance certain large-scale projects independently. Syndicated loans are one of the most significant forms of large-scale project funding that banks frequently require for these projects. Syndicated loans can be defined as an auxiliary financing instrument with multiple lenders for borrowers with complex requirements, such as funding large-scale green and sustainability-linked projects. Syndicated loans come in the reporting of other banks, which also determine the loan parameters and allocate the credit risk exposures of the borrower to commodities when implementing or investing in large-scale projects or investments (Cai et al. 2018). Therefore, banks can make safer or secure loans for large-scale financing projects that have helped develop more systematic risk management ratios in debt markets.

The process commences when a borrower has mobilised funds through a syndicated loan. This means the borrower chooses one of the many accounts that manages and coordinates all aspects of a loan. After the lead, the partner determines in detail the amount of funding, terms of return on the funding, and other relevant factors (Godlewski 2008). A lead bank seeks construction participation from other financial tools, such as commercial banks, sophisticated investors, or other agencies and finalises the financing framework with the participants. After all concerned have understood, the borrower signs the loan agreement, along with the lead bank and all lenders. After signing the loan agreement, the arranged lenders provide credits to the borrower according to the specified arrangement.

The lead bank monitors the client's adherence to obligations on the loan amount throughout the loan. The borrower fulfils the loan payments in the agreed amount of time, along with any applied interest and repayment of the principal amount (Yılmaz 2003). The process of syndicated loan is demonstrated in Fig. 4.21.

The use of syndicated loans has its advantages and disadvantages. As a positive, they help corporations reach out to even higher amounts since there are more than two lenders. When many lenders are involved, lenders have less risk exposure. If one lender goes into financial trouble or defaults, other lenders remain, so the borrower's outlook is not too bleak (Wild 2004). Specific borrower needs can be easily addressed with syndicated loans. For example, the borrower can even finance green, social, or sustainability-linked projects with the regulated flexibility offered by the loan.

Conversely, it has also been established that syndicated loans have some drawbacks. Since they are more complex and bigger, syndicated loans need more extensive talks and a longer documentation process than bilateral loans. It can be even more complicated as each debtor has different needs and expectations that may need to be addressed. Furthermore, there are instances where syndicated loans can be offered at a higher

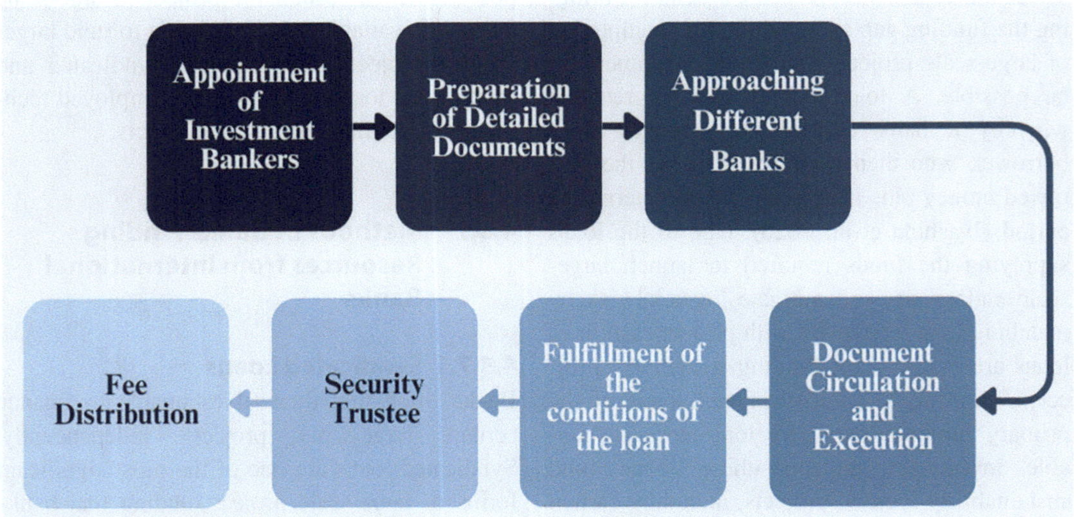

Fig. 4.21 Process of a syndicated loan

rate as they involve more than one lender (Wild 2004).

Syndicated loans can be given from both commercials and banks. For instance, 21 banks have provided an approximately EUR 2.3 billion syndicated loan to The Dardanelles Bridge, also known as the 1915 Çanakkale Bridge, which was constructed to connect the European and Asian sides of the Dardanelles. The Dardanelles Bridge syndicated loan is not only financed by Turkish banks such as Akbank and Türkiye İş Bankası but also by international banks such as Industrial and Commercial Bank of China (ICBC) and Korea Development Bank (KDB). This loan, consisting of 8 different loan tranches, including Export Credit Agency (ECA) and Islamic financing methods, has a maturity of 15 years and a grace period of 5 years (1915canakkale 2024). Dardanelles Bridge syndicated loan is one of the significant examples of commercial syndicated loans.

As an example of syndicated bank loans, İşbank's sustainability-linked loan exceeding USD 1 billion in 2024 is a prime example. İşbank secured a 367-day sustainability-linked syndicated loan of EUR 562.7 million and USD 442 million to support foreign trade financing. İşbank's syndicated loan's sustainability performance criteria prioritise providing financial loans to small- and medium-sized female entrepreneurs and smallholder farmers. The agreement was signed by 48 banks from 21 countries in the Americas, Europe, the Middle East, and Asia, with 18 new banks joining (İşbank 2024). İşbank's collaboration with stakeholders has resulted in a new syndicated loan that would finance clients' foreign trade transactions while supporting Turkish agriculture and women's societal and economic roles.

Besides all this, there is also a specialised form of loan within the financial landscape similar to syndicated loans, whose name is club loans. A club loan is a loan where lenders collectively provide funds to a borrower. In club loans, the number of lenders is less than in syndicated loans, and the loan size is smaller than in syndicated loans. These lenders typically have a direct relationship with the borrower and have an equal say in the club loan terms and conditions (Nowak et al. 2018). Because of the smaller number of lenders, club loans are more flexible than syndicated loans, and club loan fees and costs are fewer. For example, a company requires USD 30 million to invest in a sustainability initiative. Instead of pursuing a large-sized syndicated loan, the corporation approaches three of its main banking partners. These banks establish a club loan, each providing an equal portion of about USD ten million. They get together and agree on loan conditions. Once approved, the loan is used by the business to invest in sustainability projects, with the banks working together to keep an eye on the loan's performance and the business's compliance with the terms of the agreement.

4.3.7.2 Bilateral and Multilateral Loans

Based on the terms and structures of the loans, loans that play a major role in funding sustainable development projects and preserving economic equilibrium can be categorised as bilateral or multilateral. The main difference between bilateral and multilateral loans is the number of lenders. Bilateral loans involve only one lender and one borrower, while multilateral loans combine multiple lending sources for a single borrower. Examples of borrowers include individuals and businesses such as SMEs, while examples of lenders include banks and financial institutions. Because only one lender is involved and just one lender and one borrower negotiate the terms and conditions directly, bilateral loan conditions are usually more flexible than multilateral loans, costs are much lower, and the size of the bilateral loan is generally smaller in dimension compared to the multilateral loan (Barthelmess and Langlois 2020). Bilateral loans are an essential financing mechanism for businesses and those wanting to meet their needs. It is a common financial tool to apply for bilateral loans to finance important expenses such as home, car, education, and health.

Unlike bilateral loans, multilateral loans involve multiple lenders; thus, negotiating a multilateral loan is more complex than a bilateral loan due to the involvement of various lenders. Because of various lenders, multilateral loan terms and conditions are stricter than bilateral loans, and fees are generally higher than bilateral loans (Odedokun 2004). Despite higher costs and more stringent conditions, multilateral loans can fund large-scale projects without a single lender. Thanks to these characteristics of multilateral loans, this type of loan has become an important resource for implementing large-scale projects that promote long-term development and economic stability.

To illustrate the distinction between bilateral and multilateral loans, consider the scenario of a small business owner seeking loan capital for operational expansion. If a bank agrees to extend a USD 100,000 loan at a 5% annual interest rate with a five-year repayment schedule and the small business owner agrees with just one bank, this exemplifies a bilateral loan function because the bank acts as a single lender. The business owner who must repay the principal amount and interest during the agreed-upon period acts as a single borrower. Only two parties are in this agreement, and all the terms are formed within the framework of the bargaining agreement between the two parties. However, if the same business needed a larger-sized loan and it was difficult for a single bank to provide this amount, the loan could be provided multilaterally. In this case, multiple banks would come together to provide credit, and the loan required by the business could be provided in this way, called a multilateral loan.

4.3.7.3 Commercial Loans

Commercial loans are agreements whereby financial institutions provide or undertake to provide access to capital to public or private entities so that they may become able to maintain or expand their operations. In the ordinary course of commercial life, public or private entities may attempt to expand their geographic scope by establishing a branch or acquiring real property, broaden their consumer reach by investing in marketing operations, and improve their production capacity and quality by acquiring means and technologies of production. Also, maintaining business operations may be challenging, resulting in bankruptcy when unexpected circumstances occur, such as financial crises, pandemics, natural disasters, etc. Whatever the internal or external circumstances, financial institutions may enable public or private entities to reach their goals while directing them towards more sustainable options.

The process for obtaining commercial loans should be evaluated to grasp how public or private entities may be directed towards providing more sustainable goods and services. Although the documents requested by financial institutions may vary, the applicants may typically be required to submit their financial reports and documents indicating that they are active in business. After calculating the transaction risk, financial institutions may refuse the application outright or determine individual agreement terms, especially regarding interest rates and repayment terms. Public or private entities whose project minimises or eliminate environmental risks may benefit from advantageous interest rates and repayment terms.

Among public or private entities, micro-, small-, and medium-sized enterprises (MSMEs) play an important role in the green economy (OECD 2024a). In this sense, the European Bank for Reconstruction and Development provided Sparkasse Bank Makedonija in North Macedonia with a EUR 23 million loan so that it could onlend to MSMEs enterprises to support their investments associated with energy efficiency and renewable energy (Mirbabaeva 2022). Commercial loans also play a key role in supporting environmentally sustainable transportation. In 2022, for example, the European Investment Bank provided Cabify, which is a Spanish company operating an online transportation platform, with a EUR 40 million loan to support its initiative to decarbonise the fleet in Spain by 2025 by acquiring electric vehicles and setting up the infrastructure for charging these vehicles (EIB 2022).

4.3.7.4 Retail Loans

Other than commercial entities, consumers may also need financial support in the ordinary course of life. Indeed, consumers have fundamental needs to live a life worthy of human dignity, such as a safe house, accessible means of transportation to commute from home to work and vice versa, educational materials, and so on (Woodhall 1995; Luong 2010).

Retail loans offer huge opportunities to nudge consumers into adopting more sustainable consumption habits. Considering the volume and value of transactions, it may not be conceivable for financial institutions to offer a differentiated scheme for each personal loan and credit card expenditure to direct consumers to more sustainable options. However, the mortgage may play a pivotal role. According to the European Banking Authority (2023), for example, 79% of all loans to households in the EU were allocated in the form of mortgages in 2022. In 2016, the Dutch Development Bank granted Banco Promerica a USD 30 million loan so that it may launch a green mortgage program by offering lower interest rates and longer repayment periods (IFC 2019a). Likewise, the European Bank for Reconstruction and Development provided Sparkasse Bank Makedonija in North Macedonia with a EUR five million loan so that it on-lent funds to consumers who invest in residential green buildings and technologies improving the energy efficiency of residential buildings (Rozanova 2020).

4.3.7.5 Challenges and Opportunities in Loans for Sustainable Firms

Loans are one of the most important financial instruments for funding sustainable projects. Syndicated loans enable multiple financial institutions to share the risks of large-scale green and sustainability-linked project financing. While bilateral loans target smaller loan sizes involving only one lender and one borrower, multilateral loans are more complex processes for larger loan sizes, combining multiple lending sources for a single borrower. Commercial loans support the green economy by steering commercial entities during the continuation and expansion of business operations. Finally, retail loans include consumers in sustainable finance, especially when they invest in real estate and use credit cards. The summary of the loan types is presented in Fig. 4.22.

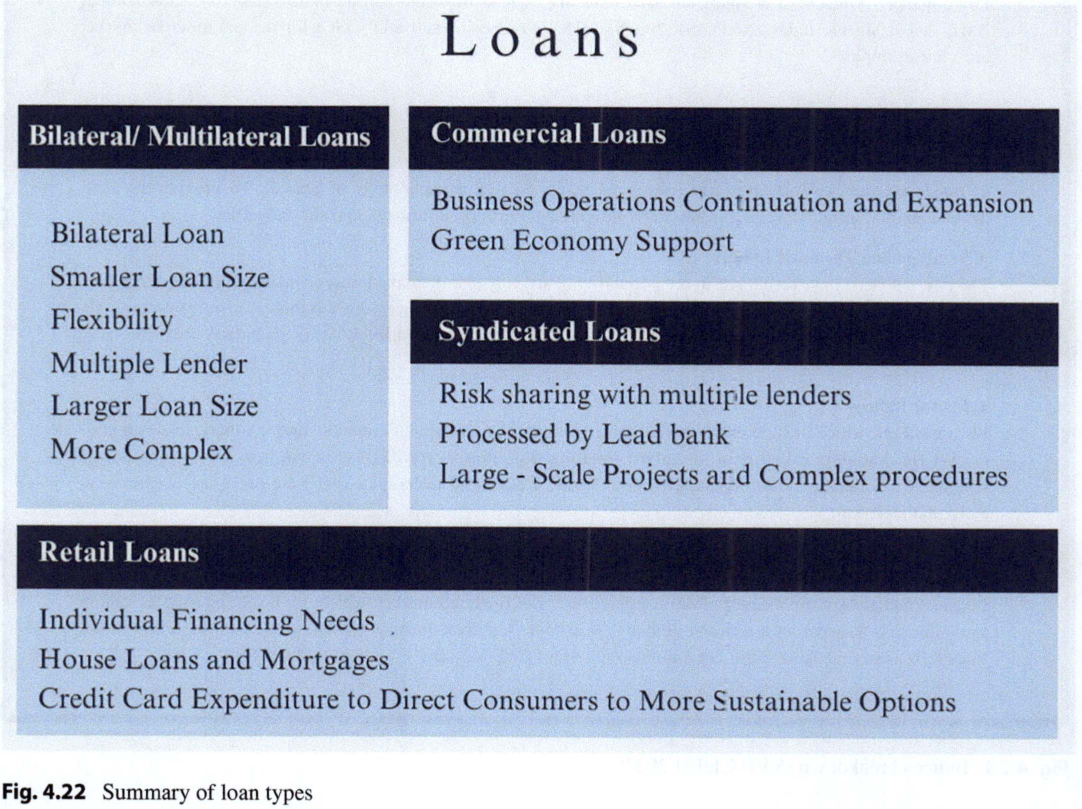

Fig. 4.22 Summary of loan types

To sum up, there are many types of loans that could assist to finance and refinance of sustainability projects and business. For ESG-compliant operations, tools including commercial loans, invoice financing, guarantees, letters of credit, supply chain financing (SCF), and syndicated loans manage cash flow, reduce financial risks, and offer liquidity. SCF mechanisms, for example, promote environmentally and socially responsible company practices by establishing a value chain. Performance-linked loans direct businesses towards sustainable objectives by forming financial arrangements based on ESG criteria. Furthermore, commercial and personal loans encourage businesses and individuals to make sustainable investments, while syndicated and multi-party loans help large-scale initiatives. Banks and other financial institutions may play a vital role in promoting sustainable growth and a greener economy, thanks to this all-encompassing financial strategy.

4.3.8 Sustainability Indices

A market, asset class, market sector, or investment strategy's performance could be represented and measured by an index, a collection of stocks, derivatives, or other financial instruments. An index is a selection of any set of observable securities in a market segment that is statistically representative. One example of a large-cap representation of the US equities market is the well-known S&P 500. The index level, or numerical value, fluctuates in response to changes in the overall value of the securities that comprise the index (S&P Global 2024b).

Indices can represent different countries, sectors, and themes. For example, the MSCI US Index is an index that symbolises a country's equity market. Conversely, indices like the MSCI US Healthcare Index symbolise the sector. Figure 4.23 outlines the main components of indices.

1. Broad Market Indices
Large market segments or a significant asset class are tracked by broad market indices. S&P 500, S&P Global BMI (Broad Market Index), S&P/ASX 200, S&P/TSX Composite, and S&P GSCI for the commodities market are a few examples.

2. Sector Indices
Sector indices follow market groupings based on industries, breaking down large markets into more specialized areas like technology and health care. These indices have the ability to track at an even more precise level, giving investors exposure to software businesses in particular as well as technology in general. To provide the most detailed market insights, sector variants delve into industry groups, industries, and sub-industries.

3. Strategy and Thematic Indices
The past ten years have seen a rise in the popularity of strategy and thematic indices, which are intended to reflect an investment strategy or target a specific market niche. Thematic indices, such as those for infrastructure, clean energy, or biotech stocks, typically reflect certain market niches or themes. Strategy indices frequently use unconventional techniques or focus on the basics.

4. Factor Indices
Factor indices, which include value, momentum, low volatility, and quality indices, among others, focus on non-market risk variables rather than the return resulting from market risk. Investors can now access components separately or in combination, opening them a world of advanced techniques that were previously exclusive to active management.

5. Fixed Income Indices
Fixed income indices are meant to follow different parts of the bond market, which is reinventing itself as a more accessible and transparent asset class globally these days. The fixed income market, along with the indexes that track it, is widely recognized for its advantages in terms of yield, diversity, and reduced volatility.

Fig. 4.23 Indices breakdown (S&P Global 2024b)

With the growing interest in indices, investors are looking for new ways to use index-linked investment products such as ETFs to invest in capital markets. Sustainability indices are tools used to assess a company's ESG responsibility. Their score will increase proportionately to how much consideration they give these factors as they grow their company. Beatriz Fernández, an environmentalist and professor at the Instituto Superior de Medio Ambiente, states that the purpose of sustainability indexes is to inform institutional and retail investors about the importance of a company's ESG responsibility, in addition to financial results, when making decisions to buy shares (BBVA 2019).

One further advantage is that the indices allow users to monitor the market. The asset classes of bonds and stocks are both tracked by indicators. One of the indices most frequently cited in US markets is the S&P 500. Professional investors, therefore, usually use S&P performance as a benchmark to assess all or a portion of their performance. Investors anticipate returns that exceed the corresponding market indices for actively managed funds (such as mutual funds), given that these funds typically have more significant costs than index funds. The S&P 500 index is frequently used as a stand-in for the whole stock market (RobinHood 2020).

4.3.8.1 Benefits of the Indices for Sustainable Finance

For several reasons, "indices" are essential tools and strategies in sustainable finance. They are crucial devices for assessing how well sustainable financing products are performing. By giving investors a reliable benchmark, they assist investors in evaluating how well their sustainable investments perform in contrast to other sustainable options or traditional investments. Investors can use benchmarks given by indexes such as the Bloomberg Barclays MSCI Green Bond Index to evaluate the performance of their green bond investments relative to the market (Bloomberg Barclays Indices 2021). Investors must evaluate their portfolios with the knowledge to guarantee that their sustainable investments yield competitive returns; this comparison is crucial (Sustainable Finance Working Group 2022).

Regarding sustainable finance, using indices for performance benchmarking involves several critical considerations when making investment decisions. First and foremost is the necessity of performing a comparative analysis, in which investors assess several investment vehicles simultaneously. This comparison method is fundamental in sustainable finance, where complex factors like social governance and environmental effects can cause returns to be evaluated very differently from traditional investments (Statman 2005). The ability of benchmarking indices to enable thorough risk assessment in sustainable finance is essential to their usefulness (Christoffersen 1998). These indices provide investors with a tool to evaluate companies' sustainability performance, enabling them to make informed investment choices based on environmental, social, and governance (ESG) factors (Rekenthaler 2024).

In the evolving landscape of sustainable finance, performance benchmarking indices such as the Bloomberg Barclays MSCI Green Bond Index are crucial for informed investment decision-making (Statman 2002). The Bloomberg Barclays MSCI Green Bond Index tracks the performance of green bonds issued to fund environmentally beneficial projects, ensuring they meet stringent criteria aligning with sustainability and financial goals (Bloomberg Barclays Indices 2021). This alignment enhances the credibility and attractiveness of green bonds. Furthermore, these indices provide historical performance data, aiding in risk assessment by helping investors understand the long-term stability and volatility of returns (Christoffersen 1998). They also offer insights into market trends, identifying outperforming and underperforming sectors, thus guiding strategic positioning within the investment landscape (Fama and French 1993).

Conversely, indices are transparent and standardised way to track the success and effects of different sustainable investments and ensure they meet goals (Steuer and Tröger 2022). They also foster accountability by creating a clear baseline that guarantees funds are spent as intended and are not misused or misrepresented. Sustaining credibility and confidence in the market depends on this responsibility (Global Impact Investing Network 2023). Then, by independently verifying and monitoring important ESG factors, indices bolster investor confidence by confirming the legitimacy of financial products that make sustainability claims (S&P Dow Jones Indices 2024). This transparency also supports market integrity by lowering the possibility of "greenwashing" and guaranteeing that only investments fulfilling stringent sustainability standards are featured (Global Impact Investing Network 2023). By providing standardised metrics that satisfy disclosure and reporting requirements, indices support regulatory compliance and promote sustainability through standardised practices (Giamporcaro and Gond 2024). For example, the S&P Green Bond Index tracks the global green bond market, including only bonds that meet specific criteria for funding eco-friendly initiatives, providing investors with a clear benchmark for assessing the authenticity and profitability of green bond investments (S&P Dow Jones Indices 2024).

Finding opportunities through indicators is essential for investors looking to find companies and projects that excel at sustainable practices. Investors can use these indicators to match their portfolios with ESG goals and spot possible investment opportunities (Jones and Felps 2013). Thanks to this data-driven strategy, investors may identify businesses that value sustainability and are well-positioned to profit from new trends and market demands. Investing in sustainable enterprises and initiatives also has the potential to yield long-term financial gains and promote positive ESG impacts (Flammer 2015).

Additionally, investors can make better-informed selections. This strategy improves portfolio management techniques through efficient resource allocation and may eventually maximise returns for investors (DeMiguel et al. 2009). With the help of these indices, investors can find and back businesses with robust sustainability policies, guaranteeing that their financial investments have a good influence on the environment and society. ESG indices provide investors with a standard to measure corporate performance on sustainability indicators, enabling them to make more responsible and knowledgeable investment decisions (Kotsantonis and Serafeim 2019). Financial returns from investing in ESG indices can be competitive, offering advantages for ethics and the economy (Clark et al. 2015).

Furthermore, according to a wealth of data, businesses that implement strong ESG practices typically outperform their competitors in the long run. Over eighteen years, it has been demonstrated that companies with high sustainability outperform their peers regarding stock market returns and accounting performance (Eccles et al. 2014). Furthermore, most research supports a positive relationship between ESG criteria and corporate financial performance, highlighting the financial benefits of integrating sustainable practices into business strategies, according to a thorough meta-analysis of over 2000 empirical studies (Friede et al. 2015).

Moreover, indices contribute to market consistency and dependability by providing explicit standards for what qualifies as a green bond, social bond, or other sustainable finance product. This standardisation improves accountability because businesses and issuers can be compared to their counterparts and held to predetermined benchmarks. An important function of indices is to provide investors with a clear set of criteria that directs them towards sustainable financial products that meet their ESG goals. These indices set out particular criteria for various kinds of sustainable finance instruments, including green bonds, which could mandate that projects that get funding concentrate on pollution control, energy efficiency, or renewable energy. By following these guidelines, issuers can ensure their financial products meet industry-accepted sustainability standards. This will give investors dependable and transparent choices for chasing financial returns while supporting socially and environmentally conscious projects (Scholtens and Kang 2013). Investors navigating the world of sustainable finance instruments must prioritise consistency in evaluation, which is made easier by standardised indices. These indices guarantee fair comparisons by creating consistent standards for evaluating various sustainable financing tools. To give investors clarity and transparency, green bond indices, for example, can require that funded projects concentrate on energy efficiency, renewable energy, or pollution avoidance. This uniformity improves the validity and significance of comparisons across various sustainable financing instruments by empowering investors to make well-informed selections by evaluating options according to the same standards (ICMA 2021a).

Index-based standardisation is essential to improving the sustainable finance market's dependability. Indices ensure that the financial goods they contain follow strict sustainability requirements by setting explicit benchmarks. By doing this, the possibility of businesses deceiving customers into believing their products are sustainable is decreased. Investors can feel secure knowing that the goods included in these indices meet strict sustainability standards. Index providers and companies are also held responsible for their sustainability claims because their performance is monitored over time, and they have to fulfil certain requirements to be included. This accountability framework ensures that businesses and initiatives follow their sustainability pledges (Larsen 2019). Indices facilitate the comparison of companies and issuers against their peers. This comparison is valuable for investors and other stakeholders who want to identify leaders in sustainability practices. It also encourages competition among companies to improve their ESG performance to meet or exceed industry standards (The World Bank 2021). For instance, the S&P Green Bond Index tracks green bonds that meet specific criteria for environmental benefits. This index provides a clear and standardised benchmark for what constitutes a green bond, helping to ensure that only bonds with genuine ecological environmental benefits are included. By setting these standards, the index enhances the accountability and reliability of green bonds in the market (S&P Dow Jones Indices 2024).

4.3.8.2 Mechanisms and Practices in Developing Sustainable Finance Indices

ESG-compliant investments are difficult to evaluate and monitor, and indexes, particularly, are vital for sustainable finance. By serving as benchmarks for sustainable investment portfolios, these indices—which include the FTSE4Good and Dow Jones Sustainability Indices—assist investors in coordinating their strategies with sustainability objectives. S&P Global (2024) states that the Dow Jones Sustainability Indices assess economic, environmental, and social variables using about twenty-four categories. The results are scores ranging from 0 to 100 in each area and an overall score. Similar to this, the FTSE4Good Index provides a total score ranging from 0 to 5 in addition to specific scores for important areas such as corporate governance, the fight against corruption, human rights, labour rights, environmental management, and labour standards in the supply chain (FTSE Russell 2021). These indices support long-term ESG advantages, as well as responsible investing.

Creating and utilising sustainable finance indices includes gathering precise and thorough ESG data, using reliable techniques to rank and assess businesses, and updating the indexes frequently to consider the most recent developments in sustainability. Only businesses making financial decisions based on ESG criteria are included, thanks to the thorough study and selection procedure (Do Good People n.d.). This procedure is necessary for the indexes to continue serving as trustworthy and pertinent standards for sustainable investing. These indexes, which are widely used and significantly impact promoting sustainability in the financial sector, are essential tools for investors looking to align their portfolios with morally and environmentally responsible company practices as sustainable investing gains traction. The following are the main components that go into developing and utilising sustainable finance indices.

4.3.8.3 Development of ESG Criteria

ESG standards are benchmarks for an organisation's practices that investors with a social conscience use to evaluate possible investments. Environmental criteria consider a company's performance as a steward of the environment; social criteria examine its management of relationships with suppliers, customers, and communities; and governance criteria cover leadership, executive compensation, internal controls, and shareholder rights (Global Reporting Initiative 2024).

In recent years, integrating ESG criteria into investment decisions has gained significant traction globally. Studies have shown that companies with strong ESG performance tend to exhibit better long-term financial performance and lower risk profiles. For instance, an Oxford University and Arabesque Partners meta-analysis found that 88% of reviewed studies demonstrated a positive correlation between ESG factors and financial performance (Clark et al. 2015).

Moreover, the rise of sustainable finance has led to the development of various ESG rating agencies and data providers, offering investors a wide range of ESG-related information to inform their investment decisions. These agencies assess and rate companies based on their performance across ESG factors, providing valuable insights into a company's sustainability practices (Kotsantonis and Serafeim 2019).

4.3.8.4 Selection of Index Constituents

Selecting index constituents based on ESG criteria involves thoroughly analysing and evaluating companies' sustainability practices. This selection process often incorporates various data sources, including ESG ratings, corporate sustainability reports, and third-party assessments. One common approach is using quantitative screens to identify companies with the highest ESG scores within each industry sector. These screens may consider factors such as carbon emissions, diversity and inclusion metrics, labour practices, and board diversity. Companies that meet or exceed predefined thresholds for these criteria are included in the index, while those that fall short are excluded. Some sustainable indices also employ qualitative assessments to holistically evaluate companies' ESG performance. This may involve reviewing company policies, initiatives, and controversies to assess their commitment to sustainability beyond numerical metrics (MSCI ESG Research 2024).

4.3.8.5 Weighting of Constituents

A combination of characteristics, market capitalisation, and ESG scores may determine the index weighting of firms. Certain indices may use equal weighting to highlight smaller businesses that excel at sustainability. Understanding how ESG variables affect investment decisions largely depends on the weighting scheme used by sustainable indices. The emphasis placed on market capitalisation compared to sustainability performance varies between weighting schemes (MSCI ESG Research 2024).

(a) ESG Score-Based Weighting: This method assigns weight to organisations based on their ESG ratings or scores. The index's heavier weighting of companies with higher ESG scores reflects stronger sustainability performance. Because it gives preference to businesses with excellent ESG practices, this approach brings the index closer to sustainability goals.
(b) Market Capitalisation Weighting: Conventionally, indices weighted according to market capitalisation give companies a certain amount of weight based on their market value; larger companies have higher weights. Although these indices might consider ESG aspects, sustainability variables indirectly impact index composition and influence market dynamics.
(c) Combination Weighting: The constituent weights of certain indices are established by combining market capitalisation and ESG scores. This approach seeks to strike a balance between market representation and sustainability goals by more sophisticatedly evaluating the contributions made by the companies to the index.
(d) Equal Weighting: Some indices use equal weighting, meaning that every component firm has the same weight, irrespective of its ESG score or market value. This approach lowers the concentration risk by promoting diversity and highlighting the significance of smaller businesses that excel in sustainability.

Using a suitable weighting scheme, sustainable indices can successfully reflect investors' preferences for ESG integration while staying in line with larger market trends.

4.3.8.6 Performance Measurement and Reporting

Performance measurement typically involves evaluating key metrics such as total return, volatility, and tracking error relative to a benchmark index. Sustainable indices may incorporate additional performance indicators specific to ESG objectives, such as carbon footprint reduction, diversity metrics, or alignment with global sustainability goals. For instance, the Global Reporting Initiative (GRI) provides a framework for sustainability reporting, offering guidelines for companies and indices to disclose relevant ESG metrics transparently (Global Reporting Initiative 2024).

Through engagement in activities like Principles for Responsible Investment (PRI), clients can enhance accountability in the investment space. By reporting the integral performance, sustainable finance indices encompass more than just the reports; investors plan and actualise their sustainability goals (Giamporcaro and Gond 2024). By providing comprehensive performance measurement and reporting, sustainable finance indices enable investors to make informed decisions and track progress towards their sustainability objectives.

In conclusion, sustainability indices are key tools that provide a comprehensive framework for assessing the performance and impact of investments against ESG criteria. These indices aid informed decision-making by allowing investors to benchmark sustainable investments, promote market transparency, and increase accountability. They support risk management by emphasising non-market risk factors and facilitating the alignment of portfolios with sustainability objectives. Developing ESG criteria, carefully selecting and weighing index constituents, and robust performance measurement and reporting ensure these indices reflect sustainable practices. Engagement and governance further enhance corporate responsibility, while adherence to regulatory and market standards underpins their credibility and reliability. Overall, sustainability indices play an important role in advancing sustainable finance by balancing ethical considerations with financial performance, promoting long-term value creation, and enhancing the integrity of capital markets.

4.3.9 Venture Capital and Private Equity in Sustainable Finance Mechanism

Sustainable finance mechanisms refer to financial instruments, strategies, and practices that integrate ESG considerations into investment decisions and capital allocation processes, creating opportunities for individual investors and invested entities (Migliorelli 2021). These mechanisms promote sustainable development by directing capital towards activities and projects that generate positive social, environmental, and financial outcomes. Cumming et al. (2023) emphasise that venture capital (VC) and private equity (PE) are two critical platforms for financing businesses at the startup and scaling stages, respectively. VC and PE provide expertise and risk capital, promoting innovative and sustainable business models across economies on a global scale.

4.3.9.1 Venture Capital (VC)

Venture capital (VC) is defined as financing provided to early-stage or emerging companies with high growth potential, typically in exchange for an ownership stake. According to Cumming et al. (2023), startups, early-stage businesses, and emerging companies with higher growth potential can obtain funding from VC. Noteworthy venture capital is often provided by institutional investors, such as VC firms or angel investors, who are willing to take on high levels of risk in exchange for the potential for significant returns on their investment (Cumming et al. 2023). This type of funding is commonly sought by entrepreneurs and small businesses with limited opportunities to access traditional forms of financing, such as bank loans or public markets. For instance, in 2023, the USA witnessed the lowest 30% reduction in venture capital (Market Research Future 2024).

On the contrary, VC in the energy storage sector reported growth due to a focus on clean energy to achieve zero emissions by 2050. Notably, the clean energy sector has demonstrated the potential to create the highest venture capital returns in 2024 (Market Research Future 2024). The energy trends are reflected in Fig. 4.24. As apparent in the funding trends, sustainability constitutes the greatest share as the most funded sector in 2023.

Therefore, sustainable investing offers an opportunity to transform financial markets towards accountability regarding social and ecological issues. Nonetheless, it is not the only form of investment available to enterprises, as they can resort to traditional approaches based on profit maximisation. The main difference between conventional investing and sustainable investing is that the former is purely based on profit maximisation. At the same time, the latter prioritises ethical and social standards above profit maximisation.

In addition to garnering interest, sustainable businesses are also obtaining significant funding from venture capitalists, representing a paradigm-shifting move in the direction of impact investing day after day. For example, Bay Area Coffee Supplier is a Double Bottom Line Venture Capital (DBL) partner promoting the "Electrify Everything" movement. The startup unveiled a smaller model of an electric coffee roaster that can be used in retail shops. In this case, the sustainable perspective involves increasing reliance on clean energy to limit business impact on the environment. Given its simplicity, there is no need for local permits, gas lines, training, or construction to set it up. This remains an important part of the growing frontier in the wider "electrify everything" movement for buildings, appliances, cars, and more.

VC drives innovation and supports sustainable finance mechanisms (Barber et al. 2021). VC investors contribute to developing and scaling sustainable solutions across various sectors, such as renewable energy, clean technology, and envi-

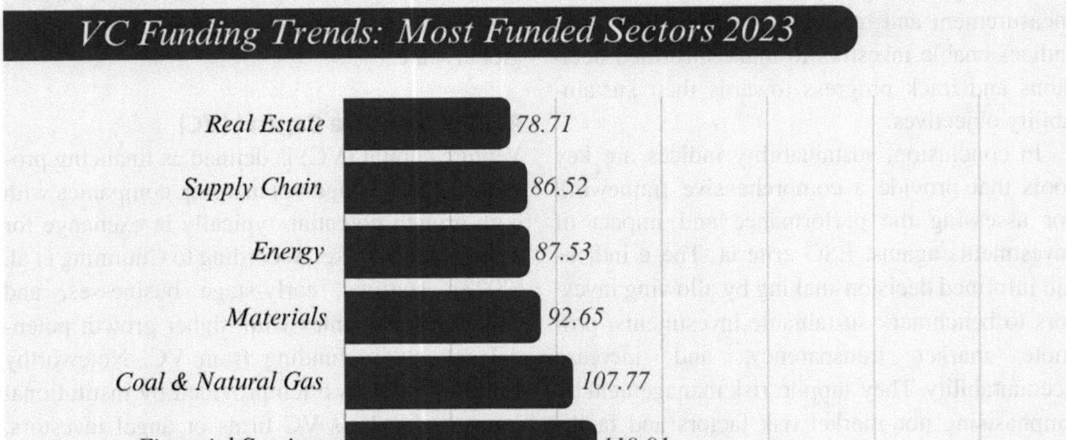

Fig. 4.24 VC funding trends: Most funded sectors 2023 (Market Research Future 2024)

ronmental conservation. These investments fuel the growth of environmentally friendly businesses and incentivise entrepreneurs to develop innovative products and services that address pressing sustainability challenges (Agrawal and Hockerts 2021). Further, VC firms increasingly consider ESG criteria in their investment decisions, aligning their portfolios with SDGs and driving positive social and environmental impacts. As a result, VC serves as a catalyst for sustainable finance, channelling capital towards ventures that generate financial returns and create meaningful and lasting benefits for society and the planet.

Additionally, Impossible Foods received over USD 1.6 billion in funding in 2023 from venture capitalists such as Khosla and Horizon Ventures to support their bioengineering business and enhance the production of plant-based meat products in the USA (Beyond Meat 2024). This implies that both VC and PE not only engage in businesses that are solely for profit but also are interested in the sustainability perspective of their impacts. For example, Barber et al. (2021) pointed out that PE and VC funders engage in sustainable finance mechanisms, such as impact investing. Notably, impact investing is a strategy that aims to generate specific social or environmental effects and financial returns.

Companies turn to venture capital firms to invest in green projects and support the implementation of sustainable business models, in addition to traditional financing options such as loans from banks and family members. VC investment in farming-based startups has increased. This reflects a growing interest in agricultural innovation and sustainable food production. These startups are leveraging technology to address various challenges in agriculture, such as improving crop yields, reducing environmental impact, and enhancing supply chain efficiency.

According to the Commission on Sustainable Agriculture Intensification (CoSAI), the realisation of the UN SDG on food and agriculture by 2030 requires USD 320 billion in annual investment. PE/VC is one of the funders necessary for sustainable agriculture. The desired financial resources are yet to be achieved. Nonetheless, considerable efforts have been made towards sustainability objectives, especially in the Global South. For example, between 2010 and 2019, over USD 2 billion was invested in agricultural innovation (CoSAI 2021). As much as the Q1 VC has mainly focused on AI, agriculture remains one of the sectors attracting investor interest.

Notably, this chapter highlights and explains three main roles of venture capital in sustainable firms: resource provision, innovation, and attracting investments. Firms provide the necessary resources to sustain startups. VC firms must offer the needed human resources and expertise, sometimes resulting in long-term follow-ups and support (KPMG 2023). Venture capitalists must have a profound understanding of what a sustainable business model means—this involves investing vast amounts over a long period despite the high likelihood of a collapse.

Conversely, VCs play a vital role in commercialising breakthrough sustainable technologies, and innovations are necessary for developing a new wave of VC firms (CoSAI 2021). These companies rely on extras such as mentorship, tips, and networks from VCs. Investing in innovations that solve provable problems, the business can get money from the market when commercialising it. Lastly, attract corporate investment to support sustainable innovation. For instance, PowerVault designs and manufactures smart energy storage systems (solar and grid electricity). The UK-based company raised USD 5.4 million in equity crowdfunding (CoSAI 2021).

In summary, sustainability has become a basic business principle for VCs. Their role in providing necessary resources, supporting innovation, and attracting corporate investments continues to have a considerable influence on ESG. Therefore, VCs effectively support viable enterprises; they should emphasise meticulous due diligence, a far supported vision, and prioritise true innovation over short-lived fashions. Keeping a broad portfolio of companies across different industries and stages also helps to reduce risks. If done right, venture capital supports sustainable startups and technologies.

4.3.9.2 Private Equity (PE)

Private equity (PE) is a category of investments that occurs by directly injecting funds into private companies or assets that cannot be listed on stock markets (Wood and Wright 2009). The funds are raised permanently from captive investors, including pension funds, endowments, and affluent individuals intending to establish equity in "investable" companies. Over a definite timeline, usually between three and seven years, these firms aim to increase the value of their portfolio companies before executing an exit strategy to regain their invested funds. Existing methods often include selling the firm to a different investor, floating out the company through an IPO, or merging it with another company (Levis 2011). PE investments are characterised by a longer investment period and greater control in management than investments in publicly traded companies, thereby enhancing the chances of social and economic changes. Nevertheless, studies indicate their effects, particularly in the nascent phases (Wood and Wright 2009).

ESG is redirecting finance towards companies and projects while prioritising ethical and social considerations. Although the social impact has been less concentrated, they are fully financed, capable, and have the resources to execute the sustainability strategies of their companies through passive and active investments and intervention (Ewens and Farre-Mensa 2020). Since private equity investors do not only use the funds but also adopt the principles of long-term value and sustainability while actively participating in the portfolio companies, they assist in promoting such practices.

In this way, PE can work towards sustainable development and ensure that the investments that create social and environmental impact and financial returns are supported (Wood and Wright 2009). According to Harvard Business Review, 90% of the Global Private Equity initiatives factor ESG into their decisions, while 77% use it for partner selection (HBR 2022). For example, the Dutch pension investor APG reportedly invested USD 36 billion in 75 general partners (GPs). They published 37% responsible financing of these investments, indicating an increased focus on ESG (HBR 2022). Further, Carlyle is one of the private equity firms leveraging their capital to support sustainable businesses and projects, contributing to environmental sustainability and social impact while seeking financial returns. Other PE firms that focus on ESG, including Galway Sustainable Capital, Encourage Capital, and CrossBoundary, seek to address social and environmental issues that affect society.

Notably, PE is essential in each investment phase to create value in a sustainable firm. In pre-deal, PE ensures deal evaluation and due diligence using risk management and value-preservation lenses to conduct forward-looking assessments. During the tenure, PE ensures that ESG is incorporated into value-creation plans through climate risk assessments and baseline emission measurements, among other approaches. This makes sustainable firms attractive to buyers focusing on ESG-enhancing businesses. Lastly, PE firms provide ESG reports during the exit phase to enhance the sales process and maximise value for sustainable firms (KPMG 2023).

4.3.9.3 IPOs of Sustainable Products and Service Providers

Providing initial public offerings (IPOs) for providers of sustainable products and services entails listing companies with EGS disciplines on the stock market. In the report on sustainability measures, while planning to go public, more companies are increasingly going public, attracting demands for financial reforms, corporate governance systems, and sustainable products and services (PwC 2011), while Husain and Nair (2011) argue that these IPOs raise capital from public investors so that such sustainable businesses can enable growth by funding any of expanding or scaling up production or R&D for new sustainable solutions. IPOs in sustainable sectors are expected to mobilise funds that help the concerned companies scale their sustainability objectives while leading to improved visibility and the participation of sustainable business models among the investors and the public, thus triggering a shift to the desired sustainable future.

Over the past few years, IPOs of companies that deal with sustainable products and services have launched, amazing many investors. There are also forecasts that ESG-related assets under management will reach 33.9 trillion in 2026 with 12.9 CAGR growth worldwide (PwC 2024). Since these businesses are often innovators in renewable energy, waste management, and synthesising sustainable consumer goods, they use public markets to enhance their expansion and reach. For instance, Beyond Meat is a company that aims to cut the needless environmental damage incurred in traditional meat production by making burgers from plants. The 2019 IPO of Beyond Meat at USD 25 per share fundamentally changed the organisation, as the funding was significant, and there was evidence that food alternatives would continue to be in demand, which only increased when it went public (Beyond Meat 2024). It emphasised that the market acceptance of plant-based products was possible, and this fostered other sustainable food companies to seek stock market listings.

Another interesting IPO is of Tesla in electric vehicles and the clean energy market. Even though the IPO of Tesla was back in 2010, the significance it has created to the market of sustainable products is noteworthy. For example, Tesla went public with a price of 17 dollars per share and started trading shares at 19 dollars. By the end of that fiscal year, the company was closing shares at 23.89 dollars, a gain of 40.5% (Divine and Davis 2022). Today, Tesla shares are synonymous with the electric vehicle brand, which propelled the use of EV batteries and improved renewable energy integration. Their market performance has vindicated the premise that a business that builds around sustainability can be profit-generating while also being part of the transformation towards clean energy.

When harnessing the power of venture capital, private equity, or even initial public offering of shares within the confines of investing in sustainability, investors and companies face unique challenges. One of the key problems has to do with the way impact is measured. Without a global standard for monitoring sustainability, assessing and comparing investments becomes much more achievable. It is also related to information asymmetries where businesses overstress on their turn and how green they are trying to be to convince investors, thus misallocating investment. However, even with increased awareness, acceptance in the market is challenging, as not all customers prioritise sustainability. Such factors negatively influence the market potential for the provision of sustainable products and services, and in considering the sustainability of business ventures, there is always an element of optimism bias among traditional investors, which may lead to inadequate funding (Lin 2022).

Despite these factors, there are still some possibilities for success, especially in sustainable venture capital (SVC), where there is a need to target young startups that are more likely to achieve market fit. There is an increasing trend in consumer awareness and demand for sustainability in the products they use, which supports expansion in this area of business (Holtslag et al. 2021). More companies are integrating sustainability into their core strategies, creating a larger market for sustainable technologies and services. Regulatory support, in the form of incentives, grants, and subsidies, helps reduce financial risks and encourages investment in sustainability. Internal enablers, such as top management support, and external enablers, including government support, offer valuable opportunities for SVC (Antarciuc et al. 2018).

4.3.10 Carbon Pricing

Grasping the concept of carbon pricing involves appreciating its role as a strategic tool in the broader variety of sustainable development. At its core, this mechanism subtly pushes enterprises to consider the environmental footprint of their operations, encouraging a shift towards more sustainable practices. This model does not just depend on its financial aspect; it is about creating a consciousness where the Earth's health becomes a critical parameter in the blueprint of progress. It balances economics and ecology, guiding humanity towards a future where their actions harmoniously align with the Earth's well-being. Carbon pricing is a straightforward yet effective idea. Assume that every business or person that releases carbon dioxide (CO_2) into the atmosphere is also responsible for a hidden environmental cost, such as air pollution or climate change. Currently, this expense is not included in the cost of the goods or services bought or used, like petrol for cars or electricity for homes. The method of assigning a cost to carbon emissions is known as carbon pricing. This encourages people to adopt more environmentally friendly practices and reduce their carbon footprint. It serves as a reminder that adding carbon to the atmosphere has actual costs that impact the environment. This approach is based on the economic idea of external factors, which says that when the environment is affected in ways not included in the cost of something, there should be rules to address it (Bowen 2011). Fundamentally, carbon pricing manifests in two distinct methods: carbon taxation and emissions trading (Boyce 2018). Carbon taxation involves directly assigning a monetary value to carbon emissions, achieved through a tax placed on the carbon content found in fossil fuels. This approach provides predictability but lacks certainty in emission reductions, as carbon taxation provides a clear cost for carbon emissions, making its financial impact predictable for businesses and individuals. However, it doesn't ensure a specific reduction in emissions because it imposes a price rather than a strict emissions limit. The efficacy of a carbon tax in reducing emissions largely depends on the tax level and the willingness of emitters to pay the tax versus reducing their carbon output. This approach contrasts with cap-and-trade systems, which offer certainty in emissions reduction by setting a firm cap but introduce variability in the cost of emissions permits (Metcalf and Weisbach 2009). Conversely, cap-and-trade systems set a maximum allowable level of emissions while allowing the trading of emission allowances. This market-based approach ensures a specific emission target is met and allows for flexibility in achieving it.

Globally, carbon pricing strategies show remarkable diversity, reflecting each region's unique economic and environmental contexts. The European Union's Emissions Trading System (the EU ETS) exemplifies a successful cap-and-trade model, demonstrating how regional collaboration can effectively reduce emissions. This contrasts with Sweden's robust carbon tax approach, highlighting direct taxation methods' environmental and economic sustainability. Canada's hybrid model in North America blends a federal carbon pricing benchmark with provincial flexibility, showing an approach that allows diverse regional circumstances. Conversely, countries like South Africa and Chile are exploring carbon pricing among their unique developmental challenges in the developing world, signalling a growing global acknowledgement of its importance in climate change mitigation strategies.

The modern environmental structure highlights the urgency of integrating carbon valuation into climate change mitigation strategies. This concept is rooted in assigning a tangible value to carbon emissions, transforming the abstract concept of environmental impact into a measurable economic entity. This transition is essential to balance industrial growth with ecological sustainability. Moreover, introducing the carbon value proposition has significant effects on the market. It essentially changes the economic system of businesses and consumers, encouraging the adoption of cleaner technologies and sustainable practices. This economic incentive is crucial in encouraging a transition towards a low-carbon economy, potentially triggering innovation in green technology sectors.

Conventional environmental policy instruments, universal technology, and performance norms are insufficient for decarbonisation due to most countries' prevalence and diversity of greenhouse gases (GHGs). As a result, the focus has shifted steadily towards instruments based on markets, such as carbon pricing mechanisms (Aldy and Stavins 2012). Realising "polluter-pays", emitters should be responsible for covering the external costs resulting from their pollution. Using legislative tools for carbon pricing is potentially the most cost-effective approach to cutting greenhouse gas emissions (Narassimhan et al. 2017). Meanwhile, carbon pricing has evolved from a cutting-edge policy implemented solely by pioneer nations to a worldwide standard that all responsible states must follow (Thisted and Thisted 2019).

Carbon pricing instruments take on various shapes and sizes, each with unique characteristics and effects on emission reductions. Besides, global analysis of the carbon market reveals that different nations, states, and regions design their carbon pricing schemes based on their unique circumstances, requirements, and political objectives. As a result, different countries use various methods and approaches for pricing carbon (Stavins 2007; The Climate Reality Project 2017; Carbon Pricing Leadership Coalition 2020). This section presents the most frequently used concepts for carbon pricing: carbon tax, carbon market mechanisms (cap and trade and baseline and credit), voluntary carbon market, and internal (shadow) carbon pricing via case studies and comparisons. In Fig. 4.25, essential concepts of carbon pricing mechanisms are given.

4.3.10.1 Carbon Tax

One of the carbon pricing instruments is carbon taxation. The idea of emissions taxation, which forms the basis of the concept, is an old concept. Economists have been working to create a better model for the complexity of this concept since 1920 (McKitrick 2016). Conversely, other environmental tax concepts have a long history, especially since 1970, when environmental taxes have been used more frequently. These taxes aim to reduce environmental problems (Kettner-Marx and Kletzan-Slamanig 2018). Likewise, carbon taxes have been a tool that economists have used for a long time (Baranzini et al. 2000).

A carbon tax is a charge levied on emitters that functions similarly to a pollution tax. Businesses that use carbon-based fuels, such as coal, oil, petrol, and natural gas, are subject to the levy. A carbon tax puts a price on carbon directly by establishing an explicit tax rate on greenhouse gas emissions or, more frequently, on the carbon content of fossil fuels (Tsai 2020). Finland, which started carbon taxation in 1990, was followed by Denmark, the Netherlands, Sweden, Norway, Italy, New Zealand, Switzerland, and

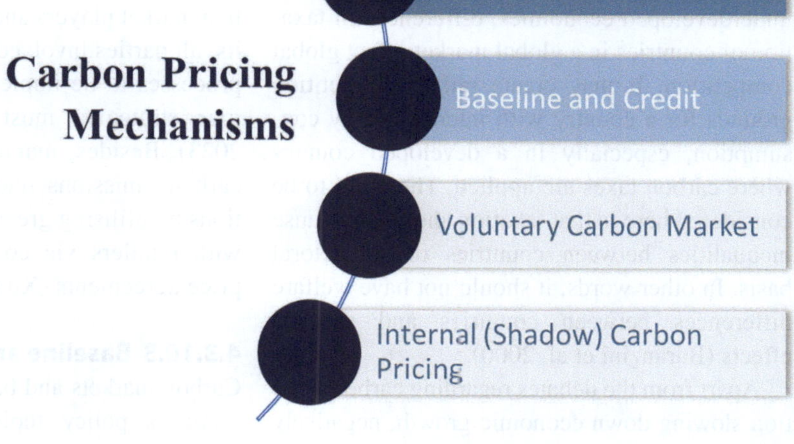

Fig. 4.25 Overview of carbon pricing mechanisms (Küfeoğlu 2024a)

Canada. Energy security and emissions reduction movements have also driven some countries to begin carbon taxation. As a result, applications have started in France, Japan, and China (Lin and Li 2011).

Some criticisms are essential to mention here to understand the mechanism better. The main purpose of the carbon tax is to price the amount of carbon released. However, the market determines whether carbon emissions will decrease or not. Therefore, one might argue that the primary goal of the carbon tax is to subject carbon emitters to a payment rather than reducing greenhouse gas emissions. This situation is related to the emission elasticity of the sector. The market does not have emission elasticity if there is no emission reduction due to taxation in the sector (McKitrick 2016). For example, if a product does not lose demand even if its price increases in a certain market, it can be said that the tax does no more than create more costs for the consumer and producer. Therefore, the tax will not be effective if more costly production occurs in the sectors instead of emission reduction (Lin and Li 2011). Another topic of discussion is that carbon taxes may cause some adversities. The adverse features of carbon taxation are that it increases operating costs in the short term, increases the prices of products, harms competition, especially in sectors where energy is used more intensively, and negatively affects economic growth (Lin and Li 2011). In addition, due to the difference in the existence and scope of carbon taxes in developed, developing, and underdeveloped economies, differences in taxation of countries in a global market affect global competition. It may create unfair competition grounds for a country with intense energy consumption, especially in a developed country where carbon taxes are applied. The point to be considered here is that taxation should not cause inequalities between countries on a sectoral basis. In other words, it should not have welfare differences between countries and sectoral effects (Baranzini et al. 2000).

Apart from the debates regarding carbon taxation slowing down economic growth, negatively affecting welfare, and being more costly in some sectors, it also has positive effects. It might considerably impact the reduction of energy consumption, increasing efficiency, and directing people to use clean energy sources (Lin and Li 2011). For example, the study by Hájek et al. (2019) states that by increasing the carbon tax by an average of one euro per ton, annual greenhouse gas emissions per person would decrease by 11.58 kg.

4.3.10.2 Cap and Trade

To ensure environmental success, cap-and-trade regimes are recommended to be established at thresholds involving reductions in emissions through practical emissions control measures (Mcallister 2012). The cap-and-trade system limits the amount of greenhouse gas emissions that are permitted for specific emissions. Each allowance, which makes up the entire cap, allows a corporation to release one ton of emissions. The government gives allowances to businesses either free or via an auction (Aldy and Stavins 2012). If emissions exceed this cap, polluters are compelled to make payments for their carbon emissions, which forces businesses to decarbonise. In the cap-and-trade system, increasingly stringent caps are being gradually implemented to reduce greenhouse gas emissions (Prindle et al. 2010). Moreover, since supply and demand determine the price, polluters purchase and sell permits that enable them to emit just a specific quantity (Wagner 2013). Since systems entail quota distribution to carbon market players and the trade of carbon credits, all parties involved must cooperate for these processes to be implemented, and their behaviour strategies must be considered (Ernest 2023). Besides, manufacturers can lower their carbon emissions under cap-and-trade regulations by utilising green technology and working with retailers via cost-sharing and wholesale price agreements (Xu et al. 2017).

4.3.10.3 Baseline and Credit

Carbon markets and baseline-and-credit systems serve as policy tools for mitigating climate change. The design of these marketplaces is critical to ensuring economic emission reductions while maintaining environmental integrity (Betz et al. 2022). The baseline-and-credit model states that activities that produce emissions are assigned an emissions intensity based on their comparison to a baseline. In contrast, activities that attain emissions intensities below the baseline are awarded credits. Such credits must be purchased by activities whose emissions intensities are above the baseline (Buckley et al. 2007). Participants are incentivised to identify lower emission production processes by the possibility of obtaining credits for lowering emissions relative to baseline and the pressure to avoid purchasing permits for emissions exceeding baseline (Streck 2020).

Cap-and-trade and baseline-and-credit schemes are different compliance markets. An incentive is given for a course of action in a baseline-and-credit scheme that is thought to lower emissions from a reference scenario. To make up for emissions whose mitigation would be more expensive than the cost of the credit, the units produced—known as "credits"—represent emissions reductions and are typically sold to purchasers in compliance or voluntary markets (World Bank Group 2016). Furthermore, a maximum amount for emissions is defined in a cap-and-trade system, and emission permits are either sold at auction or given away without charge per predetermined standards (Centre For Tax Policy and Administration 2008). Regulations imposing a carbon tax or cap-and-trade both lower the release of carbon and revenue, but how best to apply them relies on the rates of taxes and emission trading costs; neither strategy has a distinct benefit (Xu et al. 2016). In Fig. 3.2, similarities and differences of carbon tax, cap and trade, and baseline and credit are proposed in the positive and negative impacts of emission reduction.

4.3.10.4 Voluntary Carbon Market

In the voluntary carbon market (VCM), people or organisations purchase carbon credits produced by independent, non-regulatory crediting schemes and standards to offset their carbon footprint (Gould 2023). These credits are usually made by programs that either remove carbon from the atmosphere through afforestation or carbon capture and storage or reduce or avoid carbon emissions through energy efficiency, renewable energy, and reforestation. When an organisation measures its carbon footprint, it can buy these credits to offset the emissions it cannot internally eliminate. This allows the organisation to achieve specified sustainability targets or claim "carbon neutrality" (Goldstein et al. 2020).

Since transactions occur outside of any official mandate, the decentralised nature of the VCM sets it apart. Direct negotiations or carbon credit exchanges between buyers and sellers set pricing, which is influenced by the state of the market, the legitimacy of the project, and other elements like carbon credit type and quality. These markets offer a dynamic and adaptable framework for financing climate-friendly initiatives, facilitating wider involvement in worldwide climate action beyond legally required entities (Blaufelder et al. 2021) (Fig. 4.26).

The European Union's Renewable Energy Project and the China Reforestation and Afforestation Project are prime examples of noteworthy regional initiatives addressing climate change. This initiative aims to improve biodiversity, reduce soil erosion, and trap carbon dioxide. The resulting carbon credits will be sold to businesses to offset emissions and help China achieve its larger environmental goals (Zheng and Dutton 2024). Parallel to this, the EU Wind Power Project entails the installation of wind turbines in Germany and Spain, among other EU nations. By generating sustainable energy, this project lessens reliance on fossil fuels and significantly lowers CO_2 emissions. The project's carbon credits enable businesses to offset their carbon footprints, facilitating the shift to a low-carbon economy and bolstering regional energy security (European Commission 2024c).

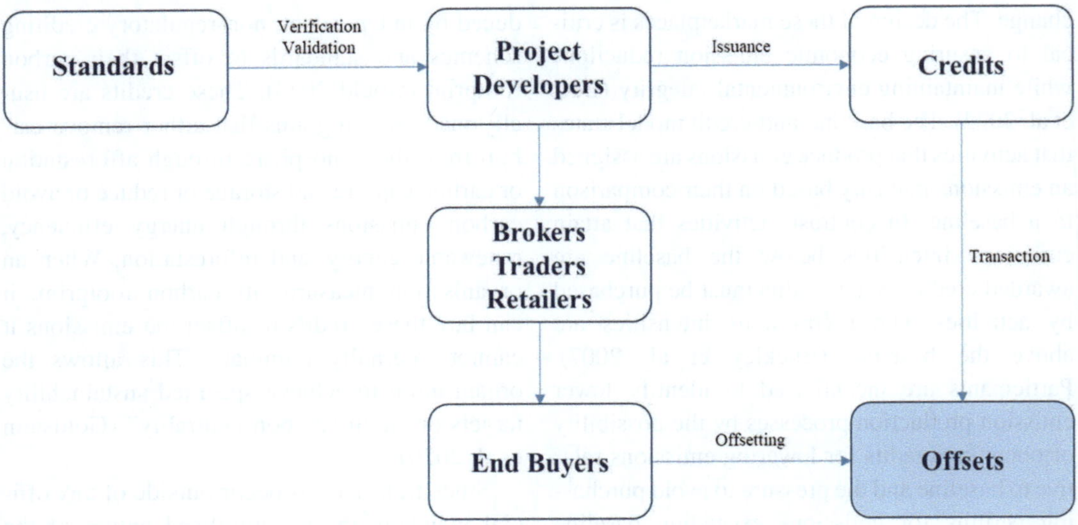

Fig. 4.26 Summary of voluntary carbon market mechanism (Küfeoğlu 2024a)

Reflecting the growing interest of the public and corporate sectors in addressing climate change, the VCM has expanded quickly in recent years. The predicted value of the worldwide VCM in 2021 was close to USD 2 billion, and it has the potential to become much higher as more businesses commit to net-zero goals and investors see the financial advantages of using sustainable finance (Adams et al. 2021). Increased community resilience, reduced poverty, and better biodiversity are just a few of the social, environmental, and economic co-benefits that result from the wide range of initiatives that this market supports (Goldstein et al. 2020).

4.3.10.5 Internal (Shadow) Carbon Pricing

Internal carbon pricing is a voluntary way for businesses to internalise the implicit (real or projected) price of carbon within different rules and regulations, regardless of whether all or a portion of an organisation's activities are not presently covered by external carbon laws (Bento and Gianfrate 2020). To meet the objectives of their carbon strategies, a growing number of businesses have started using the technique of pricing carbon for internal decision-making in the last few years (Dawes et al. 2023). Increasing public consciousness of the requirement for sustainable behaviour, the resulting social and legal pressure, and the financial incentive to lower risks or costs by enhancing processes and goods are the primary drivers behind organisations' implementation of internal carbon pricing (Gorbach et al. 2022). Many companies apply internal carbon pricing to manage their decarbonisation pathway. Hereby, according to CDP's most recent statistics on internal carbon pricing from 2020, growth is still occurring globally (CDP 2021). Currently, 853 firms have disclosed that they employ internal pricing on carbon, representing a 43% rise since 2018 (CDP 2021). For instance, Microsoft committed to being carbon-neutral as a corporation in 2012. Microsoft decided to impose a carbon charge that divides the expense of being carbon-neutral across its main business groups to achieve that aim. They invest in initiatives to cut carbon emissions, recycle electronic rubbish, produce green electricity, improve building efficiency, and offset emissions through global community programs to offset those emissions. Based on its emissions and the carbon price for the current year, each business unit makes a proportionate payment to the carbon fund. Every business unit, where possible, actively cuts emissions connected to power (e.g., through efficiency initiatives). Renewable power purchase agreements and renewable energy certificates are two exam-

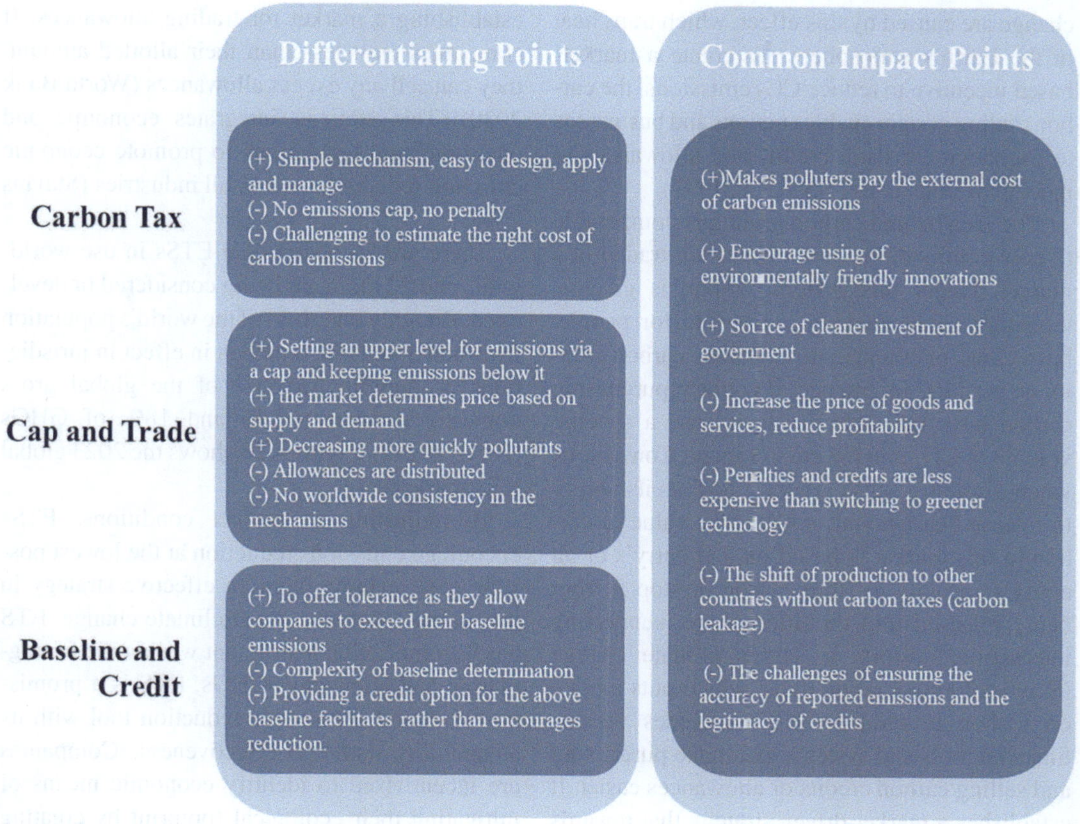

Fig. 4.27 Common and differentiating points of favoured carbon pricing mechanisms (Baranzini et al. 2017; Buckley 2004; Goulder and Schein 2013)

ples of green power investments that help achieve carbon neutrality for actual electricity-related emissions. Microsoft then looks to a portfolio of community investments in carbon offset projects to mitigate emissions from air travel, fuel use (such as backup generators), and any electricity-based emissions that cannot be mitigated through qualified green power investments (United Nations Climate Change 2023) (Fig. 4.27).

4.3.11 Carbon Trade

The idea of the carbon trade was born in direct response to an international agreement named the Kyoto Protocol (UNFCCC 2024a). The Kyoto Protocol marked a significant milestone in global efforts to combat climate change by setting legally binding targets for 37 industrialised countries, requiring them to reduce their greenhouse gas emissions during the commitment period from 2008 to 2012. By imposing legally enforceable objectives on 38 developed nations to cut their greenhouse gas emissions during the commitment period from 2008 to 2012, the Kyoto Protocol represented a critical turning point in the global effort to tackle climate change (UNFCCC 2024b). These goals attempted to reduce emissions to an average of 5.2% below 1990 levels (United Nations 1997). The fundamental problem that the Kyoto Protocol attempts to solve is carbon, a chemical element plentifully stored in fossil fuels, including coal, oil, and natural gas. These fossil fuels produce carbon dioxide (CO_2), a strong greenhouse gas that greatly intensifies the greenhouse effect and enters the atmosphere as it burns (National Geographic Society 2024). Global warming and climate change are caused by this effect, which traps heat in the Earth's atmosphere. To create a market-based incentive to reduce CO_2 emissions, the carbon trading system enables nations and businesses to exchange emission credits and allowances to meet their objectives (UNFCCC 1997).

The idea behind carbon trading is comparable to how commodities or securities are traded in a market. Under this system, carbon is given a monetary value, making it possible for people, businesses, or countries to exchange carbon credits or permits. In essence, a country purchasing carbon buys the authority to release a specific volume of CO_2 into the environment. Conversely, when a country sells carbon, it forfeits its ability to release that amount of CO_2. The value of carbon in this market is based on a country's or an entity's capacity to store carbon or stop it from being released into the atmosphere, supporting international efforts to lessen climate change (Stavins 2008). Carbon trade, which puts a price on carbon to reduce GHGs, includes various financial tools and systems that make purchasing and selling carbon credits or allowances easier. It establishes a market-driven strategy that rewards pollution reduction by placing a monetary value on emissions (Kenton 2024).

Understanding the mechanics of carbon trade products and services is vital to grasp their role in mitigating the negative impact of climate change. To encourage decreased GHG emissions, these goods and services function inside structured carbon markets through various financial techniques (Kossoy et al. 2015). Furthermore, there are three notable systems in which carbon trading operates: emission trading systems (ETS), voluntary carbon markets, and carbon removal market.

4.3.11.1 Emission Trading System (ETS)

One of the market-based tools for regulating and lowering GHG emissions is an emission trading system (ETS), a mechanism on which global initiatives to mitigate climate change heavily rely. Enabling entities to sell emission allowances within the specified limit operates by imposing a cap or restriction on the total emissions permitted. The ETS encourages enterprises to cut emissions by setting a cap on overall emissions and establishing a market for trading allowances. If companies emit less than their allotted amount, they can sell any excess allowances (World Bank 2020). This strategy integrates economic and environmental regulations to promote economic emission reductions across all industries (Stavins 2008).

There are 36 operational ETSs in use worldwide, and 22 more are being considered or developed. Roughly one-third of the world's population lives under an ETS, which is in effect in jurisdictions accounting for 58% of the global gross domestic product (GDP) and 18% of GHGs (ICAP 2024a). Figure 4.28 shows the 2023 global ETS markets.

By adjusting to market conditions, ETSs encourage emissions reduction at the lowest possible cost, making them an effective strategy in the global effort to combat climate change. ETS has become a vital instrument worldwide for regulating and alleviating GHGs. ETS is a promising worldwide emissions reduction tool with its adaptability and cost-effectiveness. Companies are incentivised to identify economic means of mitigating their ecological footprint by creating an ETS. Allowances, carbon emissions, and market conditions come together to form an evolving framework that encourages cuts in emissions at the lowest feasible cost. Because of its flexibility, affordability, and efficacy, ETS has long been regarded as the most promising tool for reducing greenhouse gas emissions worldwide (Küfeoğlu 2024a). Considering this knowledge, it is crucial to highlight the European Union ETS and United Kingdom ETS as trailblazing examples.

European Union (EU) ETS

"Cap and trade" is the guiding philosophy of the EU ETS. The maximum quantity of GHGs that installations and operators covered by the system are permitted to emit is called the "cap". By lowering this cap annually, the EU may gradually reduce its emissions to meet its climate target. The EU ETS has allowed European power and industrial facilities to lower their emissions by approximately 47% by 2023 when compared to 2005 levels (European Commission 2024b). EU ETS mechanism is shown in Fig. 4.29.

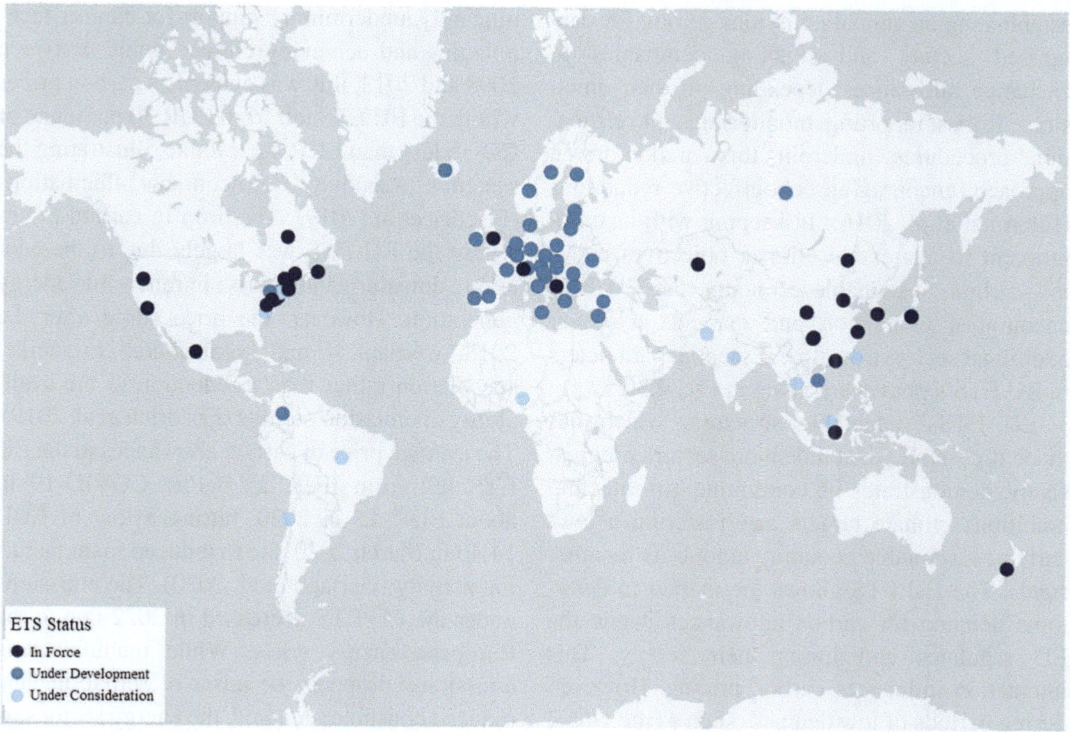

Fig. 4.28 Global emissions trading systems map (ICAP 2023a; Küfeoğlu 2024a)

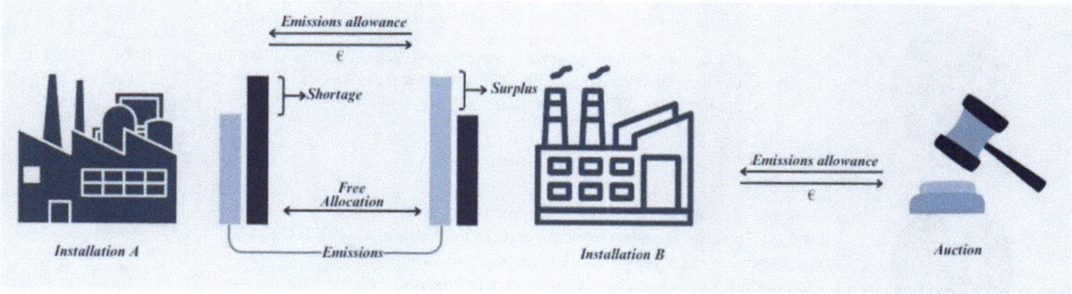

Fig. 4.29 EU ETS mechanism (Küfeoğlu 2024a; European Commission 2015)

Figure 4.29 shows the EU ETS mechanisms (Küfeoğlu 2024a). The EU ETS operates in this manner: Installation B does not have enough free allowances to offset its emissions, so it can only comply with the cap by buying permits from Installation A or through the auction. If their emissions exceed the free allowances given at the start of the year, a participant may keep any extra allowances they acquire from others or at auctions. Participants can also decide to hold onto their allowances for future use. Because there is a demand from participants for whom it is more costly to make cuts than for other participants, and because the supply is limited or capped, allowances are valuable.

The EU ETS is a key component of the EU's policy to cut GHG emissions and combat climate change (European Union 2015). The system's issues have worsened as the EU aspires to reduce emissions by 55% by 2030 and achieve net-zero emissions by 2050 (European Commission 2024c). It uses a cap-and-trade mechanism by establishing an annual emissions ceiling for designated sectors and enabling companies to exchange allowances depending on their emissions. Robust reporting, monitoring, and verification procedures underpin this market-driven approach, encouraging cost-effective reductions (Ellerman et al. 2016). In keeping with its commitment to aggressive climate objectives and a low-carbon, sustainable economy, the EU ETS encourages innovation and expands to cover additional sectors over time. Essential parameters in EU ETS's phases are given in Fig. 4.30.

EU ETSs' adaptable structure, which has made it possible to include more sectors progressively, demonstrates its commitment to meeting ambitious climate targets and fostering a low-carbon, sustainable economy among its member states. The EU ETS allows the market to determine demand for carbon allowances, while the EU regulates and limits their supply. This interaction influences carbon pricing. However, during periods of low demand, such as the global financial crisis 2010, carbon prices can drop significantly, undermining support for cleaner technologies and achieving climate goals. Between 2008 and 2013, this was evident as carbon prices within the EU ETS fell from EUR 30 per tonne of CO_2 to less than EUR 5 per tonne, illustrating the system's sensitivity to economic fluctuations (Fischer et al. 2019). The drop in carbon prices within the EU ETS was largely due to the economic downturn and the rise in renewable energy generation. However, the price surge after the 2018 overhaul is mainly attributed to market speculation rather than a reduction in the availability of emission permits (Friedrich et al. 2019). The average price of carbon allowances in the EU ETS fell from EUR 25 before COVID-19 to about EUR 15 in 2020, hitting a low of EUR 14.40 in March 2020 due to reduced manufacturing activity (Gerlagh et al. 2020). The emissions under the EU ETS decreased in 2022 due to the European energy crisis. While manufacturing emissions dropped because of inflation and reduced consumer demand, the energy sector saw a slight increase in emissions due to higher coal

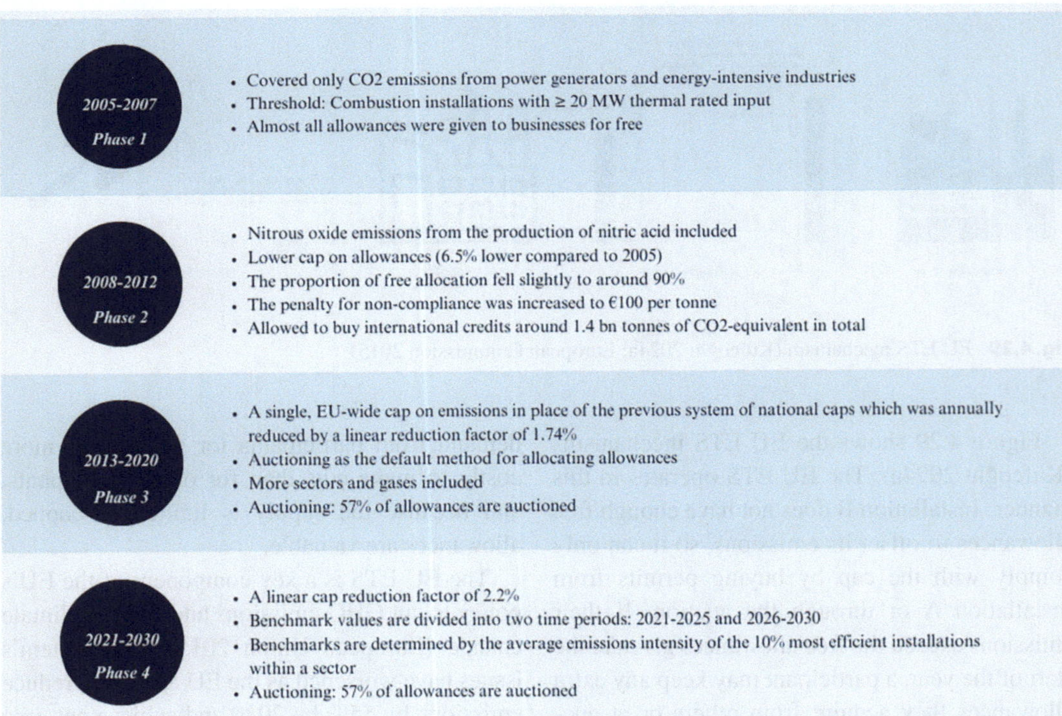

Fig. 4.30 Phases of EU ETS (Küfeoğlu 2024a; ICAP 2023)

use (European Council 2023). The average price of EU ETS credits in 2023 was between EUR 80 and EUR 100 per tonne of CO_2. Factors such as current regulatory changes, market expectations for upcoming carbon laws, and economic conditions affecting industrial emissions all influenced prices. The cost of EU ETS credits has fallen significantly in 2024, now trading between EUR 60 and EUR 80 per tonne of CO_2, a notable decrease from the higher prices seen in 2023. This decline has been attributed to economic slowdown, increased allowance supply, and policy modifications. EU ETS price change in years is given in Fig. 4.31.

In conclusion, the EU ETS is essential to the EU's comprehensive climate change policy. The EU ETS limits greenhouse gas emissions and promotes a market-driven strategy for their reduction by implementing a cap-and-trade system. Significant reductions in emissions from key industries such as power providers, huge energy consumers, and the aviation sector have been made possible by this approach. The EU ETS' Phase IV changes reflect the program's progress and emphasise its commitment to achieving ambitious climate targets through continuous adaptation. Notwithstanding its achievements, the EU ETS must contend with issues like volatile carbon prices brought on by outside forces like market speculation and economic downturns. These problems demonstrate how difficult it is to oversee a market-based system for reducing emissions and how constant improvement is required to guarantee its efficacy (Küfeoğlu 2024a).

The future of the EU ETS is shaped by increasingly ambitious climate goals, sectoral expansion, and evolving policy frameworks. More aggressive reductions in emissions caps are being driven by the EU's goal of achieving climate neutrality by 2050 and the "Fit for 55" package, which aims to reduce GHG emissions by 55% by 2030. This is causing a tightening of the supply of allowances and increased carbon prices (European Commission 2024c). In addition, a rival system (ETS2) for buildings and road transportation is being introduced, and the EU ETS is broadening its purview to include new industries, including sea transport. While the Carbon Border Adjustment Mechanism (CBAM), scheduled to go into effect in 2026, will combat carbon leakage by imposing carbon fees on imports from regions with less rigorous rules,

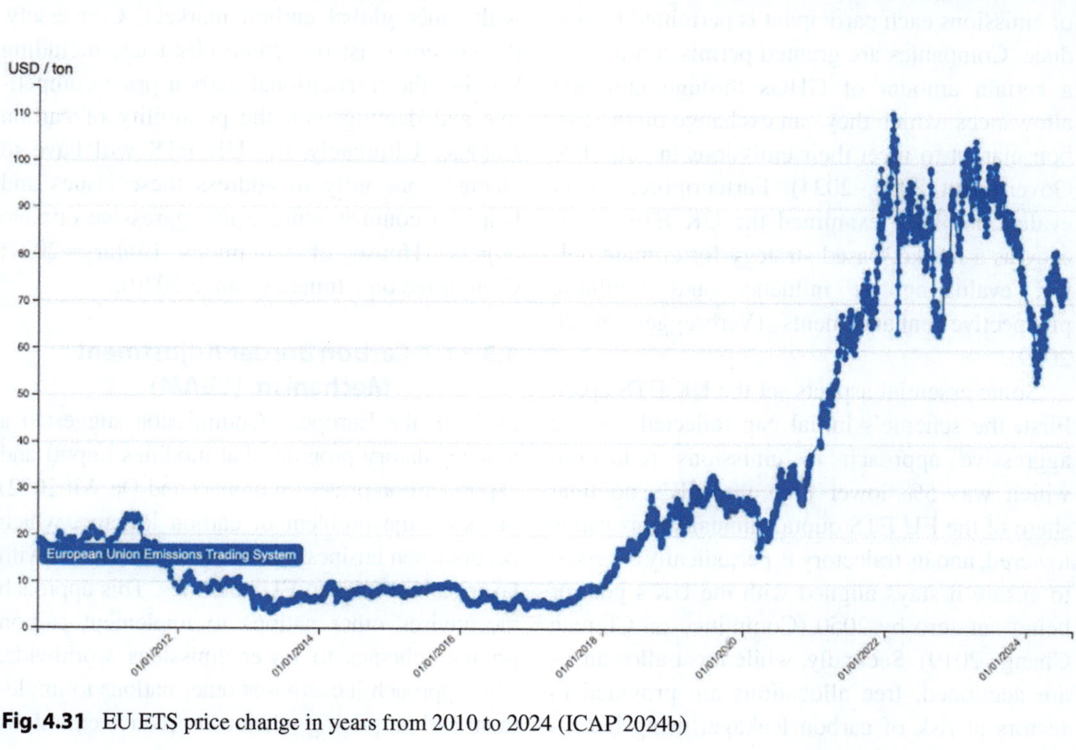

Fig. 4.31 EU ETS price change in years from 2010 to 2024 (ICAP 2024b)

higher carbon prices are anticipated to encourage businesses to adopt cleaner technology (European Council 2023). Furthermore, with possible connections to other markets and an expanding role in establishing global norms, the EU ETS is becoming increasingly seen as a worldwide model for carbon pricing. Further modifications are expected to maintain the system's efficacy and durability as it adjusts to shifting economic, technological, and geopolitical settings (Carbon Market Watch 2024a).

The United Kingdom (UK) ETS

Introduced in January 2021, the UK Emissions Trading Scheme (UK ETS) is a crucial market-based mechanism designed to control GHG emissions inside the UK. After Brexit, the UK left the European Union Emissions Trading System (EU ETS). The UK ETS was designed to meet the country's unique climate goals and economic circumstances (House of Commons Library 2021). It includes important industries, including heavy industry, power generation, and domestic aviation, which produce almost one-third of the UK's emissions. The "cap and trade" model set a maximum limit on the total amount of emissions each participant is permitted to produce. Companies are granted permission to emit a certain amount of GHGs through emission allowances, which they can exchange on the carbon market to meet their emissions targets (UK Government 2023, 2024). Furthermore, recent evaluations have examined the UK ETS's efficacy as a market-based strategy for climate policy, evaluating its influence and outlining prospective enhancements (Verbruggen et al. 2019).

Some essential aspects set the UK ETS apart. First, the scheme's initial cap reflected a more aggressive approach to emissions reduction, which was 5% lower than the UK's nominal share of the EU ETS quota. Annually, this cap is lowered, and its trajectory is periodically assessed to ensure it stays aligned with the UK's goal of being net zero by 2050 (Committee on Climate Change 2019). Secondly, while most allowances are auctioned, free allocations are provided to sectors at risk of carbon leakage, using benchmarks that reflect the performance of the most efficient installations. This ensures that while emissions are being reduced, the competitiveness of UK industries is preserved. The scheme's Carbon Price Support mechanism also sets a minimum price for carbon allowances, providing a stable and predictable market environment. The UK government can adjust this price floor, potentially leading to higher carbon prices than those in the EU ETS, reinforcing the UK's commitment to leading in climate action (Committee on Climate Change 2019).

Simultaneously, the UK ETS incorporates market stability features to prevent excessive price volatility. One such mechanism is a cost containment mechanism, which releases additional permits if prices rise too high. This strategy supports long-term investments in emissions reduction technologies by ensuring the carbon market is predictable and stable (Committee on Climate Change 2019). Though the UK ETS and the EU ETS are quite similar, they are built with the ability to deviate from the EU framework in the future. This can be increasing the program's coverage to new industries, further reducing the emissions cap, or even establishing connections with other global carbon markets. Conversely, the system must overcome obstacles, including keeping the international carbon price competitive and dealing with the possibility of carbon leakage. Ultimately, the UK ETS will have to change constantly to address these issues and help the country achieve its aggressive climate targets (House of Commons Library 2021; Committee on Climate Change 2020).

4.3.11.2 Carbon Border Adjustment Mechanism (CBAM)

In 2019, the European Commission suggested a new regulatory program that modifies import and export carbon prices (Schippers and De Wit 2022) to tackle the problem of carbon leakage, which occurs when businesses relocate from nations with laxer carbon rules to EU countries. This approach incentivises other nations to implement carbon pricing schemes to lower emissions worldwide. This approach incentivises other nations to implement carbon pricing schemes to reduce emissions worldwide. Moreover, by placing a price on carbon for imported products based on their carbon content, it aims to level the playing field (Quick 2020). A new regulatory program called the Carbon Border Adjustment Mechanism (CBAM) aims to incentivise firms outside the EU to reduce their emissions as an essential part of the Fit for 55 package, which is a set of recommendations meant to update and modernise current EU law to bring it into compliance with the climate goals, particularly the goal of lowering net greenhouse gas emissions by at least 55% by 2030. On December 13, 2022, the European Parliament and the European Council signed a temporary political agreement on CBAM. The European Council then formally enacted the legislation in April 2023 (European Commission 2021b; European Council 2023).

Implementing CBAM in an open climate alliance to support domestic mitigation efforts would be beneficial and be phased in gradually to mitigate political resistance from local and foreign companies (Jakob 2023). The new regulations initially apply to certain goods within the cement, electricity, fertiliser, aluminium, steel, and hydrogen sectors and some upstream and downstream products (primarily iron, steel, and aluminium) imported from non-EU nations. Except for those that take part in or are connected to the EU ETS (currently Iceland, Norway, Liechtenstein, Switzerland, and five other small territories), CBAM covers imports of covered commodities from non-EU nations. The EU Commission intends to broaden CBAM's applicability by 2030 to include all industries covered by EU carbon trading (European Council 2023).

The companies subject to CBAM have a reporting duty without financial liabilities from October 1, 2023, until December 31, 2025. Importers must document both direct and indirect emissions from the manufacture of imported items. EU importers require quarterly CBAM reports outlining imported amounts, embedded emissions, and paid carbon taxes. Importing CBAM items will be restricted to registered declarants only as of January 1, 2026. On the same day certificate trading starts, importers must buy enough emission permits to cover the embedded emissions from imports. Financial penalties may be imposed if there is a discrepancy between imported embedded emissions and purchased emission permits, which is determined by an annual CBAM statement. Organisations willing to import products made outside of the EU must buy certifications that match the amount of carbon emissions those products' manufacture caused. The price of CBAM certificates will be determined by the European Commission using the average weekly price of ETS auctions as a guide (European Commission 2023a). Accordingly, CBAM certificates will be based on the ETS, guaranteeing that the cost of CBAM certificates will be as near to the cost of ETS permits as feasible while simultaneously maintaining the administrative authorities' ability to govern the system intact (The Conference Board 2022). In Fig. 4.32, the CBAM working scheme is illustrated.

The free distribution of EU ETS allowances for industries covered by CBAM will gradually end between 2026 and 2034 as part of the CBAM

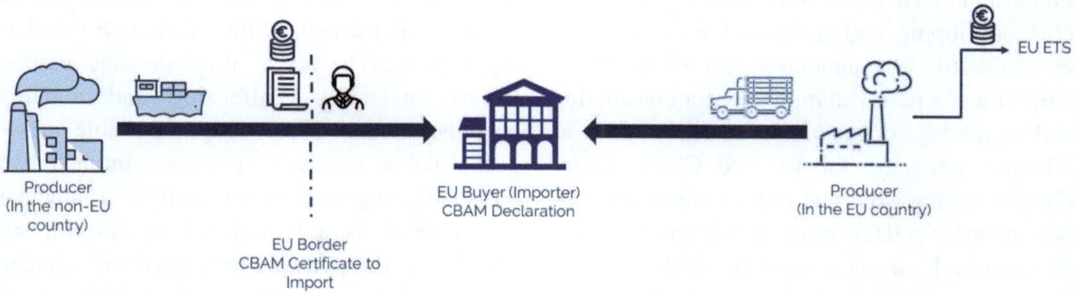

Fig. 4.32 CBAM working scheme (The Conference Board 2022)

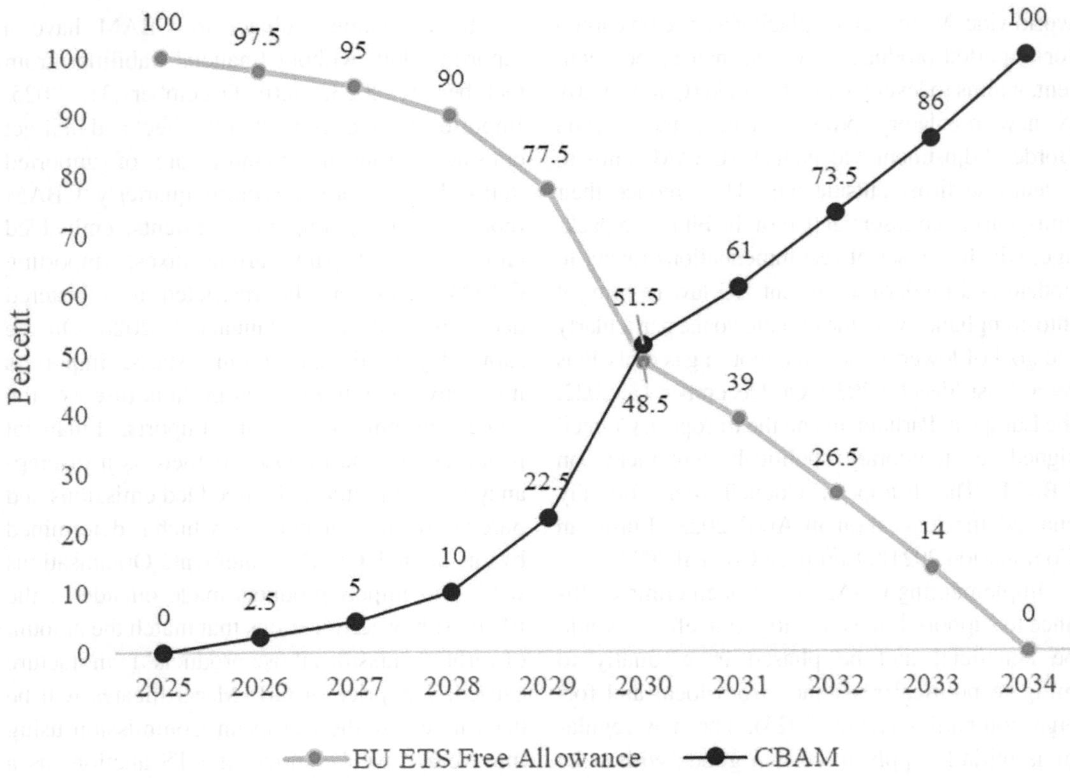

Fig. 4.33 Over the 2025–2034 period, phase out the EU ETS free allowances percentage and phase in the CBAM percentage (ICAP 2023e)

phase-in strategy. Free allocation will be phased down gradually at first, then more quickly as the period occurs. Additionally, it will immediately align with the CBAM phase-in, meaning that only the share of emissions not subjected to free allocation within the EU ETS will be covered by CBAM throughout the transitional period (ICAP 2023e). Figure 4.33 portrays the phase-out of EU ETS-free allowances and CBAM phase-in.

CBAM can potentially accelerate the transition to a more energy-efficient economy that benefits developing and developed nations (Bueb et al. 2016). The announcement of the EU's CBAM was a powerful motivator for considering carbon pricing for several jurisdictions, such as Ukraine, Uruguay, Taiwan, and China. Under CBAM, asking exporters to show their real carbon intensity collects more significant economic advantages. It enhances legal possibilities, raises the danger of evasive techniques, and adds an administrative burden (Mehling and Ritz 2023).

In broad terms, the EU has taken a significant step in tackling the issue of carbon leakage and advancing worldwide emission reductions with the adoption of CBAM. CBAM seeks to encourage countries worldwide to implement carbon pricing plans by modifying import and export carbon prices. The phased method, which targets specific industries first and then expands gradually, is a practical tactic to lessen political opposition. The reporting requirements and monetary fines for inconsistencies show a commitment to openness and accountability. Although mandating exporters to reveal their genuine carbon intensity under CBAM offers legal and economic benefits, it also acknowledges possible drawbacks such as evasive tactics and administrative costs. Aligning trade policy with environmental sustainability is a remarkable endeavour, and CBAM has the potential to further both climate goals and a more environmentally friendly global financial system.

4.3.11.3 Carbon Removal Market

A subset of the larger carbon market, the carbon removal industry focuses on processes and technologies that extract CO_2 from the atmosphere and store it for a long time. Carbon removal deals directly with removing existing CO_2 from the atmosphere, unlike carbon offsetting, which usually entails reducing or preventing future emissions (e.g., through renewable energy projects or forest preservation). This distinction is critical since global efforts are being made to actively reduce previous emissions and the rate at which new greenhouse gases are added (IPCC 2018).

The variety of carbon removal technologies and processes includes engineered methods like direct air and bioenergy with carbon capture and storage and nature-based options like reforestation and soil carbon sequestration. By permanently storing CO_2 to keep it from re-entering the atmosphere, usually through geological storage or long-lasting goods, each technique contributes to offsetting carbon emissions (IPCC 2018).

The market for carbon removal has emerged as a new front in the battle against climate change in response to the growing need for alternatives to emission reductions. The market is being driven by the increasing realisation that lowering emissions alone won't meet net-zero emissions targets by the middle of the century. The IPCC states that even with aggressive reductions, a certain amount of carbon removal will be required to offset inevitable emissions from difficult-to-abate industries, including agriculture, cement manufacturing, and aviation (IPCC 2018). Thus, removing carbon is seen as a key component in the shift to a global economy that is robust to climate change.

The market for carbon removal plays a significant role in propelling advancements in this field through large-scale corporate activities. To encourage high-quality carbon removal projects, for example, Google has announced that it will invest USD 35 million in the carbon credit market (Green Earth 2024). Similarly, Apple's carbon removal strategy strongly emphasises using natural solutions to offset its carbon footprint, such as better land management techniques and forest restoration (Apple 2023). These instances show how digital companies are using their power to push the limits of sustainable innovation and quicken the expansion of the carbon removal industry.

The carbon removal market is expanding quickly due to legislative initiatives, technological advancements, and increased corporate commitment to climate change. Key innovations that lower costs and improve scalability include direct air capture, bioenergy with carbon capture and storage, and nature-based solutions (Fuss et al. 2018a, b). Large-scale carbon removal will prevent global warming (IPCC 2018). It is also crucial to provide policy support through programs like international carbon market agreements, tax credits, and carbon pricing (Honegger and Reiner 2018). Furthermore, due to businesses investing in offsets and technology to satisfy their climate targets, corporate net-zero pledges drive up demand for carbon removal solutions (Cames et al. 2016). However, several major barriers remain, including expensive costs, energy requirements, and uncertainty around long-term carbon storage (Bui et al. 2018). Building trust and guaranteeing the efficacy of carbon removal initiatives will depend on addressing these issues through enhanced market mechanisms, standardisation, and verification frameworks (Schneider and La Hoz Theuer 2019).

Before investigating the table outlining the various carbon market systems—ETS, voluntary carbon markets, and carbon removal markets—it's crucial to realise that each has unique strengths and weaknesses. Although reducing greenhouse gas emissions and promoting carbon removal are the goals of these mechanisms, their methods and efficacy vary. Table 4.2 shows the strengths and weaknesses of these systems.

After analysing its advantages and disadvantages, it is critical to look at each carbon market system's projected future paths and market expectations. It can be helpful to comprehend the changing dynamics and the future trends that these markets will likely follow to learn how to adjust to new possibilities and difficulties. However, for this, it is first necessary to examine the recent behaviour of the markets. In 2024,

Table 4.2 Existing strengths and weaknesses of carbon trade systems

	Strengths	Weaknesses
Emission trading systems (ETS)	Cost-Effectiveness: Because the ETS permits flexibility in achieving emission reductions, businesses can cut emissions, where doing so will result in the lowest costs. This may result in more effective reductions and overall cost savings (Böning et al. 2023) Incentive for Innovation: The ETS, by putting a price on carbon, provides businesses with a financial incentive to develop and invest in low-carbon technologies (Böhringer et al. 2017) Market-Driven Flexibility: The ETS's market mechanism helps maintain price stability and guarantees that emissions targets are met by adjusting the cost of emissions by supply and demand (Tietenberg 2006)	Price Volatility: Market dynamics can cause ETS to see large price swings, which could make carbon costs less predictable for companies (Burtraw et al. 2000) Complexity and Administrative Costs: To assure efficacy, the system may be difficult to develop and administer, requiring significant administrative resources and monitoring (Matos et al. 2022) Leakage Risk: Businesses may move to areas with laxer laws, which could result in "carbon leakage" and prevent overall world emissions from declining (Aldy and Stavins 2012)
Voluntary carbon markets (VCMs)	Flexibility and Innovation: VCM offers flexibility for projects that may not be covered by mandatory regulations, fostering a wide range of innovative carbon reduction projects (Hamilton et al. 2010) Consumer Engagement: VCM allows individuals and companies to offset their carbon footprint voluntarily, which can raise awareness and drive broader climate action (Battocletti et al. 2023) Additionality: Many voluntary projects aim for additionality, delivering carbon reductions that would not have occurred otherwise (Schneider 2007)	Problems with Quality Control: The inconsistent quality of carbon credits may jeopardise the market's legitimacy. Monitoring and verification practices can vary (Borgmann et al. 2024) Lack of Standardisation: Inefficiencies and market fragmentation can result from a lack of regulatory control and universal standards (Bumpus and Liverman 2008) Limited Scale: The voluntary market's influence on global emissions reductions may be limited because it is smaller than compliance markets (Drew et al. 2023)
Carbon removal market	Direct Effect on Atmospheric CO_2: The goal of carbon removal from the atmosphere can directly lower atmospheric CO_2 levels and possibly offset emissions that are difficult to offset (Smith et al. 2015) Helps Long-Term Objectives: CRM helps long-term climate objectives by addressing residual and possibly reversing historical emissions (Low et al. 2024) Diverse Strategies: The market offers many strategies to remove carbon dioxide, such as reforestation and soil carbon sequestration (Chazdon et al. 2016)	High Prices and Uncertain Technology: Several carbon removal methods are either very expensive or still in development, which may limit their applicability (Minx et al. 2018). Scalability Issues: There are considerable obstacles to scaling carbon removal technology to the necessary level for a discernible climatic impact (Geden et al. 2018) Possible Environmental Trade-Offs: Certain carbon removal techniques may have unanticipated negative effects on the environment or society, such as changes to biodiversity or land use (Lackner et al. 2018)

many regions have had limited movement on adopting carbon pricing. Nonetheless, there are encouraging indications that middle-income nations are starting to use these approaches more aggressively. This change is important because, even after ten years of steady increases in carbon prices, overall costs are still too high to result in appreciable emissions reductions. The continued insufficiency of carbon pricing emphasises the necessity for more comprehensive actions. Positively, revenue from carbon pricing has risen to all-time highs, indicating that more people are realising how powerful this tool can be in influencing climate policy and raising money (World Bank 2024a).

This spike in revenue has sparked the growth of new adaptable designs and methodologies tailoring carbon pricing schemes to different national contexts and particular requirements. This flexibility shows how the system may

Table 4.3 Strengths and weaknesses of carbon trade systems according to market expectations

Carbon market system	Market expectation
Emission trading systems (ETS)	Regulatory Support: Stricter regulations are anticipated to increase carbon prices (Carbon Market Watch 2024b) Price Volatility: Although prices could fluctuate, they should eventually stabilise (Carbon Market Watch 2024b) Market Expansion: Growth is anticipated as more areas and industries implement ETS frameworks (Carbon Market Watch 2024b) Innovation Incentives: These are anticipated to stimulate innovation in methods and technologies to lower emissions (Carbon Market Watch 2024b)
Voluntary carbon markets (VCMs)	Growth Potential: Significant growth is predicted as companies and individuals look to reduce their carbon footprints (Ecosystem Marketplace 2023) Quality and Transparency: Improved standards and verification procedures are anticipated to guarantee credit integrity (Ecosystem Marketplace 2023) Diversification: A wider range of initiatives, such as reforestation and renewable energy, are anticipated to arise (Ecosystem Marketplace 2023) Price Trends: As the market develops, prices should standardise, and liquidity should rise (Ecosystem Marketplace 2023)
Carbon removal markets	Emerging Sector: As removal technologies advance, growth is anticipated (Mannion et al. 2023) Frameworks for Regulations and Policies: Rules conducive to growth are essential (Mannion et al. 2023) Technological Advancements: Carbon capture and storage developments are anticipated to propel market growth (Mannion et al. 2023) Price and Demand: As technologies advance, prices are anticipated to rise first and then decline (Mannion et al. 2023)

change to accommodate new situations. However, the moves of the carbon credit markets have been uneven, showing a combination of advancements and disappointments. Given the current quiet market circumstances and waning trust, there is an urgent need for actions targeted at regaining the integrity and credibility of these markets. These obstacles and adjustments illustrate the intricate interactions between market dynamics, policy innovations, and carbon pricing methods. These interactions will significantly impact global carbon pricing initiatives' future efficacy and scope (World Bank 2024b). Based on the behaviour of the carbon markets 2024, the market expectations of these systems can be seen in Table 4.3.

4.3.12 Risk and Insurance

Sustainable finance must consider both risk and insurance in creating financial systems that are resilient and responsive to global challenges, such as climate change, loss of biodiversity, or social injustice. There is a growing awareness to address ESG criteria not only at the level of business strategies but also in risk modelling and underwriting, as such challenges are systemic and have a long-term impact. Also, the economic principles of sustainable insurance contribute to the development of capital funding for environmentally and socially responsible projects, leading to the reduction of carbon footprint. Insurers have a significant role in advancing sustainable development by offering products, like green bonds or loans for renewable energy projects, that support environmentally responsible behaviour.

The engagement of suppliable risks with sustainable finance principles, particularly risk management and insurance, stems from comprehensive stakeholder engagement, from policyholders to regulatory authorities, to mitigate the risks and harness the opportunities presented by sustainability. Such an approach is useful in fostering a robust financial ecosystem conducive to economic development and social welfare in the long run. It is evidenced that embedding sustainability within the risk management and insurance process protects against future uncertainties and creates avenues for new value. As the economy is ever more complex with structural, operational, and competitive industry changes, practising sustainable methods is becoming an organic business requirement (Swiss Re Institute 2020).

4.3.12.1 Climate Risk

Climate risk's broad socio-economic and environmental effects make it a crucial concern for global development. The World Bank underlines that extreme weather events, water shortages, and food insecurity—all of which affect more developing nations—are made worse by climate change. These effects exacerbate poverty and instability by destroying infrastructure, upsetting livelihoods, and raising health risks (World Bank 2024a). The World Bank will provide USD 38.6 billion in climate finance in 2023 to assist efforts towards adaptation and mitigation, indicating the substantial economic costs involved. By constructing sturdy infrastructure and encouraging sustainable habits, this investment seeks to increase resilience (World Bank 2024b).

Risks related to climate change have a significant human cost, impacting hundreds of millions of people annually. For example, Vietnam is sensitive to weather extremes and sea-level rise, which can lead to economic disruption and relocation. In the same way, 600 million or more people lack access to electricity in sub-Saharan Africa, highlighting the pressing necessity of renewable energy projects to enhance resilience and quality of life (World Bank 2024a). These difficulties highlight the need for concerted international efforts to properly manage and reduce climate hazards to promote sustainable development and the elimination of poverty.

Climate risks, one of the major elements in the portfolio of sustainable finance, relate to the risk of loss and adverse impacts of climate-related events. They can be classified as physical risks (e.g., hurricanes, floods, and sea-level rise), transition risks (e.g., climate policies and development of low-carbon technologies), and liability risks (e.g., environment-related legal claims). The IPCC highlights that due to climate change, the frequency of extreme weather events is greatly increased owing to global warming, which requires proper risk management and insurance systems to be deployed (IPCC 2018). Furthermore, the Task Force on Climate-Related Financial Disclosures (TCFD) points out that such risks must be included in the overall financial decision-making process to avert economic catastrophe in the future (FSB-TCFD 2017).

Climate change and its associated risks demand a combination of climate scenarios, stress testing, and other methodologies from banks to predict the institution's risk. The Global Network of Financial Systems for Greening favours better practices regarding the risks posed by climate change (NGFS 2019). Insurers and risk managers also help by creating cheaper and timely climactic trigger types of insurance that automatically pay under certain conditions and encouraging practices that promote protection through discounts on premiums. This sweeping inclusion of climate issues enhances resilience while addressing climate change holistically.

Physical Risks

The coordinated impacts of climate change on foundations, human well-being, and the environment are alluded to as physical climate dangers. As a rule, these dangers are partitioned into intense and incessant categories. Extraordinary climate conditions that can result in fast and critical harm, such as typhoons, surges, and fierce blazes, illustrate intense dangers. Then again, chronic risks are related to long-term shifts, including rising ocean levels, rising temperatures, and altered precipitation designs (Küfeoğlu et al. 2014). "The recurrence and concentration of extreme climate occasions have expanded over the past few decades, driving to noteworthy financial misfortunes and human enduring", said a report distributed by the Interval Board on Climate Alter (IPCC 2021). These episodes not only upend communities but also put a weight on the soundness of the economy and open assets.

Acute physical risks are event-driven, occurring suddenly and causing immediate, often severe impacts. Such risks include heatwaves, intense flooding, storms, hurricanes, and wildfires. Heatwaves can lead to health crises, increased energy demand, and infrastructure damage (IPCC 2021). Intense rainfall, storms, or rapid snowmelt can cause rivers to overflow, resulting in property damage, transportation disruptions, and loss of life (NOAA 2021). Storms and hurricanes bring strong winds, heavy rain, and storm surges, causing widespread destruction to buildings, infrastructure, and natural ecosystems (WEF 2020). Extended dry periods combined with high temperatures increase the likelihood and severity of wildfires, destroying forests and homes and severely affecting air quality due to smoke (UNEP FI 2019). The immediate nature of acute physical risks means they can cause significant disruptions in a short period. Businesses might face sudden operational halts, supply chain interruptions, and increased emergency response and recovery costs. For instance, a severe storm can damage critical infrastructure, leading to production delays and increased insurance claims (CDSB 2019). Communities may also experience loss of life, displacement, and long-term socioeconomic impacts (EEA 2017).

Conversely, chronic physical risks are associated with primary long-term climate change trends caused by the gradual deterioration of several factors such as temperature changes, sea-level rise, shift in rain patterns, or the increasing level of ocean acidification. In particular, the increase in average temperatures may impact agriculture in terms of productivity, higher cooling demands, and alteration of ecosystems (IPCC 2021). Because meltwater from glaciers and polar seas, combined with expansion due to heat, raises sea level, gradual encroachment will inundate the land, accelerate erosion processes near the coast, and limit the amount of freshwater available (NOAA 2021). Changes in the average precipitation amount and water runoff into inland water bodies will lead to drought periods in some countries and a surplus in others, affecting water accessibility, farming, and ecosystems (WEF 2020). One of the consequences of oceans taking in increased carbon dioxide is their ionic alteration, which will directly affect the calcium carbonate-dependent marine creatures (UNEP 2021).

Nevertheless, while subtle, the chronic physical risks have a severe, long-lasting effect on the environment. Some areas must be changed based on economic conditions; for example, agricultural zones or more energy will be required to apply more loads. Some infrastructure components may no longer be adequate for new climate conditions and must be redesigned or reinforced. As an example, rising sea levels may put coastal assets at risk and require costly adaptation strategies such as sea barriers or relocation (CDSB 2019). Over time, climate changes can also damage an ecosystem, diminishing the variety of species within it and the benefits that these ecosystems deliver (FSB-TCFD 2017).

Risks of a Transition Climate

Dangers related to the move to a low-carbon economy stem from societal and financial shifts. The administrative, legitimate, mechanical, and commercial changes that go with nations' and businesses' change to feasible hones incorporate a few perils. Businesses and economies dependent on fossil fuels may confront extreme results due to moving dangers.

"The move to a low-carbon economy will require noteworthy changes in vitality generation, transportation, and industrial processes, which might lead to considerable budgetary dangers for companies that are ill-equipped", agreeing to think about by the Assignment Constraint on Climate-related Budgetary Revelations (TCFD 2020). Stricter natural laws, a development in customer tastes towards economic items, and headways in green innovation are examples of these patterns.

In the shift to a sustainable economy, financial institutions "face both risks and opportunities", according to the UNEP Finance Initiative study "Beyond the Horizon". According to UNEP FI (2020), "those that adapt can benefit from emerging opportunities, while those that fail to integrate climate considerations into their strategies may suffer significant losses". One division that illustrates the dangers well is the car industry. Ordinary automakers go up against imposing deterrents due to rising customer requests for electric vehicles and government weight to lower carbon emanations. Businesses that do not alter quickly risk losing clients and becoming unsteady fiscally.

Climate Risk Management

Risk management may be defined as the systematic identification, assessment, and management of risks that might affect the achievement of an organisation's objectives (ISO 2018). In the context of climate-related financial disclosures, such risk management is crucial in understanding and addressing the adverse economic effects of climate risk on organisations (TCFD 2024). Regarding climate-related risks, the Task Force on Climate-Related Financial Disclosures (TCFD) has provided a structured approach to managing such risks. This framework highlights the role of governance, strategy, risk management processes, and appropriate metrics and targets (FSB-TCFD 2017; UNEP FI 2021). These aspects will be elaborated on in the following sections, emphasising governance, strategy, and metrics.

Assessing Financial Impacts

An essential aspect of the management of climate-linked risks is the evaluation of their financial consequences. Institutions should assess how climate-related risks and opportunities impact revenue, costs, assets, liabilities, and equity financing according to:

- Revenues: Determining possible variations in the sales of goods and services because of risks associated with climate change.
- Expenditures: Estimating the amount needed to avoid and seize climate-related risks and opportunities and the cost of investing in new technologies and new processes.
- Assets and liabilities: Considering the likely effect of climate-related risks on the valuation of long-lived assets and the valuation of liabilities as well.
- Capital and Financing: Assessing how climate change could affect the organisation's debt and equity structure or its ability to raise new capital or roll over existing debt.

Deutsche Bank has acknowledged the significance of measuring and managing climate-related risks as a worldwide organisation. By embedding climate change risk consideration into its business as a risk management process, Deutsche Bank has, to some extent, lessened the adverse effects of climate change (Deutsche Bank 2023). The bank employs sophisticated climate change risk assessment tools to assess climate change and its effects on the bank's loan books, investment plans, and overall financial health (Deutsche Bank 2023). In addition, Deutsche Bank has made promises to deepen its sustainable finance operations and back projects and firms that promote the sustainability of the environment. This aligns with the international climate change objectives and limits the bank's exposure to vulnerable climate sectors that may become more restricted by climate policies and regulations (Deutsche Bank 2023).

As for the bank remains equipped for different scenarios and actively modifies its plans as necessary (Deutsche Bank 2022). It uses the information gleaned from anticipation and scenarios performed regularly to determine how the climate may affect its business operations in the future. This must be done for different potential future scenarios to arrive at the suitable decision that would be the most effective towards achieving the bank's objective. As such, it allows the bank to effectively manage various possible risks.

The effort is, or at least, shares a vision of being guided by key principles: transparency, engagement, and communication. "Through its engagement with stakeholders, Deutsche Bank has also disclosed its climate-related risks and opportunities in line with TCFD recommendations" (TCFD 2023). Unfortunately, the impact does not stop at financial exposure. Instead of that, risk management efforts helped improve Deutsche Bank's brand perception while fostering growth that is also sustainable in the long run (Deutsche Bank 2023).

The World Economic Forum stresses that financial stability should go hand in climate risk management only to render useful outcomes (WEF 2022b). The orientation around climate proves useful, even having been verified by the European Central Bank (ECB 2023). There seems to be another aspect of the argument that the ECB highlights very well. A secure framework not only helps with the regulatory requirements but also, due to the changing nature of the landscape, helps stay ahead in the disruptive environment. The International Monetary Fund (IMF) also notes that integrating climate risks into financial strategies helps institutions mitigate potential losses and capitalise on emerging opportunities in the green economy (IMF 2022).

4.3.12.2 Integrated Climate Risk Assessment: A Comprehensive Approach

Integrating Physical and Transition Risk Assessment

Scenario analysis explains climate-related risks and opportunities by examining assumptions, limitations, and hypothetical outcomes. They test conventional assumptions by highlighting key factors affecting the future. Comprehensive methodologies are available in this regard. However, these tools consider and evaluate physical, transition, or liability risks independently and differently. It is necessary to adopt a systematic approach to climate risk assessment because the effects of these risks on the financial sector are systematic. These interactions must be considered to fully assess climate risk exposure and develop a comprehensive climate strategy (CISL 2022).

Growing Integration of Climate Risks in Assessment Tools

Since 2021, the tool market has witnessed significant developments. There is a growing demand for solutions that can address and evaluate physical and transition risks together. This trend has led to consolidating services among tool providers with different specialisations. The impact of this shift is also evident in the business consultancy and accounting sectors, highlighting the urgency and importance of integrated climate risk assessment (UNEPFI 2023).

Addressing the Interconnected Challenges

As the Cambridge Institute for Sustainability Leadership (2022) emphasises, a systematic approach to sustainability is beneficial and critical for the financial sector. While awareness of the interconnectedness of nature, humans, and climate challenges is increasing, integrated methods are still in their infancy. This underlines the need for a comprehensive approach to address these challenges effectively (CISL 2022).

Integrated Climate Risk Assessment Principles and Steps

Although integrated approaches are still in their early stages, detailed studies in this context are included in the "Climate Tango" report presented by the Cambridge Institute for Sustainability Leadership. CISL has identified five core principles and three steps. These principles and steps aim to evaluate and manage climate risks holistically. It has also been stated that this is important for the sustainability of the financial sector.

Businesses must integrate climate considerations into their strategic planning to manage transition risks. This includes conducting thorough climate risk assessments, investing in sustainable technologies, and advocating for supportive policy changes. Financial institutions play a crucial role by shifting investments from high-carbon activities to sustainable initiatives, facilitating the transition to a low-carbon economy (McGrady 2024). The Task Force on Climate-Related Financial Disclosures (TCFD) provides a framework for companies to disclose climate-related risks and opportunities, helping stakeholders make informed decisions (FSB-TCFD 2024).

4.3.12.3 Insurance

Climate change due to human actions increases the frequency and intensity of extreme weather events globally (European Environment Agency 2021). These climate hazards affect different regions, economic sectors, and societies in various ways. While heavy rainfall and other extreme weather events are becoming more likely in some areas, others may experience intense heat extremes, wildfires, and droughts. Moreover, many social and economic sectors, including tourism, the labour market, health, and agriculture, are threatened by climate change. Poor and vulnerable populations in developing countries are heavily affected by the outcomes of climate change since they have less ability to fight these outcomes. These outcomes increase poverty's frequency and severity; thus, development progress becomes slower. Climate change harms agricultural productivity, water and food security, infrastructure, and value chains among poor and vulnerable populations, and it undermines their ability to recover and compensate (InsuResilience 2020). Thus, the insurance sector is particularly impacted, as increasing the frequency and severity of extreme events could lead to higher premiums. In the long term, this may result in some people being unable to afford insurance, thereby increasing social inequalities (European Commission 2024a). Climate risk insurance, one of the risk transfer solutions, is one of the most effective practices because it initially prevents these impacts (InsuResilience 2020).

4.3.12.4 Impact of Climate Change on the Insurance Industry

Climate change has profoundly impacted global supply chains and infrastructure, hurting firms' capital structures, household wealth, and government funding. As a result, the insurance sector has also been significantly affected. By protecting assets, earnings, and lives against unexpected financial shocks, insurance aims to reduce the uncertainty of daily life. However, there has been more uncertainty lately (Swiss Re Institute 2024a).

Economic, societal, and physical climate changes have increased uncertainty significantly in the past 30 years (Swiss Re Institute 2024a). Furthermore, according to the expectations, it is not likely to decrease in the future due to globalisation, political tensions, demographics, and climate change. Higher economic uncertainty and increased risk awareness of households and businesses in parallel could result in higher non-life insurance premium volumes, especially in the long term (Swiss Re Institute 2024a).

According to Munich Re (2024a), increasing risk of wildfires and bushfires has become an essential issue for the insurance industry in recent years. For instance, in 2024, thousands of acres burned, and many evacuated due to a wind-driven fire close to Tracy, California, despite a relatively mild fire season in 2023. As a result of this, other states experienced insurance catastrophes. Increased risks have led some insurers to stop providing coverage or tighten their terms and raise premiums. This pushed many insureds out of the standard market and into the surplus lines (E&S) market. The Insurance Information Institute (2024) describes the surplus lines market as a niche sector comprising highly specialised insurers. These insurers take on risks that licensed (admitted) companies either refuse to cover or will insure only at a significantly higher cost, with extensive exclusions, or with very high deductibles (Insurance Information Institute 2024). The E&S market is vital for coverage in high-risk wildfire areas, but it is built to handle only some types of risk (Munich Re 2024b).

According to the World Economic Forum (2023b), from 2020 to 2021, the protection gap grew from USD 117 billion to USD 161 billion. As a result, retreat risk occurred for insurers in some regions of natural catastrophe coverage (WEF 2023b). Of the 66 natural disasters that resulted in at least a billion dollars in damages in 2023, 63 were weather-related. However, insurance did not cover 60% of weather and climate-related losses in 2023. AON (2024) defines the protection gap as the difference between total damage and losses covered by insurers, emphasising it as a critical reference point for identifying opportunities. Additionally, they emphasise the significant potential to close this protection gap, provide clients with tangible benefits, and enhance the industry's capacity to help society respond to extreme weather and climate hazards.

USD 380 billion is the total loss of natural disasters globally in 2023 (AON 2024). Moreover, this was the eighth time global losses exceeded USD 300 billion and were 22% higher than the long-term average. In 2023, insurers globally provided coverage totalling USD 118 billion. However, after seeing the lowest gap in 2022, the protection gap rose to an average of 69% (AON 2024). However, this is uneven across the world. While most disaster-related losses are covered in the USA, the majority of losses in the APAC and EMEA regions, as well as other parts of the USA, remain uninsured. Closing the protection gap will continue to be both a challenge and a significant opportunity for specific regions.

Climate change is compelling the insurance sector to be innovative and adaptable in responding to the needs and challenges of climate adaptation. Insurance, particularly for property assets, is a relatively short-term product, renewed every 1–5 years. However, physical and financial protection requires long-term strategies. Integration into risk management frameworks will enhance the insurance sector's preparedness and resilience against climate disasters by identifying and mitigating climate risks (Jarzabkowski et al. 2019).

4.3.12.5 Climate Risk Insurance

Schaefer and Waters (2016) define climate risk insurance as insurance products that cover losses from extreme weather events, which are becoming more severe and frequent due to climate change. The negative impacts of disasters on the economy can be reduced with the successful implementation of climate risk insurance, which creates incentives for preventive measures. Similarly, InsuResilience Global Partnership (2020) explains climate risk insurance as a collection of instruments that protect financial risk transfer against extreme weather events, which have become more frequent and intense due to climate change. This system protects groups from regional entities to governments and individuals against loss of life, livelihood, or assets. These groups receive immediate and effective payments if insured under this mechanism. This system is particularly important for the poor and vulnerable people in developing countries because they will be disproportionately affected by these extreme weather events. The impacts of climate change threaten existing infrastructure and value chains, especially reducing the capacity of these groups in these countries to cope and recover. As a risk transfer solution, climate risk insurance prevents or mitigates the negative impacts of climate change. Climate risk insurance works through two primary mechanisms:

Direct Parametric Insurance

In parametric direct insurance, policyholders are households, agricultural businesses, market vendors, or companies such as financial institutions. NGOs, banks, farmers' cooperatives, or local insurance companies typically offer policies at the regional level as retail products. These policies provide compensation rights based on predefined parameters against extreme weather events. For example, payouts can be used for purposes such as continuing to pay school fees or purchasing food to compensate for harvest losses. Additionally, they can help increase savings, credit efficiency, and investments in higher-yield activities (InsuResilience Global Partnership 2020).

Indirect Insurance

In indirect insurance, however, the policyholders are municipalities or governments. It can be applied at a macro-level to governments and a meso-level to microfinance institutions, NGOs, and credit unions. The policyholders act as intermediaries in the payment process for those affected by extreme weather events. Thus, the target group benefits indirectly. For example, food aid and infrastructure reconstruction funded by insurance payouts after a drought can help many people in a short period through indirect insurance (InsuResilience Global Partnership 2020).

Africa is vulnerable to the outcomes of the climate crisis, which affects food and water security, economies, and agricultural production. By 2040, the total GDP of the continent is expected to reduce by around 2% and 4%, while reductions in rain-fed farm yields could be seen at 50%. In this context, interest in climate risk insurance increases for the vulnerable populations most. For instance, the Rural Resilience Initiative (R4) and Oxfam want to help farmers with food insecurity. The purpose is to prevent selling their assets and encourage them to invest in profitable agricultural activities. This microinsurance initiative provides coverage either through affordable premium payments or by participating in community projects. It provides benefits, especially for women's decision-making abilities and access to credit. These efforts have improved women's empowerment and inclusivity in countries like Ethiopia and Senegal. Another important example is the Index-Based Livestock Insurance (IBLI), which offers drought insurance to livestock-rearing communities in Kenya and Ethiopia. This insurance, activated by satellite data, has shown significant benefits, such as reducing food consumption decline and preventing the sale of assets during droughts among pastoralists. However, it has been observed that IBLI often struggles to reach the most vulnerable populations, mainly benefiting those who are just above the poverty line. Both the R4 and IBLI programs highlight the potential to protect vulnerable communities in Africa from the increasing effects of climate change. However, there are challenges in making these insurance models sustainable and ensuring they reach the most disadvantaged groups (ActAlliance 2020).

Bangladesh faces significant climate-related risks, with flooding threatening about 70% of the population. The agricultural sector, which employs two-thirds of the population, is especially vulnerable. In that context, Green Delta Insurance Company (GDIC), supported by the International Finance Corporation, introduced Weather Index-Based Crop Insurance (WIBCI) in 2015. This insurance protects crops like rice, tomatoes, and potatoes from weather-related risks such as wet spells, extreme rainfall, and temperature fluctuations. From 2016 to 2020, GDIC's WIBCI program expanded fast, growing by approximately 400% annually and covering around 50,000 farmers under WIBCI and 100,000 farmers across all climate risk products. The five-year overall claims ratio is 48%, and the claims incidence rate is 45%. This shows that almost half of the farmers received payouts. To make it more accessible, GDIC makes partnerships with local communities and sells products through group policies with partner organisations to reduce the risk of misselling and to promote renewals. Partnerships with NGOs are essential, representing over 70% of premiums and 80% of the customer base. Although the average premium per client has decreased, GDIC is committed to providing affordable coverage, including low-cost options for the poorest farmers. With donor support for infrastructure and capacity building, GDIC aims to insure 500,000 farmers by 2025 (Miliman 2022).

MENA is the most water-stressed region in the world, representing around 6% of the world's population but less than 2% of its freshwater supply. Water availability per capita will be reduced by half by 2050, severely impacting the agricultural sector, which consumes about 85% of the water in many MENA countries. The Advancing Climate Risk Insurance Plus (ACRI+) initiative aims to provide financial and strategic capacities for MSMEs to cope with climate risk, focusing on Morocco. ACRI+ is under the "Promotion of Integrated Climate Risk Management and Transfer" project funded by the German Federal Ministry of Environment, Nature Conservation, and Nuclear Safety. Over 97% of the products available in Morocco are crop insurance, with just over 2.9% attributed to hail insurance and 0.1% livestock mortality insurance. Morocco claims the highest rate of insurance market penetration in Africa. In 2015, the insurance industry generated a USD 3 billion US turnover, of which USD 41 million contributed to climate insurance products. Although the state subsidises the insurance sector by 90%, it is the people who face the greatest challenges. The primary issue is that awareness of these products remains low. Moreover, despite the diverse needs of MSMEs, there are no personalised products (ActAlliance 2020).

UNEP FI Principles for Sustainable Insurance

Sustainable insurance, as defined by (UNEP FI 2024), involves strategically conducting all insurance value chain activities and stakeholder interactions responsibly and with a future-orientated approach. Sustainable insurance seeks to minimise risk, create new approaches, strengthen business outcomes, and support *the* ecological, social, and economic sustainability they empha-

Table 4.4 Principles for sustainable insurance

Principle 1	Focuses on embedding ESG issues into decision-making processes, from company strategy and risk management to product development and investment practices. It emphasises integrating ESG factors at all levels, including recruitment, training, and claims management
Principle 2	Encourages collaboration with clients and business partners to raise awareness, manage risk, and develop solutions related to ESG issues. This involves dialogue with clients, suppliers, insurers, reinsurers, and intermediaries and promoting the adoption of these principles across the industry
Principle 3	Advocates working with governments, regulators, and other stakeholders to promote widespread action on ESG issues, supporting policy frameworks that enable risk reduction and innovation. It also calls for dialogue with various stakeholders, including NGOs, academia, and the media, to foster research, education, and public awareness
Principle 4	Emphasises the importance of accountability and transparency, urging companies to regularly disclose their progress in managing ESG issues through public reporting and stakeholder engagement

sise. The UNEP FI Principles for sustainable insurance provide a global framework for the insurance industry to address ESG risks and opportunities. Table 4.4 shows four principles as follows.

As climate-related natural disaster events increase and intensify in frequency, the insurance industry is in an essential position. It, therefore, calls for innovative and highly flexible solutions in the insurance sector to protect vulnerable populations from financial loss resulting from these events. Thus, direct and indirect climate risk insurance have emerged as the most essential tools for helping secure the most vulnerable populations while promoting community resilience. Nevertheless, the sector faces important barriers, such as growing protection gaps and the need to target the most disadvantaged groups effectively. The successful implementation of climate risk insurance can reduce the negative economic impacts of disasters and encourage preventive measures and long-term sustainability. Collaboration between stakeholders and alignment to innovative climate risk insurance practices and sustainable insurance principles will be crucial as the industry develops solutions to climate challenges and ensures fair protection for all.

4.3.13 Islamic Structures

Islamic finance is pivotal in the modern financial landscape by offering a values-based approach that emphasises ethical behaviour, transparency, predictability, fairness, and social responsibility (ICMA, IsDB and LSEG 2024). Its structures ensure that economic activities align with societal welfare and sustainable development objectives. Thus, Islamic finance is a natural partner to green finance initiatives promoting sustainability and responsible resource use. It structures require transactions to be supported by tangible assets and rely on the sharing of profits and/or losses among the provider and receiver of funds (Khalil et al. 2002). Islamic structures appear in various forms and are not subject to the numerus clausus principle, meaning that new structures may be established according to the developments in the financial sector. Islamic finance asset growth between 2016 and 2022 and 2027 projection are illustrated in Fig. 4.34.

As shown in Fig. 4.31, the total value of assets managed within the scope of Islamic structures reached USD 4.5 billion in 2022 and is expected to exceed USD 6 billion in 2027 (ICD and LSEG 2023). The Sukuk issuance globally reached USD 168,4 billion 2023, with an expectation to increase slightly to USD 170 billion in 2024 (S&P Global 2024a). Conversely, the sustainable Sukuk issuance reached nearly USD 11 billion in 2023 and exceeded USD 2 billion in the first quarter of 2024, as indicated in Fig. 4.31 (S&P Global 2024a).

Sukuk are the most common Islamic financial structures, typically in the form of asset-based

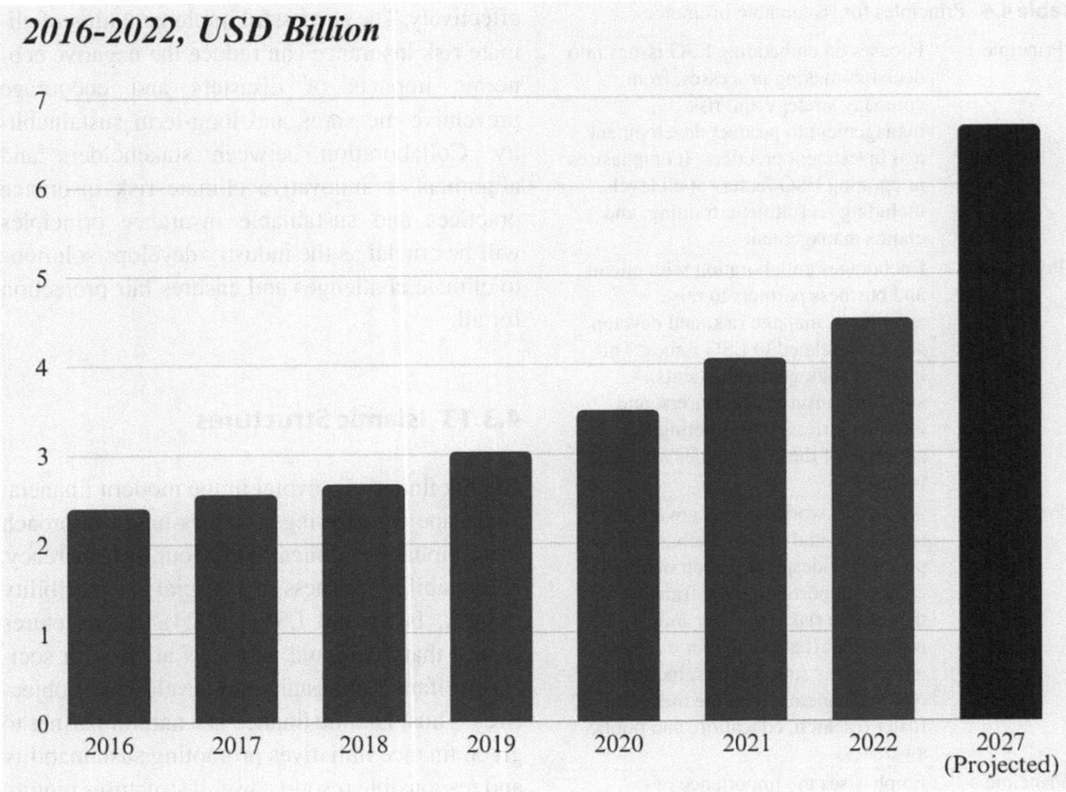

Fig. 4.34 Islamic finance assets growth (ICD and LSEG 2023)

Sukuk (ICMA, IsDB and LSEG 2024). Sukuk issuers distribute profits or revenues from tangible assets, such as rental income, to Sukuk holders in these structures. Sukuk holders own a share of the property rights over the underlying tangible assets, which must comply with Islamic law (ICMA, IsDB and LSEG 2024). Sukuk can be broadly categorised into asset-based and asset-backed structures. Asset-based Sukuk is most prevalent, where investors have a beneficial interest in the underlying assets but not direct ownership; the risk is largely linked to the creditworthiness of the issuer rather than the asset itself (Jobst et al. 2008). In contrast, asset-backed Sukuk offers investors direct ownership of the underlying assets, providing additional security and aligning more closely with the principles of risk-sharing central to Islamic finance (Haneef 2010).

Sukuk issuance has grown significantly due to its alignment with ethical investment principles, making it an attractive option for investors seeking stable and socially responsible opportunities. Furthermore, the rise of green Sukuk—Sukuk structured to finance environmentally sustainable projects—has highlighted the role of Islamic finance in promoting green transformation. Green Sukuk aligns closely with Islamic and green finance objectives by funding projects that exclude harmful industries and focus on sustainable development (World Bank 2019b). As a result, green Sukuk has become a critical tool in financing renewable energy, energy efficiency, and other green initiatives, driving the convergence of ethical and sustainable investment frameworks (Liu and Lai 2021).

Other than Sukuk, Mudarabah and Musharakah are contractual structures that enable investors to establish partnerships (Jais et al. 2019). The difference between Mudarabah and Musharakah lies in the contribution devoted to the partnership, the role in the management, and the share of profit and loss. Indeed, a partnership constitutes a Mudarabah when one partner grants financial capital and the other human capital (i.e., management experience), all partners share profits according to the ratio indicated in the contract, but the partners who provide funds bear losses (Jais et al. 2019; Iqbal and Llewellyn 2002; Khalil et al. 2002; ICMA, IsDB and LSEG 2024). Mudarabah is extensively used in Islamic banking for investment management, where the bank acts as the entrepreneur (Mudarib) and the depositor as the capital provider (Rab-ul-Mal). This structure allows investors to participate in profitable ventures without the direct responsibility of managing the investment (El-Gamal 2006). However, investors form Musharakah when all partners contribute capital in cash or kind, participate in the management, and share both profit and loss (Jais et al. 2019; Iqbal and Llewellyn 2002; ICMA, IsDB and LSEG 2024). Here, partners allocate profits based on the predetermined ratios indicated in their partnership contract, while losses are divided according to each partner's proportion of the total capital (Iqbal and Llewellyn 2002; ICMA, IsDB and LSEG 2024). Musharakah provides flexibility as it allows for both temporary and long-term partnerships. It is ideal for large-scale infrastructure and real estate projects, where joint venture arrangements can optimise resource use and management (Zamir and Mirakhor 2007).

Murabahah is a sale contract in which lenders sell assets to borrowers at a predetermined price, which includes a certain profit margin added to the original purchase price (Iqbal and Llewellyn 2002; ICMA, IsDB and LSEG 2024). In the application for Murabahah financing, borrowers indicate which assets they want to purchase (Iqbal and Llewellyn 2002). Lenders act as intermediaries and purchase these assets from third-party sellers to resell them to borrowers (Iqbal and Llewellyn 2002; ICMA, IsDB and LSEG 2024). Murabahah is highly favoured in trade financing and asset acquisition due to its straightforward structure, which involves a fixed profit margin agreed upon upfront. This transparency appeals to both parties as it eliminates uncertainty, aligning with the risk-averse nature of Islamic finance (Ahmed 2011). In consumer finance, Murabahah contracts enable individuals and businesses to acquire various assets, from vehicles to machinery, without engaging in interest-bearing loans, thereby adhering to Sharia principles (Siddiqi 2006). Borrowers pay the predetermined price to lenders either in a lump sum or in instalments; however, the ownership (and risks attached to the ownership of assets) is transferred upon delivery (Iqbal and Llewellyn 2002). For example, an individual who aims to purchase a house may apply to an Islamic bank, and if accepted, the bank purchases the house indicated in the application. The bank then sells the house to the individual at a price calculated by adding a certain profit margin to the original purchase price paid by the bank.

Takaful is a type of insurance contract where a group of participants agrees to contribute to a pool of funds, which is used to support members who suffer losses due to predetermined risks, thereby sharing the risks collectively rather than transferring them to a third-party insurance company (Iqbal and Llewellyn 2002). This cooperative model aligns with the principles of mutual assistance and shared responsibility central to Islamic finance, as it operates based on shared risk and profit. Takaful participants contribute to the fund and act as both the insurer and the insured, making it distinct from conventional insurance, which is profit-driven and often involves elements of uncertainty and gambling prohibited in Islamic law (Khan and Bhatti 2008). Takaful can be structured in various forms, such as Mudharabah (profit-sharing) or Wakalah (agency) models, offering flexibility in managing funds and distributing profits. Ali (2016) states that the Mudharabah model fosters a collective benefit and a sense of community by sharing any excess funds generated from investments between the participants and the Takaful operator by a pre-arranged ratio. Particularly in areas with sizable Muslim populations, takaful has experienced tremendous growth as a moral substitute for traditional insurance. It is increasingly used to cover many risks, including health, life, and property, providing Sharia-compliant protection to individuals and businesses (Billah 2007). Islamic finance instruments are summarised in Table 4.5.

Islamic finance structures align with green finance practices, considering both systems aim to prevent societal (i.e., environmental) damages and promote ethical and responsible use of resources (Liu and Lai 2021). Indeed, the SDGs and Islamic law are estimated to converge in over 95% of their objectives (UKIFC 2023). This synergy allows them to operate in tandem with each other and enormously accelerate the green transformation (ICMA, IsDB and LSEG 2024). As to asset-based structures, for example, the eligibility of the underlying tangible assets required for the issuance of green sukuk would not cause a contradiction, given that both green finance and Islamic finance tend to exclude the same type of assets. Despite this synergy, Islamic structures have not completely merged with the green finance system, considering they offer predictability and transparency as Islamic law requires.

Table 4.5 Summary of Islamic finance instruments

Structures	Key characteristics
Sukuk	Sukuk is a contractual structure that represents a share of an underlying asset. Sukuk issuers pay profits or revenues from tangible assets, such as rental income, to Sukuk holders. Sukuk holders own a share of the property rights over the underlying tangible assets, the eligibility of which is determined according to Islamic law
Mudarabah	Mudarabah is a partnership contract whereby one partner grants financial capital and the other human capital (i.e., management experience). All partners share profits according to the ratio indicated in the contract. Only the financial capital providers bear losses
Musharakah	Musharakah is a partnership contract whereby all partners contribute capital in cash or kind and participate in the management. Profits are allocated among partners based on the predetermined ratios indicated in the partnership contract. Losses are divided according to each partner's proportion of the total capital
Murabahah	Murabahah is a sale contract between lenders (financial institutions) and borrowers. Borrowers determine which assets to be purchased. Lenders purchase these assets from third-party sellers to resell them to borrowers. Borrowers pay the predetermined price, which includes a certain profit margin added to the original purchase price, to lenders either in a lump sum or in instalments
Takaful	Takaful is a type of insurance contract whereby a group of participants undertake to contribute to a pool of funds in exchange for receiving financial support when they suffer from damage because of a predetermined risk, meaning that they collectively share the risks rather than transferring them to insurance companies

This huge potential has already been reflected in the financial sector in that the issuance of sustainable Sukuk shows an upward trend, albeit slowly. According to the DIFC, UKIFC, GEFI, and LSE (2023), the green Sukuk issuance can reach USD 30 billion to USD 50 billion by 2025. Particularly, in 2024, the Dubai Islamic Bank issued a USD 1 billion sustainable sukuk under the Wakala structure. The proceeds of the Sukuk were intended to allow financing of eligible sustainable initiatives like green building projects, alternative energy, and clean transport (ICMA, IsDB, and LSEG 2024). Also, in 2023, DP World Crescent Ltd. Sukuk Mokhuerra structured a sustainable one worth USD 1.5 billion. The proceeds of this Sukuk were to be applied to project categories, including clean transportation, energy efficiency and renewable energy, and green buildings (ICMA, IsDB and LSEG 2024). Finally, the Islamic Development Bank issued a sustainable Sukuk worth USD 2 billion in 2024, intending to use the proceeds in sustainable initiatives in member countries, including foiling food insecurity, poverty alleviation, climate action, and resilience building (IsDB 2024).

References

1915canakkale (2024) Proje Finansmanı. https://www.1915canakkale.com/hakkinda/proje-finansmani. Accessed 24 May 2024

Abate G, Basile I, Ferrari P (2021) The level of sustainability and mutual fund performance in Europe:

an empirical analysis using ESG ratings. Corp Soc Respons Environ Manag 28(5):1446–1455

ActAlliance (2020) Climate risk insurance and risk financing in the context of climate justice. https://actalliance.org/publications/climate-risk-insurance-and-risk-financing-in-the-context-of-climate-justice/. Accessed 15 Aug 2024

Adams T, Winters B, Nazareth A, Carney M (2021) Taskforce on scaling voluntary carbon markets. Institute of International Finance (IIF). https://www.iif.com/Portals/1/Files/TSVCM_Report.pdf. Accessed 12 Oct 2024

Agrawal A, Hockerts K (2021) Impact investing: review and research agenda. J Small Bus Entrep 33(2):153–181

Ahmed H (2011) Product development in Islamic banks. Edinburgh University Press, Edinburgh

Aldy JE, Stavins RN (2012) The promise and problems of pricing carbon. J Environ Dev 21(2):152–180

Ali AJ (2016) Fundamentals of Islamic finance and banking. Wiley, Hoboken, NJ

Alyasa-Gan S, Che-Yahya N (2022) Intended use of IPO proceeds and survival of listed companies in Malaysia. J Risk Financ Manag 15(3):145

Amari J (2024) Japan's first sovereign climate transition bonds debut. Glob Financ Mag. https://gfmag.com/economics-policy-regulation/japans-first-sovereign-climate-transition-bonds-debut/. Accessed 22 Aug 2024

Antarciuc E, Zhu Q, Almarri J, Zhao S, Feng Y, Agyemang M (2018) Sustainable venture capital investments: an enabler investigation. Sustainability 10(4):1204. https://www.mdpi.com/2071-1050/10/4/1204. Accessed 12 Oct 2024

AON (2024) Climate and catastrophe insight. https://www.aon.com/en/insights/reports/climate-and-catastrophe-report. Accessed 20 Aug 2024

Apple (2023). Apple expands innovative Restore Fund for carbon removal. https://www.apple.com/uk/newsroom/2023/04/apple-expands-innovative-restore-fund-for-carbonremoval/#:~:text=opens%20in%20new%20window,nature%2Dbased%20carbon%20removal%20offsets

Asian Development Bank (2023) Climate finance landscape of Asia and the Pacific. ADB, Manila. https://doi.org/10.22617/TCS230305-2. Accessed 6 Sept 2024

Auerbach AJ, Hassett KA (2003) On the marginal source of investment funds. J Public Econ 87(1):205–232

Baranzini A, Goldemberg J, Speck S (2000) A future for carbon taxes. Ecol Econ 32(3):395–412

Baranzini A, van den Bergh JCJM, Carattini S, Howarth RB, Padilla E, Roca J (2017) Carbon pricing in climate policy: seven reasons, complementary instruments, and political economy considerations. WIREs Clim Change 8:e462. https://doi.org/10.1002/wcc.462

Barber BM, Morse A, Yasuda A (2021) Impact investing. J Financ Econ 139(1):162–185

Barthelmess B, Langlois J (2020) SME financing in MENA: a quantitative and qualitative analysis of multilateral and bilateral development lenders' intermediated lending practices. Rev Middle East Econ Financ 16(3):20200013

Battocletti V, Enriques L, Romano A (2023) The voluntary carbon market: market failures and policy implications. European Corporate Governance Institute (ECGI) Working Paper Series in Law.. https://www.ecgi.global/sites/default/files/working_papers/documents/thevoluntarycarbonmarket.pdf. Accessed 12 Oct 2024

BBVA (2019) BBVA listed again in Dow Jones Sustainability Index. https://www.bbva.com/en/sustainability/bbva-listed-again-in-dow-jones-sustainability-index/. Accessed 13 Jun 2024

Ben-David I, Franzoni F, Moussawi R (2017) Exchange-traded funds. Annu Rev Financ Econ 9(1):169–189

Bento N, Gianfrate G (2020) Determinants of internal carbon pricing. Energy Policy 143:111499. https://doi.org/10.1016/j.enpol.2020.111499

Betz R et al (2022) The carbon market challenge. Cambridge University Press, Cambridge. https://doi.org/10.1017/9781009216500

Beyond Meat (2024) Beyond meat announces pricing of initial public offering. https://investors.beyondmeat.com/news-releases/news-release-details/beyond-meatr-Kannounces-pricing-initial-public-offering#:~:text=LOS%20ANGELES%20%2C%20May%2001%2C%202019,public%20of%20%2425.00%20per%20share. Accessed 12 Oct 2024

Billah MM (2007) Applied Takaful and modern insurance: law and practice. Sweet & Maxwell Asia, Selangor

BlackRock (2024) Understanding repurchase agreements. https://www.blackrock.com/cash/en-us/insight-and-education/understanding-repurchase-agreements. Accessed 15 Oct 2024

Blaufelder C, Levy C, Mannion P, Pinner D (2021) A blueprint for scaling voluntary carbon markets to meet the climate challenge. McKinsey & Company. https://www.mckinsey.com/capabilities/sustainability/our-insights/a-blueprint-for-scaling-voluntary-carbon-markets-to-meet-the-climate-challenge. Accessed 12 Oct 2024

Bloomberg Barclays Indices (2021) Bloomberg barclays MSCI Global Green Bond Index. https://www.msci.com/documents/10199/242721/Barclays_MSCI_Green_Bond_Index.pdf/6e4d942a-0ce4-4e70-9aff-d7643e1bde96. Accessed 22 May 2024

BNP Paribas (2024) The Republic of Iceland issues the first sovereign gender bond. https://cib.bnpparibas/the-republic-of-iceland-issues-first-sovereign-gender-bond/. Accessed 22 Aug 2024

Boffo R, Patalano R (2020) ESG Investing: practices, progress and challenges. OECD, Paris. www.oecd.org/finance/ESG-Investing-Practices-Progress-and-Challenges.pdf. Accessed 12 Apr 2024

Böhringer C, Garcia-Muros X, Cazcarro I, Arto I (2017) The efficiency cost of protective measures in climate policy. Energy Policy 104:446–454. https://www.sciencedirect.com/science/article/abs/pii/S0301421517300071. Accessed 12 Oct 2024

Böning J, Di Nino V, Folger T (2023) Benefits and costs of the ETS in the EU. A lesson learned for the CBAM design. In: European Central Bank Working Paper Series. https://www.ecb.europa.eu/pub/pdf/scpwps/ecb.wp2764~3ff8cb597b.en.pdf. Accessed 12 Oct 2024

Borgmann M, Gierds J, Fischedick M, Henning HM, Matthies E, Pittel K, Renn J, Sauer DU, Spiecker I (2024) What is the voluntary carbon market—and what contribution does it make to climate action? In: Discussion Paper of the Academies' Project "Energy Systems of the Future". Series on "Energy Systems of the Future" (ESYS). https://doi.org/10.48669/esys_2024-6. https://www.leopoldina.org/fileadmin/redaktion/Publikationen/Nationale_Empfehlungen/2024_Impuls_ESYS_Kohlenstoffmarkt-EN.pdf. Accessed 12 Oct 2024

Borsa Istanbul (2024) Sustainability Indices. https://www.borsaistanbul.com/en/sayfa/2227/sustainability-indices. Accessed 13 Oct 2024

Bowen A (2011) The case for carbon pricing. Policy Brief, Grantham Research

Boyce JK (2018) Carbon pricing: effectiveness and equity. Ecol Econ 150:52–61. https://doi.org/10.1016/j.ecolecon.2018.03.030

Brigham EF, Ehrhardt MC (2016) Financial management: theory and practice, 15th edn. Cengage Learning. https://www.examplelink.com/brigham-financial-management. Accessed 10 Sept 2024

Bryan K (2023) Sustainability bond market stumbles as investors get picky. Financial Times. https://www.ft.com/content/309a703a-3a5f-420e-afea-38a142a2f21a. Accessed 16 May 2024

Buckley NJ (2004) Short-run implications of cap-and-trade versus baseline-and-credit emission trading plans: experimental evidence. Department of Economics, McMaster University

Buckley NJ, Müller R, Mestelman S (2007) Baseline-and-credit emission permit trading: experimental evidence under variable output capacity. Routledge eBooks, pp 31–50. https://doi.org/10.4324/9780203935361-10

Bueb J, Hanania LR, Clézio AL (2016) Border adjustment mechanisms: elements for economic, legal, and political analysis (working paper series). https://doi.org/10.35188/unu-wider/2016/063-8

Bui M, Adjiman et al (2018) Carbon capture and storage (CCS): the way forward. Energy Environ Sci 11(5):1062–1176. https://pubs.rsc.org/en/content/articlelanding/2018/ee/c7ee02342a. Accessed 12 Oct 2024

Bumpus AG, Liverman DM (2008) Accumulation by decarbonization and the governance of carbon offsets. Econ Geogr 84(2):127–155. https://doi.org/10.1111/j.1944-8287.2008.tb00401.x. https://www.researchgate.net/publication/227654068_Accumulation_by_Decarbonization_and_the_Governance_of_Carbon_Offsets. Accessed 12 Oct 2024

Burtraw D, Krupnick A, Markandya A (2000) The ancillary benefits and costs of climate change mitigation: a conceptual framework. OECD, Paris. https://doi.org/10.1787/9789264188129-en. https://www.oecd-ilibrary.org/environment/ancillary-benefits-and-costs-of-greenhouse-gas-mitigation_9789264188129-en. Accessed 12 Oct 2024

Cai J, Eidam F, Saunders A, Steffen S (2018). Loan syndication structures and price collusion. Available at SSRN 3250817. Accessed 25 May 2024

Cames M, Harthan RO, Füssler J, Lazarus M, Lee CM, Erickson P, Spalding-Fecher R (2016) Öko Institut e.V. https://climate.ec.europa.eu/system/files/2017-04/clean_dev_mechanism_en.pdf. Accessed 12 Oct 2024

Carbon Market Watch (2024a) EU ETS 101: a beginner's guide to the EU's emissions trading system. https://carbonmarketwatch.org/publications/eu-ets-101-a-beginners-guide-to-the-eus-emissions-trading--system-2024-update/. Accessed 12 Oct 2024

Carbon Market Watch (2024b) ETS impact on climate policy. https://carbonmarketwatch.org/?s=ETS+Impact+on+Climate+Policy. Accessed 12 Oct 2024

Carbon Pricing Leadership Coalition (2020) Defining carbon pricing policy and instruments. https://www.carbonpricingleadership.org/blogs/2020/4/20/defining-carbon-pricing-policy-and-instrumentsCarbonPricingRevenues092016.Pdf. Accessed 5 Jan 2024

CDP (2021) Putting a price on carbon: the state of internal carbon pricing by corporates globally

CDSB (2019) Physical risks and opportunities: guidance for companies on identifying and disclosing physical risks and opportunities from climate change

Centre For Tax Policy and Administration (2008) Environmentally related taxes and tradable permit systems in practice

Chazdon RL et al (2016) Carbon sequestration potential of second-growth forest regeneration in the Latin American tropics. Sci Adv 2(5):e1501639. https://doi.org/10.1126/sciadv.1501639. Accessed 12 Oct 2024

Christoffersen PF (1998) Evaluating interval forecasts. Int Econ Rev 39(4):841–862. https://www.jstor.org/stable/2527341. Accessed 3 Jun 2024

CISL (2022). Sustainability Leadership. Available at: https://www.cisl.cam.ac.uk/ (Accessed 12 October 2024)

Clark GL, Feiner A, Viehs M (2015) From the stockholder to the stakeholder: how sustainability can drive financial outperformance. Oxford University and Arabesque Partners. https://arabesque.com/research/From_the_stockholder_to_the_stakeholder_web.pdf. Accessed 5 Jun 2024

Climate Bond Initiative (2022) Climate bonds initiative Green Bond Database methodology. https://www.climatebonds.net/market/green-bond-database-methodology. Accessed 4 Apr 2024

Climate Bonds Initiative (2023) Green Bond Market Summary 2023. https://www.examplelink.com/climate-bonds-market-summary. Accessed 10 Sept 2024

Climate Bonds Initiative (2024) Sustainable Debt Market Summary Q1 2024. https://www.climatebonds.net/files/reports/cbi_mr_q1_2024_01e_1.pdf. Accessed 7 Sept 2024

Committee on Climate Change (2019) The future of carbon pricing in the UK. Committee on Climate Change, London. https://www.theccc.org.uk/wp-content/uploads/2019/08/Vivid-Economics-The-Future-of-Carbon-Pricing-in-the-UK.pdf. Accessed 12 Oct 2024

Committee on Climate Change (2020) The sixth carbon budget: the UK'S path to net zero. Committee on Climate Change, London. https://www.theccc.org.uk/wp-content/uploads/2020/12/The-Sixth-Carbon-Budget-The-UKs-path-to-Net-Zero.pdf. Accessed 12 Oct 2024

CoSAI (2021) Case Study: Finance for Sustainable Agricultural Intensification. https://www.iwmi.cgiar.org/archive/cosai/sites/default/files/CoSAI_Finance%20Case%20Study_0/index.pdf. Accessed 12 Oct 2024

Cumming D, Kumar S, Lim WM, Pandey N (2023) Mapping the venture capital and private equity research: a bibliometric review and future research agenda. Small Bus Econ 61(1):173–221

Danske Bank (2021) Green Bond Impact Report 2020. https://danskebank.com/-/media/danske-bank-com/pdf/investor-relations/debt/green-bonds/green-bond-impact-report-2020.pdf?rev=29488d2d49b84857a181b0e885730f5e&hash=702A3583BA1D03B6D0D3356343C2D43E. Accessed 16 May 2024

Dawes A, McGeadyand C, Majkut J (2023) Voluntary carbon markets: a review of global initiatives and evolving models. CSIS

DeMiguel V, Garlappi L, Uppal R (2009) Optimal versus naive diversification: how inefficient is the 1/N portfolio strategy? Rev Financ Stud 22(5):1915–1953. http://faculty.london.edu/avmiguel/DeMiguel-Garlappi-Uppal-RFS.pdf. Accessed 3 Jun 2024

Deutsche Bank (2021) Deutsche Bank executes first green repo financing. https://www.db.com/news/detail/20210910-deutsche-bank-executes-first-green-repo-financing?language_id=1. Accessed 15 Oct 2024

Deutsche Bank (2022) Scenario analysis and climate strategy. https://www.db.com/scenario-analysis. Accessed 20 Oct 2024

Deutsche Bank (2023) Green deposits service and use of funds. https://corporates.db.com/files/documents/legal-resources/green-deposit/Deutsche_Bank_Corporate_Bank_Green_Deposit_Annex_May_2021_EN.PDF. Accessed 15 Oct 2024

Divine J, Davis A (2022) How much would $10,000 invested in Tesla Stock at IPO Be Worth Today?. https://money.usnews.com/investing/articles/if-invested-tesla-stock-ipo-worth-today#:~:text=Tesla%20went%20public%20on%20June,U.S.%20listed%20stocks%20%26%20ETFs*. Accessed 12 Oct 2024

Do Good People (n.d.) The Main Sustainability Indexes. https://www.dogoodpeople.com/csr-trends/esg-criteria/the-main-sustainability-indexes/. Accessed 8 Jun 2024

Dorleans M (2018) Financial incentives for companies' engagement in work based learning. Eastern Partnership Annual Regional Forum on Work-based Learning in VET

Drew A, Nicholson A, Gakuya J, Mahabir M., Ward R (2023) Voluntary carbon markets: opportunities, challenges and the role of insurers. Milliman. https://www.milliman.com/-/media/milliman/pdfs/2023-articles/12-1-23-voluntary-carbon-markets.ashx. Accessed 12 Oct 2024

Dubai International Financial Centre, Islamic Finance Council UK, Global Ethical Finance Initiative and London Stock Exchange (2023) Green and sustainability Sukuk Update 2023: financing a sustainable future. https://www.lseg.com/en/data-analytics/resources/reports/green-sustainability-sukuk-update-2023-report. Accessed 10 Aug 2024

ECB (2023) Climate risk and financial stability. https://www.ecb.europa.eu/climate-risk. Accessed 20 Oct 2024

Eccles RG, Serafeim G (2013) The performance frontier: innovating for a sustainable strategy. Harv Bus Rev. https://hbr.org/2013/05/the-performance-frontier-innovating-for-a-sustainable-strategy. Accessed 16 May 2024

Eccles RG, Ioannou I, Serafeim G (2014) The impact of corporate sustainability on organizational processes and performance. Manag Sci 60(11):2835–2857. https://www.hbs.edu/ris/Publication%20Files/SSRN-id1964011_6791edac-7daa-4603-a220-4a0c6c7a3f7a.pdf. Accessed 3 Jun 2024

Ecosystem Marketplace (2023) 2023 State of the Voluntary Carbon Markets Report. https://www.ecosystemmarketplace.com/publications/state-of-the-voluntary-carbon-market-report-2023/. Accessed 12 Oct 2024

EEA (2017) climate change, impacts and vulnerability in Europe 2016: an indicator-based report. EEA

EIB (2022). EU Provides EUR 40 Million Loan to Support Sustainable Transport. European Investment Bank. (Accessed: 12 October 2024)

El-Gamal MA (2006) Islamic finance: law, economics, and practice. Cambridge University Press, Cambridge

Ellerman AD, Marcantonini C, Zaklan A (2016) The European Union emissions trading system: ten years and counting. Rev Environ Econ Policy 10(1):89–107. https://doi.org/10.1093/reep/rev014

Elton EJ, Gruber MJ (2013) Mutual funds. In: Handbook of the economics of finance, vol 2. Elsevier, Amsterdam, pp 1011–1061

Elton EJ, Gruber MJ, Brown SJ, Goetzmann WN (2019) Modern portfolio theory and investment analysis, 9th edn. Wiley, Hoboken, NJ. https://www.examplelink.com/elton-modern-portfolio. Accessed 10 Sept 2024

Equitable Bank (2024) Equitable Bank completes the first-ever Social Covered Bond issued by a Canadian bank. https://eqb.investorroom.com/2024-04-23-Equitable-Bank-completes-the-first-ever-Social-Covered-Bond-issued-by-a-Canadian-bank. Accessed 22 Aug 2024

Ernest (2023). How Carbon Markets Can Move Beyond Offsets to Achieve Net Zero. Available at: https://www.ey.com/en_gl (Accessed: 25 September 2023)

ETF Database (2024) Alternative Energy Equities ETFs. https://etfdb.com/etfdb-category/alternative-energy-equities/. Accessed 13 Oct 2024

European Banking Authority (2023) EBA consumer trends report 2022/23. https://www.eba.europa.eu/sites/default/files/document_library/Publications/Reports/2023/1054879/Consumer%20Trends%20Report%202022-2023.pdf. Accessed 25 May 2024

European Central Bank (2024) Investment fund statistics. https://www.ecb.europa.eu/stats/financial_corporations/investment_funds/html/index.en.html. Accessed 13 Oct 2024

European Commission (2015) Detailed questions and answers on the proposal to revise the EU emissions trading system. https://ec.europa.eu/commission/presscorner/detail/en/MEMO_15_5352. Accessed 12 Oct 2024

European Commission (2021b) European green deal: Commission proposes transformation of EU economy and society to meet climate ambitions. https://ec.europa.eu/commission/presscorner/detail/en/IP_21_3541. Accessed 7 Jan 2024

European Commission (2022). European Green Deal: Commission Proposes Transformation of EU Economy and Society to Meet Climate Ambitions. Available at: https://ec.europa.eu/commission/presscorner/detail/en/IP_21_3541 (Accessed: 7 Jan 2024)

European Commission (2023a) Report from the commission to the European Parliament and the council: on the functioning of the European carbon market in 2022 pursuant to articles 10(5) and 21(2) of directive 2003/87/EC

European Commission (2024a) Overview of sustainable finance. https://finance.ec.europa.eu/sustainable-finance/overview-sustainable-finance_en. Accessed 9 Apr 2024

European Commission (2024b) What is the EU ETS? https://climate.ec.europa.eu/eu-action/eu-emissions-trading-system-eu-ets/what-eu-ets_en. Accessed 12 Oct 2024

European Commission (2024c) Fit for 55. Available at: https://www.consilium.europa.eu/en/policies/green-deal/fit-for-55/#:~:text=Fit%20for%2055%20refers%20to,line%20with%20the%202030%20goal. Accessed 12 Oct 2024

European Council (2023) Infographic—fit for 55: how does the EU intend to address the emissions outside of the EU? https://www.consilium.europa.eu/en/infographics/fit-for-55-cbam-carbon-border-adjustment-mechanism/. Accessed 7 Jan 2024

European Environment Agency (EEA) (2021) Europe's changing climate hazards—an index-based interactive EEA report. https://www.eea.europa.eu/publications/europes-changing-climate-hazards-1. Accessed 27 Jul 2024

European Investment Bank (EIB) (2023) Water Sector Orientation. https://www.eib.org/attachments/lucalli/20230016_eib_water_sector_orientation_en.pdf. Accessed 7 May 2024

European Union (2015) EU ETS Handbook. https://climate.ec.europa.eu/system/files/2017-03/ets_handbook_en.pdf. Accessed 12 Oct 2024

Ewens M, Farre-Mensa J (2020) The deregulation of the private equity markets and the decline in IPOs. Rev Financ Stud 33(12):5463–5509

Fabozzi FJ (2013a) Bond markets, analysis, and strategies, 8th edn. Pearson, London. https://www.examplelink.com/fabozzi-bond-markets. Accessed 13 Oct 2024

Fabozzi FJ (2013b) Bond markets, analysis, and strategies, 8th edn. Pearson, London. https://www.examplelink.com/fabozzi-bond-markets. Accessed 10 Sept 2024

Fama EF, French KR (1993) Common risk factors in the returns on stocks and bonds. J Financ Econ 33(1):3–56. https://www.sciencedirect.com/science/article/abs/pii/0304405X93900235. Accessed 3 Jun 2024

Fatin L (2019) Green bond issuance tops USD 200 bn milestone—New global record in green finance: latest climate bonds data. https://www.climatebonds.net/2019/10/green-bond-issuance-tops-200bn-milestone-new-global-record-green-finance-latest-climate. Accessed 16 May 2024

Federal Reserve Bank of New York (2024) Repo and Reverse Repo Agreements. https://www.newyorkfed.org/markets/domestic-market-operations/monetary-policy-implementation/repo-reverse-repo-agreements. Accessed 15 Oct 2024

Fenwick M, Joubert T, Van Wyk S, Vermeulen E (2022) ESG as a business model for SMEs. European Corporate Governance Institute-Law Working Paper (642)

FirstRand Bank (2022) Green deposit framework. https://www.firstrand.co.za/media/investors/policies-and-practice/pdf/FRB-green-deposit-framework.pdf. Accessed 15 Oct 2024

Fischer C et al (2019) The legal and economic case for an auction reserve price in the EU emissions trading system. Soc Sci Res Netw. https://doi.org/10.2139/ssrn.3477716

Flammer C (2015) Does corporate social responsibility lead to superior financial performance? A regression discontinuity approach. Manage Sci 61(11):2549–2568. https://doi.org/10.1287/mnsc.2014.2038. Accessed 3 Jun 2024

Forest Stewardship Council (2024) Promoting responsible forest management through certification. https://www.fsc.org. Accessed 15 Sept 2024

Friede G, Busch T, Bassen A (2015) ESG and financial performance: aggregated evidence from more than 2000 empirical studies. J Sustain Financ Invest 5(4):210–233. https://doi.org/10.1080/20430795.2015.1118917. Accessed 25 May 2024

Friedrich M, Fries S, Pahle M, Edenhofer O (2019) Understanding the explosive trend in EU ETS prices—fundamentals or speculation? https://doi.org/10.48550/arXiv.1906.10572. Accessed 12 Oct 2024

FSB-TCFD (2017) Recommendations of the task force on climate-related financial disclosures. https://www.fsb-tcfd.org/publications/final-recommendations-report/. Accessed 20 Oct 2024

FSB-TCFD (2024). Financial Stability Board Task Force on Climate-related Financial Disclosures (TCFD). Available at: https://www.fsb.org/ (Accessed: 12 October 2024)

FTSE Russell (2021) FTSE4Good Index Series. https://www.lseg.com/en/ftse-russell/indices/ftse4good. Accessed 8 Jun 2024

Fuss S et al (2018a) A framework for assessing the performance of cap-and-trade systems: insights from the European Union emissions trading system. Rev Environ Econ Policy 12(2):220–241. https://doi.org/10.1093/reep/rey010

Fuss S, Lamb WF, Callaghan MW, Hilaire J, Creutzig F, Amann T, Beringer T, Oliveira Garcia W, Hartmann J, Khanna T, Luderer G, Nemet GF, Rogelj J, Smith P, Vi JL (2018b) Negative emissions—part 2: costs, potentials and side effects. Environ Res Lett 13(6):063002. https://doi.org/10.1088/1748-9326/aabf9f. https://iopscience.iop.org/article/10.1088/1748-9326/aabf9f/meta. Accessed 12 Oct 2024

Garg A, Arora S (2024) Impact investment ecosystem critical review and recommendations. Educ Admin Theor Pract 30(5):9571–9590

Geden O, Peters GP, Scott V (2018) Targeting carbon dioxide removal in the European Union. Taylor & Francis. https://doi.org/10.1080/14693062.2018.1536600

Gerlagh R, Heijmans RJRK, Rosendahl KE (2020) COVID-19 tests the market stability reserve. Environ Resour Econ 76:855–865. https://doi.org/10.1007/s10640-020-00441-0

Getmansky M, Lee PA, Lo AW (2015) Hedge funds: a dynamic industry in transition. Annu Rev Financ Econ 7(1):483–577

Giamporcaro S, Gond JP (2024) Can We Agree on How to Regulate Sustainable Finance?. Principles for Responsible Investment (PRI). https://www.unpri.org/academic-blogs/can-we-agree-on-how-to-regulate-sustainable-finance/12323.article. Accessed 5 Jun 2024

Global Environment Facility (2022) Summary of negotiations of the Eighth Replenishment of the GEF Trust Fund. https://www.thegef.org. Accessed 7 Sept 2024

Global Impact Investing Network (2023) What you need to know about impact investing. https://thegiin.org/publication/post/about-impact-investing/. Accessed 5 Jun 2024

Global Reporting Initiative (GRI) (2024) Consolidated set of the GRI standards. https://www.globalreporting.org/how-to-use-the-gri-standards/gri-standards-english-language/. Accessed 5 Jun 2024

Godlewski CJ (2008) Banking environment, agency costs, and loan syndication: a cross-country analysis

Goldstein A, Turner WR, Spawn SA, Anderson-Teixeira K, Cook-Patton SC, Fargione J, Gibbs HK, Griscom BW, Hewson JH, Hole DG (2020) Protecting irrecoverable carbon in Earth's ecosystems. Nat Clim Chang 10:287–295. https://www.nature.com/articles/s41558-020-0738-8. Accessed 12 Oct 2024

Gompers P, Lerner J (2001) The venture capital revolution. J Econ Perspect 15(2):145–168

Gorbach OG, Kost C, Pickett CJF (2022) Review of internal carbon pricing and the development of a decision process for the identification of promising internal pricing methods for an organisation. Renew Sustain Energy Rev 154:111745. https://doi.org/10.1016/j.rser.2021.111745

Gould S (2023) Understanding the voluntary carbon markets. https://www.ifac.org/knowledge-gateway/discussion/understanding-voluntary-carbon-markets. Accessed 12 Oct 2024

Goulder LH, Schein AR (2013) Carbon taxes versus cap and trade: a critical review. Clim Change Econ 04(03):1350010. https://doi.org/10.1142/s2010007813500103

Green J (2021) Does carbon pricing reduce emissions? A review of ex-post analyses. Environ Res Lett 16(4):043004. https://doi.org/10.1088/1748-9326/abdae9

Green Earth (2024) Google's $35 Million Venture Into the Carbon Credit Market. https://www.green.earth/news/googles-35-million-venture-into-the-carbon-credit-market. Accessed 12 Oct 2024

Green Finance Institute (2023) Transition finance: new asset class or emperor's new clothes? https://www.greenfinanceinstitute.com/wp-content/uploads/2023/12/Transition-finance.pdf. Accessed 8 May 2024

Hájek M, Zimmermannová J, Helman K, Rozenský L (2019) Analysis of carbon tax efficiency in energy industries of selected EU countries. Energy Policy 134:110955

Hamilton K, Sjardin M, Peters-Stanley M, Marcello T (2010) State of the voluntary carbon markets 2010. Bloomberg New Energy Finance and Ecosystem Marketplace. https://www.forest-trends.org/wp-content/uploads/imported/final_report_2010_9-1-10_no-crop-pdf.pdf. Accessed 12 Oct 2024

Hammer B, Marcotty-Dehm N, Martin J (2024) Use of proceeds in private equity-backed initial public offerings. Available at SSRN 4747463

Haneef R (2010) From asset-backed to asset-light structures: the need to return to the roots of Sukuk. Int J Islam Middle East Financ Manag 3(3):164–175

HBR (2022) Private equity should take the lead in sustainability. https://hbr.org/2022/07/private-equity-should-take-the-lead-in-sustainability#:~:text=For%20example%2C%20the%20Dutch%20pension,a%20framework%20of%2030%20questions. Accessed 12 Oct 2024

Holtslag M, Chevrollier N, Nijhof A (2021) Impact investing and sustainable market transformations: the role of venture capital funds. Bus Ethics Environ Respons 30(4):522–537

Honegger M, Reiner D (2018) The political economy of negative emissions technologies: consequences for international policy design. Clim Policy 18(3):306–321. https://doi.org/10.1080/14693062.2017.1413322

House of Commons Library (2021) The UK Emissions Trading Scheme (UK ETS). https://researchbriefings.files.parliament.uk/documents/CBP-9212/CBP-9212.pdf. Accessed 12 Oct 2024

Hristov I, Chirico A (2019) The role of sustainability key performance indicators (KPIs) in implementing sustainable strategies. Sustainability 11(20):5742

HSBC (2023) HSBC Green bonds report. https://www.hsbc.com/investors/fixed-income-investors/green-and-sustainability-bonds. Accessed 15 May 2024

Husain & Nair (2011). Structural Loopholes in Sustainability-Linked Bonds. Available at: https://documents1.worldbank.org/curated/en/099237410062223046/pdf/IDU0e099a50307f86045a80b33 (Accessed: 12 April 2024)

ICAP (2023). EU Emissions Trading System (EU ETS). Available at: https://icapcarbonaction.com/en/ets (Accessed 26 Sept 2023)

ICAP (2023a). ICAP ETS map. https://icapcarbonaction.com/en/ets. Accessed 26 Sept 2023

ICAP (2023e) EU adopts landmark ETS reforms and new policies to meet the 2030 target image. International Carbon Action partnership. EU Emissions Trading System (EU ETS)

ICAP (2024a) Emissions trading worldwide: status report 2024. International Carbon Action Partnership, Berlin. https://icapcarbonaction.com/en/publications/emissions-trading-worldwide-2024-icap-status-report. Accessed 12 Oct 2024

ICAP (2024b) ICAP allowance price explorer. https://icapcarbonaction.com/en/ets-prices. Accessed 12 Oct 2024

ICD and LSEG (2023) Islamic Finance Development Report 2023: Navigating Uncertainty. https://solutions.lseg.com/IslamicFinance_ICD_LSEG. Accessed 20 Aug 2024

ICICI Bank (2024) What is Green Deposit—How Does Green Deposit Work. https://www.icicibank.com/blogs/fixed-deposits/what-is-green-deposit. Accessed 15 Oct 2024

ICMA (2020a) Social and sustainable bond case studies: examples from frameworks applied in capital markets. https://www.icmagroup.org/assets/documents/Regulatory/Green-Bonds/June-2020/Social-and-sustainability-bond-case-studiesJune-2020-090620.pdf. Accessed 16 May 2024

ICMA (2020b) Sustainability-linked bond principles voluntary process guidelines. https://www.icmagroup.org/assets/documents/Regulatory/Green-Bonds/June-2020/Sustainability-Linked-Bond-Principles-June-2020-171120.pdf. Accessed 20 Apr 2024

ICMA (2020c) Sustainability-linked bond principles. International Capital Market Association. ICMA. https://www.icmagroup.org/assets/documents/Regulatory/Green-Bonds/June-2020/Sustainability-Linked-Bond-Principles-June-2020-171120.pdf. Accessed 16 May 2024

ICMA (2021a) Green Bond Principles. https://www.icmagroup.org/assets/documents/Sustainable-finance/2022-updates/Green-Bond-Principles_June-2022-280622.pdf. Accessed 16 May 2024

ICMA (2022). Green Bond Principles. Available at: https://www.icmagroup.org/assets/documents/Sustainable-finance/2022-updates/Green-Bond-Principles_June-2022-280622.pdf (Accessed: 16 May 2024).

ICMA (2023) Guidance Handbook. International Capital Market Association. ICMA. https://www.icmagroup.org/assets/documents/Sustainable-finance/2023-updates/The-Principles-Guidance-Handbook-November-2023-291123.pdf. Accessed 16 May 2024

ICMA, IsDB and LSEG (2024) Guidance on Green, Social and Sustainability Sukuk. https://www.icmagroup.org/assets/documents/Sustainable-finance/ICMA-IsDB-LSEG-Guidance-on-Green-Social-and-Sustainability-Sukuk-April-2024.pdf. Accessed 20 Jul 2024

IFC (2023) Preparing for Green Bond issuances. https://www.ifc.org/content/dam/ifc/doc/2024/preparing-for-green-bond-issuances-manual-for-fis-in-lac-en.pdf. Accessed 16 May 2024

IFU (2024) BancoSol issues the first gender bond in Bolivia. https://www.ifu.dk/en/news/bancosol-issues-first-gender-bond-in-bolivia/. Accessed 22 Aug 2024

IMF (2022) Climate change and financial risk. https://www.imf.org/climate-financial-risk. Accessed 20 Oct 2024

IMF (2024) What is a repurchase agreement? https://datahelp.imf.org/knowledgebase/articles/484377-what-is-a-repurchase-agreement. Accessed 15 Oct 2024

Insurance Information Institute (2024) A firm foundation: how insurance supports the economy. https://www.iii.org/publications/a-firm-foundation-how-insurance-supports-the-economy/driving-economic-progress/surplus-lines. Accessed 25 Jul 2024

InsuResilience Global Partnership (2020) Understanding Climate Risk Insurance. https://www.insuresilience.org/resources/understanding-climate-risk-insurance. Accessed 20 Aug 2024

International Capital Market Association (ICMA) (2023a) Climate Transition Finance Handbook. https://www.icmagroup.org/assets/documents/Sustainable-finance/2023-updates/Climate-Transition-Finance-Handbook-CTFH-June-2023-220623v2.pdf. Accessed 13 Oct 2024

International Capital Market Association (ICMA) (2024c) Sustainability-linked bond principles. https://www.icmagroup.org. Accessed 12 Oct 2024

International Finance Corporation (2019a) Green Buildings. https://www.google.com/url?q=https://www.ifc.org/en/insights-reports/2019/green-buildings-report&sa=D&source=docs&ust=1712928803677741&usg=AOvVaw0z2t30u_vdQSR5zUJH-NruD. Accessed 9 Apr 2024

International Finance Corporation (2022a) Blue finance safeguarding marine ecosystems, the climate, and livelihoods. https://www.ifc.org/en/stories/2022/blue-finance-marine-ecosystems-climate-livelihoods. Accessed 13 Oct 2024

International Finance Corporation (2022b) Green Bond Framework. https://www.ifc.org/content/dam/ifc/doc/2023/IFC-Green-Bond-Framework-Jan-2023.pdf. Accessed 11 May 2024

International Finance Corporation (2022c) Guidelines for Blue Finance. https://www.ifc.org/en/insights-reports/2022/guidelines-for-blue-finance. Accessed 6 Apr 2024

International Finance Corporation (IFC) (2024) Impact investing at IFC. https://www.ifc.org/en/our-impact/impact-investing-at-ifc. Accessed 22 Aug 2024

IPCC (2018) Global warming of 1.5 °C. https://www.ipcc.ch/sr15/. Accessed 12 Oct 2024

IPCC (2021) Climate change 2021: the physical science basis. https://www.ipcc.ch/report/ar6/wg1/. Accessed 17 Aug 2024

IPSF (2023) Implementing transition finance principles—Interim report. https://finance.ec.europa.eu/document/download/f332838d-a5da-4279-8986-dbb38a212432_en?filename=231204-ipsf-transition-finance-interim-report_en.pdf. Accessed 8 May 2024

Iqbal M, Llewellyn DT (2002) Islamic banking and finance: new perspectives on profit-sharing and risk. Edward Elgar, Cheltenham

İşbank (2024) İşbank secures a sustainability-linked syndicated loan of over USD 1 billion. https://www.isbank.com.tr/en/about-us/isbank-secures-a-sustainability-linked-syndicated-loan-over-usd-1-billion. Accessed 25 May 2024

ISDA (2022) The way forward for sustainability-linked derivatives. https://www.isda.org/a/p7cgE/The-Way-Forward-for-Sustainability-linked-Derivatives.pdf. Accessed 15 Oct 2024

IsDB (2024) IsDB issues US$ 2 billion Sukuk in First Benchmark of the Year. https://www.isdb.org/news/isdb-issues-us-2-billion-sukuk-in-first-benchmark-of-the-year. Accessed 5 Aug 2024

ISO (2018) ISO 31000:2018—Risk management—Guidelines. https://www.iso.org/iso-31000-risk-management.html. Accessed 20 Oct 2024

Ivashina V, Laeven L, Moral-Benito E (2020) Loan types and the bank lending channel (No. w27056). National Bureau of Economic Research

Jais M, Sofyan F, Bacha AM (2019) Mudarabah and Musharakah as an equity financing model: issues in practice. In: Proceedings Aceh Global Conference—business, economics, and sustainable development trends, vol 2, pp 107–114

Jakob M (2023) The political economy of carbon border adjustment in the EU. Oxf Rev Econ Policy 39(1):134–146. https://doi.org/10.1093/oxrep/grac044

Jarzabkowski P, Chalkias K, Clarke D, Iyahen E, Stadtmueller D, Zwick A (2019) Insurance for climate adaptation: opportunities and limitations11. GCA

Jobst AA et al (2008) Islamic bond issuance—what sovereign debt managers need to know. IMF policy discussion paper PDP/08/3. https://www.imf.org/external/pubs/ft/pdp/2008/pdp03.pdf. Accessed 10 Sept 2024

Johnson L, Toledano P (2023) Investment incentives: a survey of policies and approaches for sustainable investment, Columbia Center on sustainable investment, October 2022. https://papers.ssrn.com/sol3/papers.cfm?abstract_id=4323885. Accessed 8 May 2024

Jones TM, Felps W (2013) Stakeholder happiness enhancement: a neo-utilitarian objective for the modern corporation. Bus Ethics Q 23(3):349–379. https://www.researchgate.net/publication/271113935_Stakeholder_Happiness_Enhancement_A_Neo-Utilitarian_Objective_for_the_Modern_Corporation. Accessed 3 Jun 2024

Kaplan SN, Sensoy BA (2015) Private equity performance: a survey. Annu Rev Financ Econ 7(1):597–614

Kenton W (2024) Carbon trade: definition, purpose, and how carbon trading works. https://www.investopedia.com/terms/c/carbontrade.asp. Accessed 12 Oct 2024

Kettner-Marx C, Kletzan-Slamanig D (2018) Carbon taxes from an economic perspective. In: WIFO Working Papers, vol 554

Khalil AAA, Rickwood C, Murinde V (2002) Evidence on agency-contractual problems in mudarabah financing operations by Islamic banks. In: Iqbal M, Llewellyn DT (eds) Islamic banking and finance: new perspectives on profit-sharing and risk. Edward Elgar, Cheltenham, pp 57–92

Khan, T. and Bhatti, M. I. (2008). Developments in Islamic banking: the case of Pakistan. Basingstoke: Palgrave Macmillan

Klein M (2016) Blending public and private finance: what lessons can be learned from IFC'S experience? EMCompass note, no. 3. World Bank Group, Washington, DC

KMRC (2022) Green Bond Market in Africa, [Online]. https://kmrc.co.ke/assets/file/KMRC-Green-Bond-Market-in-Africa%2D%2DIssue-5-October-2022-Report.pdf. Accessed 15 May 2024

Kossoy A, Peszko G, Oppermann K, Prytz N, Klein N, Blok K, Lam L, Wong L, Borkent B (2015) State and trends of carbon pricing. https://openknowledge.worldbank.org/entities/publication/344977ab-f216-5fbf-97f5-4d3119d77482. Accessed 12 Oct 2024

Kotsantonis S, Serafeim G (2019) Four things no one will tell you about ESG data. Harv Bus Rev. https://doi.org/10.1111/jacf.12346. Accessed 5 Jun 2024

KPMG (2022) Sustainability linked financing—Accounting considerations by the borrower. https://assets.kpmg.com/content/dam/kpmg/au/pdf/2022/22ru10-sustainability-linked-financing-accounting-considerations.pdf. Accessed 12 Apr 2024

KPMG (2023) ESG consideration for private equity. https://assets.kpmg.com/content/dam/kpmg/pt/pdf/private-equity-considerations-brochure.pdf. Accessed 12 Oct 2024

KPMG (2024) Credit Markets Update Q2 2024. https://corporatefinance.kpmg.com. Accessed 15 Oct 2024

Kristjanson P (2020) PROFOR special issue 22-S1-IFR. Int For Rev 22(S1):44

Küfeoğlu S (2024a) Net zero: decarbonizing the global economies. In: Sustainable development goals series. Springer, Cham. https://doi.org/10.1007/978-3-031-70322-5

Küfeoğlu S (2024b) Environmental, social, and governance. In: Net zero: decarbonizing the global economies. Sustainable development goals series. Springer, Cham. https://doi.org/10.1007/978-3-031-70322-5_2

Küfeoğlu S, Prittinen S, Lehtonen M (2014) A summary of the recent extreme weather events and their impacts on electricity. Int Rev Electr Eng 9(4):821–828

Lackner KS, Brennan S, Matter JM, Park AHA, Wright A, Zwaan B (2018) The urgency of the development of CO2 capture from ambient air. PNAS 109(33):13156. https://doi.org/10.1073/pnas.1108765109

Lalon RM (2015) Green banking: going green. Int J Econ Financ Manag Sci 3:34–42. https://doi.org/10.11648/j.ijefm.20150301.15

Larsen ML (2019) A growing toolbox of sustainable finance instruments. Green FInance and Development Center. https://greenfdc.org/a-growing-toolbox-of-sustainable-finance-instruments/. Accessed 5 Jun 2024

Levis M (2011) The performance of private equity-backed IPOs. Financ Manag 40(1):253–277

Lin L (2022) Venture capital in the rise of sustainable investment. Eur Bus Organ Law Rev 23(1):187–216

Lin B, Li X (2011) The effect of carbon tax on per capita CO2 emissions. Energy Policy 39(9):5137–5146

Liu FH, Lai KP (2021) Ecologies of green finance: Green Sukuk and development of green Islamic finance in Malaysia. Environ Plan A Econ Space 53(8):1896–1914

Loan Market Association (2023a) Green loan principles. https://www.lma.eu.com/application/files/9115/4452/5458/741_LM_Green_Loan_Principles_Booklet_V8.pdf. Accessed 6 May 2024

Loan Market Association (2023b) Sustainability-linked loan principles. https://www.lma.eu.com. Accessed 15 Oct 2024

Low S, Fritz L, Baum CM, Sovacool BK (2024) Public perceptions on carbon removal from focus groups in 22 countries. Nat Commun 15. https://www.nature.com/articles/s41467-024-47853-w. Accessed 12 Oct 2024

Luong M (2010) The financial impact of student loans. Perspect Labour Income 11(1):5–18

Mannion P, Parry E, Patel M, Ringvold E, Scott J (2023) Carbon removals: how to scale a new Gigaton industry. McKinsey Sustain. https://www.mckinsey.com/capabilities/sustainability/our-insights/carbon-removals-how-to-scale-a-new-gigaton-industry#/. Accessed 12 Oct 2024

Market Research Future (2024) Venture Capitalist Sector to see growth in sustainable sector for 2024. https://www.marketresearchfuture.com/news/venture-capitalist-sector-to-see-growth-in-sustainable-development-sector-for-2024. Accessed 12 Oct 2024

Matos S, Viardot E, Sovacool BK, Geels FW, Xiong Y (2022) Innovation and climate change: a review and introduction to the special issue. Technovation 117:102612. https://doi.org/10.1016/j.technovation.2022.102612

Mcallister L (2012) Cap and trade. Research Gate [Preprint]. https://www.researchgate.net/publication/256036280_Cap_and_Trade

McCrossan (2024). What is transition finance, and why it matters. Nico McCrossan. https://trellis.net/article/what-transition-finance-and-why-itmatters/#:~:text=Wall%20Street%20sees%20transition%20finance,Source%3A%20Shutterstock%2FVylaks

McGrady C (2024) Task Force on Climate-related Financial Disclosures (TCFD)

McGuinness PB (2019) Risk factor and use of proceeds declarations and their effects on IPO subscription, price 'fixings', liquidity and after-market returns. Eur J Financ 25(12):1122–1146

McKitrick R (2016) A practical guide to the economics of carbon pricing. SPP Res Pap 9(28)

Medina E, Caniato F, Moretto A (2023) Framing sustainable supply chain finance: how can supply chain sustainability practices and finance solutions be integrated? J Purch Supply Manag 29(3):100837

Mehling M, Ritz RA (2023) From theory to practice: determining emissions in traded goods under a border carbon adjustment. Oxf Rev Econ Policy 39(1):123–133. https://doi.org/10.1093/oxrep/grac043

Metcalf GE, Weisbach D (2009) The design of a carbon tax. Harv Environ Law Rev 33:499–556

Migliorelli M (2021) What do we mean by sustainable finance? Assessing existing frameworks and policy risks. Sustainability 13(2):975

Miliman (2022) Making climate risk microinsurance work. https://us.milliman.com/en/insight/making-climate-risk-microinsurance-work. Accessed 15 Aug 2024

Minx JC, Lamb WF, Callaghan MW, Fuss S, Hilaire J, Creutzig F, Amann T, Beringer T, Garcia WOD, Hartmann J, Khanna T, Lenzi D, Luderer G, Nemet GF, Rogelj J, Smith P, Vicente JLV, Wilcox J, Dominguez MDMZ (2018) Negative emissions—part 1: research landscape and synthesis. Environ Res Lett 13(6):063001. https://doi.org/10.1088/1748-9326/aabf9b

Mirbabaeva N (2022) EBRD and Sparkasse Bank Makedonija support small businesses in North Macedonia. European Bank for Reconstruction and Development. https://www.ebrd.com/news/2022/ebrd-and-sparkasse-bank-makedonija-support-small-businesses-in-north-macedonia.html#:~:text=The%20European%20Bank%20for%20Reconstruction,%2Dsized%20enterprises%20(MSMEs. Accessed 10 Jun 2024

Mishkin FS, Eakins SG (2018) Financial markets and institutions, 9th edn. Pearson, London. https://www.examplelink.com/mishkin-financial-markets. Accessed 10 Sept 2024

MMR (2024) Invoice factoring market report: global industry analysis and forecast (2024–2030). https://www.maximizemarketresearch.com. Accessed 15 Oct 2024

MSCI (2022) Sustainability-linked financing. https://www.msci.com/documents/1296102/28588804/MSCI_SusLinkedFinance-cfs-en.pdf/5a180f8f-0611-0171-5279-372f882a6ad1?t=1683213695301. Accessed 12 Apr 2024

MSCI (2024) What is an index. https://www.msci.com/our-solutions/indexes/making-complex-simple/what-are-indexes. Accessed 13 Jun 2024
Munich Re (2024a) Droughts and heatwaves. https://www.munichre.com/en/risks/natural-disasters/droughts-heatwaves.html. Accessed 27 July 2024
Munich Re (2024b) Wildfire protection. https://www.munichre.com/en/insights/natural-disaster-and-climate-change/wildfire-protection.html. Accessed 27 Jul 2024
Narassimhan E et al (2017) Carbon pricing in practice: a review of the evidence. Climate Policy Lab
Nasdaq (2022) Investors Want Energy Transition, Not Pure-Play "Green" Companies. https://www.nasdaq.com/articles/investors-want-energy-transition-not-pure-play-green-companies. Accessed 16 May 2024
National Geographic Society (2024) Carbon Cycle. https://www.nationalgeographic.org/encyclopedia/carbon-cycle/. Accessed: 12 Oct 2024
NGFS (2019) A call for action: climate change as a source of financial risk. NGFS
NOAA (2021) Understanding climate: climate change risk. https://www.noaa.gov. Accessed 20 Oct 2024
Nowak A, Ross A, Yencha C (2018) Small business borrowing and peer-to-peer lending: evidence from the lending club. Contemp Econ Policy 36(2):318–336
NYK (2024) Nippon Yusen Kabushiki Kaisha, NYK Secures First Transition Syndicated Loan Financing. https://www.nyk.com/english/news/2024/20240129_02.html. Accessed 9 May 2024
Odedokun M (2004) Multilateral and bilateral loans versus grants: issues and evidence. World Economy 27(2):239–263
OECD (2021) Scaling up Green, social, sustainability and sustainability-linked bond- issuances in developing countries. OECD, Paris
OECD (2022) OECD guidance on transition finance: ensuring credibility of corporate climate transition plans. Green Finance and Investment. https://doi.org/10.1787/7c68a1ee-en. Accessed 8 May 2024
OECD (2023) Mechanisms to prevent carbon lock-in in transition finance. https://www.oecd.org/environment/mechanisms-to-prevent-carbon-lock-in-in-transition-finance-d5c49358-en.htm. Accessed 9 May 2024
OECD (2024a) Financing SMEs and Entrepreneurs 2024: an OECD Scoreboard. https://www.oecd.org/publications/financing-smes-and-entrepreneurs-23065265.htm#:~:text=Financing%20SMEs%20and%20Entrepreneurs%202024%3A%20An%20OECD%20Scoreboard%20monitors%20SME,significant%20decline%20in%20SME%20lending. Accessed 10 Jun 2024
Omfif (2023) Flexibility is the main appeal of transition finance. https://www.omfif.org/2023/08/flexibility-is-the-main-appeal-of-transition-finance/
Orden R, Calonje I (2022) Sustainability-linked finance—mobilizing capital for sustainability in emerging markets. International Finance Corporation (IFC). https://www.ifc.org/content/dam/ifc/doc/mgrt/emcompass-note-110-sustainability-linked-finance-web.pdf. Accessed 12 Apr 2024
Papulová Z, Gažová A, Šlenker M, Papula J (2021) Performance measurement system: implementation process in SMEs. Sustainability 13:4794. https://doi.org/10.3390/su13094794
Parnassus Investments (2024) Sustainable investing. https://www.parnassus.com/sustainability/sustainable-investing/esg-metrics/core-equity/environmental. Accessed 13 Oct 2024
Prindle B, Zarnikau J, Allis E (2010) Barriers and policy solutions to energy efficiency as a carbon emissions reduction strategy. Elsevier eBooks, Amsterdam, pp 207–239. https://doi.org/10.1016/b978-1-85617-655-2.00008-0
PwC (2011) Factoring sustainability into IPO planning: disclosure trends reveal a changing landscape. https://www.pwc.es/es/capital-markets/assets/factoring-sustainability-into-ipo-planning-1011.pdf. Accessed 12 Oct 2024
PwC (2024) ESG-Focused Investment Report. https://www.pwc.com/id/en/media-centre/press-release/2022/english/esg-focused-institutional-investment-seen-soaring-84-to-usd-33-9-trillion-in-2026-making-up-21-5-percent-of-assets-under-management-pwc-report.html. Accessed 12 Oct 2024
Quick R (2020) Carbon border adjustment. ZEuS Zeitschrift Für Europarechtliche Studien 23(4):549–597 https://doi.org/10.5771/1435-439x-2020-4-549
Rabobank (2022) Impact report: Green loans for sustainable agriculture. Rabobank, Utrecht. https://www.rabobank.com/about-us/organization/results-and-reports/downloads. Accessed 14 Sept 2024
Regeneration.VC (2024) Portfolio. https://regeneration.vc/portfolio. Accessed 13 Oct 2024
Rekenthaler J (2024) When low risk means high risk. Morningstar Research Services LLC. https://www.morningstar.com/columns/rekenthaler-report/when-low-risk-means-high-risk. Accessed 3 Jun 2024
Research & Markets (2024) Factoring Global Market Report. https://www.researchandmarkets.com. Accessed 15 Oct 2024
Reuters (2023) Activist investor Ubben is shutting down Inclusive capital sources. https://www.reuters.com/business/finance/activist-investor-ubben-shutting-down-inclusive-capital-wsj-2023-11-29/. Accessed 13 Oct 2024
RobinHood (2020) What is a Financial Index. https://learn.robinhood.com/articles/6c0KthaqLmaqvGfoBe8AcA/what-is-a-financial-index. Accessed: 13 Jun 2024
Rozanova M (2020) EBRD to boost green buildings and mortgage finance in North Macedonia. European Bank for Reconstruction and Development, 24 February. https://www.ebrd.com/news/2020/ebrd-to-boost-green-buildings-and-mortgage-finance-in-north-macedonia.html. Accessed 10 Jun 2024
S&P Dow Jones Indices (2024) S&P Dow Jones Indices: Index Methodology. S&P Global. https://www.

spglobal.com/spdji/en/documents/methodologies/methodology-sp-green-bond-indices.pdf. Accessed 5 Jun 2024

S&P Global (2024a) Islamic Finance 2024–2025: resilient growth anticipated despite missed opportunities. https://www.spglobal.com/ratings/en/research/articles/240429-islamic-finance-2024-2025-resilient-growth-anticipated-despite-missed--opportunities-13074068. Accessed 10 Aug 2024

S&P Global (2024b) What is an Index. https://www.spglobal.com/spdji/en/research-insights/index-literacy/what-is-an-index/#:~:text=An%20index%20is%20a%20group,market%20sector%2C%20or%20investment%20strategy. Accessed 13 Jun 2024

Schaefer L, Waters E (2016) Climate risk insurance for the poor & vulnerable: how to effectively implement the pro-poor focus of InsuResilience

Schippers ML, De Wit W (2022) Proposal for a carbon border adjustment mechanism. Glob Trade Customs J 17(1):10–18. https://doi.org/10.54648/gtcj2022002

Schneider L (2007) Is the CDM fulfilling its environmental and sustainable development objectives? In: An evaluation of the CDM and options for improvement. Öko Institut e.V. https://www.monitoringmatters.org/ppdfc/cdm.pdf. Accessed 12 Oct 2024

Schneider Electric (2021) Sustainability-linked loans: a financial innovation for the future. Rueil-Malmaison: Schneider Electric. https://www.annualreports.com/HostedData/AnnualReportArchive/s/OTC_SU_2021.pdf. Accessed 13 Sept 2024

Schneider L, La Hoz Theuer S (2019) Environmental integrity of International Carbon Market mechanisms under the Paris Agreement. Clim Policy 19(3):386–400. https://doi.org/10.1080/14693062.2018.1521332. Accessed 12 Oct 2024

Schoenmaker D, Schramade W (2021) Principles of sustainable finance. Oxford University Press, Oxford

Scholtens B, Kang FC (2013) Corporate social responsibility and earnings management: evidence from Asian economies. Corp Soc Responsib Environ Manag 20(2):95–112

Schröder M (2004) The performance of socially responsible investments: investment funds and indices. Fin Mkts Portfolio Mgmt 18(2):122

SEC (2018) Investor Bulletin: Index Funds. https://www.sec.gov/resources-for-investors/investor-alerts-bulletins/ib_indexfunds. Accessed 13 Oct 2024

Shalhoob H, Hussainey K (2022) Environmental, social and governance (ESG) disclosure and the small and medium enterprises (SMEs) sustainability performance. Sustainability 15(1):200

Shell (2019) Shell signs innovative USD 10 billion revolving credit facility. https://www.shell.com/news-and-insights/newsroom/news-and-media-releases/2019/shell-signs-innovative-dollar-10-billion-revolving-credit-facility.html#vanity-aHR0cHM6Ly93d3cuc2hlbGwuY29tL21lZGlhL25ld3MtYW5kLW11ZGlhLXJlbGVhc2VzLzIwMTkvc2hlbGwtc2lnbnMtaW5ub3ZhdGl2ZS1kb2xsYXItMTAtYmlsbGlvbi1yZXZvbHZpbmctY3JlZGl0LWZhY2lsaXR5Lmh0bWw. Accessed: 9 May 2024

Shell and BCG (2022) The Voluntary Carbon Market: 2022 Insights and Trends. https://www.bp.com/content/dam/bp/business-sites/en/global/corporate/pdfs/energy-economics/energy-outlook/bp-energy-outlook-2023.pdf. Accessed 15 Oct 2024

Siddiqi MN (2006) Islamic banking and finance in theory and practice: a survey of state of the art. Islam Econ Stud 13(2):2–48

Siemens Gamesa (2019) Siemens Gamesa Pioneers the Green Foreign Exchange Hedging Market. https://www.siemensgamesa.com/global/en/home/press-releases/191030-siemens-gamesa-green-fx-with-bnp.html. Accessed 15 Oct 2024

Smith, P., Davis, et al (2015) Biophysical and economic limits to negative CO2 emissions. Nat Clim Chang volume 6, p. 42–50. https://doi.org/10.1038/nclimate2870. https://www.researchgate.net/publication/285742474_Biophysical_and_economic_limits_to_negative_CO2_emissions. Accessed 12 Oct 2024

Snam (2024) Snam successfully launched 1.5 billion euros of sustainable financing in dual tranche, the first Green Bond for 500 million euros and a Sustainability-Linked Bond (SLB) for 1 billion euros. https://www.snam.it/en/media/news-and-press-releases/comunicati-stampa/2024/Snam-sustainable-financing-Green-Bond-Sustainability-Linked-Bond.html. Accessed 12 Oct 2024

Statman M (2002) Socially responsible mutual funds. Financ Anal J 58(3):30–39

Statman M (2005) Socially responsible indexes: composition and performance. J Portfolio Manag 28(2):100–109. https://papers.ssrn.com/sol3/papers.cfm?abstract_id=705344. Accessed 3 Jun 2024

Stavins RN (2007) The relative merits of alternative instruments for environmental protection (discussion paper 07-03). Harvard Environmental Economics Program. https://scholar.harvard.edu/files/stavins/files/relative_merits_final_published_article.pdf

Stavins RN (2008) A meaningful U.S. cap-and-trade system to address climate change. Harv Environ Law Rev 32(2):293–372. https://scholar.harvard.edu/files/stavins/files/helr_cap_trade_stavins.pdf. Accessed 12 Oct 2024

Steuer S, Tröger TH (2022) The role of disclosure in Green Finance. J Financ Regul 8:1–50. https://academic.oup.com/jfr/article/8/1/1/6539733. Accessed 22 May 2024

Streck C (2020) Who owns REDD+? Carbon markets, carbon rights and entitlements to REDD+ finance. Forests 11(9):959. https://doi.org/10.3390/f11090959

Sustainable Finance Working Group (2022) 2022 G20 sustainable finance report. https://g20sfwg.org/wp-content/uploads/2022/10/2022-G20-Sustainable-

Finance-Report-2.pdf#:~:text=URL%3A%20https%3A%2F%2Fg20sfwg.org%2Fwp. Accessed 22 May 2024
Sustainability Leadership (2022). University of Cambridge Institute for Sustainability Leadership (CISL). (2022). Integrating climate and nature: the rationale for financial institutions. Cambridge, UK: University of Cambridge Institute for Sustainability Leadership. https://www.cisl.cam.ac.uk/news-and-resources/publications/integrating-climate-and-nature-rationale-financialinstitutions#:~:text=Dr%20Nina%20Seega%2C%20Research%20Director,inextricably%20linked%20on%20multiple%20levels
Swiss Re Institute (2020) The economics of climate change: no action not an option
Swiss Re Institute (2024a) Resilience Index 2024: encouraging resilience gains, but more is needed. https://www.swissre.com/institute/research/sigma-research/natural-catastrophe-insurance-global-resilience-index-2024.html. Accessed 27 Jul 2024
Swiss Re Institute (2024b) Innovative risk solutions for climate resilience. https://www.swissre.com. Accessed 15 Oct 2024
TCFD (2020) Recommendations of the task force on climate-related financial disclosures
TCFD (2023) Deutsche Banks climate-related financial disclosures. https://www.tcfdhub.org/deutsche-bank-disclosures. Accessed 20 Oct 2024
TCFD (2024). Task Force on Climate-related Financial Disclosures (TCFD). Available at: https://www.fsb.org/ (Accessed 12 October 2024)
The Business Research Company (2024) Letter of credit confirmation market report 2024. https://www.thebusinessresearchcompany.com Accessed 15 Oct 2024
The Climate Reality Project (2017) Handbook on carbon pricing instruments
The Community Foundation for the Greater Capital Region (2024) Creating a Fund. https://www.cfgcr.org/professional-advisors/options-for-giving/creating-a-fund/. Accessed 13 Oct 2024
The Conference Board (2022) EU Carbon Border Adjustment Mechanism: A primer for stakeholders. https://www.conference-board.org/topics/climate-change/EU-carbon-border-adjustment-mechanism-primer-for-stakeholders
The Motley Fool (2024) Hedge Funds vs. Mutual Funds: What's Better? https://www.fool.com/investing/how-to-invest/mutual-funds/hedge-fund-vs-mutual-fund/
The World Bank (2021) What you need to know about IFC's Green Bonds. https://www.worldbank.org/en/news/feature/2021/12/08/what-you-need-to-know-about-ifc-s-green-bonds. Accessed 5 Jun 2024
Thisted EV, Thisted RV (2019) The diffusion of carbon taxes and emission trading schemes: the emerging norm of carbon pricing. Environ Polit 29(5):804–824. https://doi.org/10.1080/09644016.2019.1661155
Tietenberg T (2006) Emissions trading principles and practice. Taylor & Francis Group. https://doi.org/10.4324/9781936331284
Tsai WH (2020) Carbon emission reduction—carbon tax, carbon trading, and carbon offset. Energies 13(22):6128. https://doi.org/10.3390/en13226128
Türkiye İş Bankası (2024) TEMA Variable Fund (821). https://www.isbank.com.tr/en/821. Accessed 13 Oct 2024
U.S. Department of Energy (2023) Homeowner's guide to the Federal tax Credit for solar photovoltaics. DOE, Washington, DC. https://www.energy.gov. Accessed 11 Sept 2024
U.S. DOE Loan Programs Office (2022) Title 17 clean energy financing program: loan guarantees for innovative energy projects. DOE, Washington, DC. https://www.energy.gov. Accessed 8 Sept 2024
U.S. Energy Information Administration (2023) Renewable energy and the impact of tax incentives. EIA, Washington, DC. https://www.eia.gov. Accessed 14 Sept 2024
U.S. Green Building Council (2024) LEED certification for sustainable buildings. https://www.usgbc.org/leed. Accessed 13 Sept 2024
U.S. Securities and Exchange Commission (2024) Investor Bulletin: Hedge Funds. https://www.sec.gov/investor/alerts/ib_hedgefunds.pdf. Accessed 13 Oct 2024
UK Government (2023) UK ETS Clamps Down on Heavy Industry Emissions. https://energyadvicehub.org/uk-ets-clamps-down-on-heavy-industry-emissions/. Accessed 12 Oct 2024
UK Government (2024) UK Emissions Trading Scheme: How it works. https://energyadvicehub.org/the-uk-emissions-trading-scheme-frequently-asked-questions/#:~:text=How%20does%20the%20UK%20ETS,a%20system%20of%20tradeable%20allowances-. Accessed 12 Oct 2024
UKIFC (2023) Global Islamic Finance and UN SDGs TaskForce Key Outputs Report. https://ukifc.com/wp-content/uploads/2023/11/GEFI526_IF_Taskforce_Report_DIGITAL-1.pdf. Accessed 10 Aug 2024
Ul Haq I, Doumbia D (2022) Structural loopholes in sustainability-linked bonds. https://documents1.worldbank.org/curated/en/099237410062223046/pdf/IDU0e099a50307f86045a80b33201d0b7057cedf.pdf. Accessed 12 Apr 2024
Ul Haq I, Doumbia D (2023) Making sustainability-linked bonds more impactful. https://www.ifc.org/en/insights-reports/2023/making-sustainability-linked-bonds-more-impactful. Accessed 12 Apr 2024
UNDESA (2024) Micro, Small, and Medium-sized Enterprises (MSMEs) and their role in achieving the sustainable development goals [Online]. https://sustainabledevelopment.un.org/content/documents/26073MSMEs_and_SDGs.pdf. Accessed 13 Apr 2024
UNEP (2021) Adaptation Gap Report 2020

UNEP FI (2019) Changing Course: a comprehensive Investor Guide to scenario-based methods for climate risk assessment, in response to the TCFD

UNEP FI (2020) Beyond the Horizon: new tools and frameworks for transition risk assessments

UNEP FI (2021) The changing role of finance in sustainability. Leadership. https://www.unepfi.org. Accessed 20 Oct 2024

UNEPFI (2023). Changing Course: A Comprehensive Investor Guide to Scenario-Based Methods for Climate Risk Assessment, in Response to the TCFD. Available at: https://www.unepfi.org/ (Accessed 12 October 2024)

UNEP FI (2024). Beyond the Horizon: New Tools and Frameworks for Transition Risk Assessments. Available at: https://www.unepfi.org/ (Accessed: 12 October 2024)

Unilever (2024) Ensuring supply chain stability through commodity futures. https://www.unilever.com/sustainability/nature/sustainable-sourcing/. Accessed 15 Oct 2024

United Nations (1997) Industrialized countries to cut greenhouse gas emissions by 5.2 percent by legally binding agreement reached by Kyoto Climate Change Conference. United Nations Meetings Coverage and Press Releases. https://press.un.org/en/1997/19971211.endev456.html. Accessed 12 Oct 2024

United Nations Climate Change (2023) Microsoft global carbon fee | global

United Nations Development Programme (UNDP) (2023) Sustainable development goals. https://www.undp.org/sustainable-development-goals. Accessed 10 Sept 2024

United Nations Framework Convention on Climate Change (UNFCCC) (1997) Kyoto Protocol to the United Nations Framework Convention on Climate Change. https://unfccc.int/documents/2409. Accessed 12 Oct 2024

United Nations Framework Convention on Climate Change (UNFCCC) (2024a) Emissions trading. https://unfccc.int/process/the-kyoto-protocol/mechanisms/emissions-trading. Accessed 12 Oct 2024

United Nations Framework Convention on Climate Change (UNFCCC) (2024b) What is the Kyoto Protocol? https://unfccc.int/kyoto_protocol. Accessed 12 Oct 2024

Vanguard (2024). Venture Capitalist Sector to See Growth in Sustainable Sector for 2024. Available at: https://www.marketresearchfuture.com/news/venture-capitalist-sector-to-see-growth-insustainable-development-sector-for-2024 (Accessed: 12 October 2024)

Verbruggen A, Laes E, Woerdman E (2019) Anatomy of emissions trading systems: what is the EU ETS? Environ Sci Policy 98:11–19. https://www.sciencedirect.com/science/article/pii/S1462901119300188. Accessed 12 Oct 2024

Wagner G (2013) Carbon cap and trade. Elsevier eBooks, Amsterdam, pp 1–5. https://doi.org/10.1016/b978-0-12-375067-9.00071-1

WBG (2024b) World Bank Group Guarantee Platform Goes Live. https://www.worldbank.org. Accessed 15 Oct 2024

WEF (2020) The Global Risks Report 2020. https://www.weforum.org/reports/the-global-risks-report-2020. Accessed 20 Oct 2024

WEF (2022b) The Global Risks Report 2022. https://www.weforum.org/global-risks-2022. Accessed 20 Oct 2024

Wild W (2004) The economic basis of syndicated lending. Doctoral dissertation, Queensland University of Technology

Wood G, Wright M (2009) Private equity: a review and synthesis. Int J Manag Rev 11(4):361–380

Woodhall M (1995) Student loans. In: Carnoy M (ed) International encyclopedia of economics of education. Elsevier, Oxford, pp 420–425

World Bank (2018) Seychelles Launches World's First Sovereign Blue Bond. https://www.worldbank.org/en/news/press-release/2018/10/29/seychelles-launches-worlds-first-sovereign-blue-bond. Accessed 12 Oct 2024

World Bank (2019a) Seychelles: Introducing the World's First Sovereign Blue Bond. https://pubdocs.worldbank.org/en/242151559930961454/Case-study-Blue-Bond-Seychelles-final-6-7-2019.pdf. Accessed 18 Apr 2024

World Bank (2019b) Green Sukuk: financing renewable energy. https://www.worldbank.org/en/news/feature/2019/04/22/green-sukuk-financing-renewable-energy. Accessed 10 Sept 2024

World Bank (2020) State and trends of carbon pricing 2020. https://openknowledge.worldbank.org/handle/10986/33809. Accessed 12 Oct 2024

World Bank (2024a) Climate change overview: development news, research, data. Accessed 20 Oct 2024

World Bank (2024b) State and trends of carbon pricing 2024. https://openknowledge.worldbank.org/server/api/core/bitstreams/253e6cdd-9631-4db2-8cc5-1d013956de15/content. Accessed 12 Oct 2024

World Bank (2024c) World development report 2024: the middle-income trap. World Bank, Washington, DC. https://www.worldbank.org. Accessed 10 Sept 2024

World Bank Group (2016) Carbon credits and additionality: past, present, and future. World Bank

World Economic Forum (2020) Measuring Stakeholder Capitalism: towards common metrics and consistent reporting of sustainable value creation. https://www3.weforum.org/docs/WEF_IBC_Measuring_Stakeholder_Capitalism_Report_2020.pdf. Accessed 16 May 2024

World Economic Forum (2023a) Global Risks Report 2023. https://www.weforum.org/publications/global-risks-report-2023/. Accessed 20 Aug 2024

World Economic Forum (2023b) Why businesses must focus on long-term value creation to drive sustainability. https://www.weforum.org/agenda/2023/02/to-make-sustainability-a-reality-businesses-must-focus-on-long-term-value-creation/. Accessed 17 April 2024

Xu X, Xu X, He P (2016) Joint production and pricing decisions for multiple products with cap-and-trade and carbon tax regulations. J Clean Prod 112:4093–4106. https://doi.org/10.1016/j.jclepro.2015.08.081

Xu X et al (2017) Supply chain coordination with green technology under cap-and-trade regulation. Int J Prod Econ 183:433–442. https://doi.org/10.1016/j.ijpe.2016.08.029

Yılmaz E (2003) Sendikasyon kredileri. Öneri Dergisi 5(19):145–154

Zamir I, Mirakhor A (2007) An introduction to Islamic finance: theory and practice. Wiley, Singapore

Zettelmeyer J, di Mauro BW, Panizzea U, Gulati M, Buchheit L, Bolton P (2022) Geneva 25: Climate and debt. Geneva Reports on the World Economy. Center for Economic Policy Research

Zheng Y, Dutton J (2024) How nature positive start-ups are helping China build a carbon neutral economy. World Economic Forum. https://www.weforum.org/agenda/2024/06/china-uplink-nature-positive-startups-carbon-neutral-economy/#:~:text=To%20tackle%20climate%20change%2C%20China,65%25%20compared%20to%202005%20levels. Accessed 12 Oct 2024

MIX
Papier aus verantwortungsvollen Quellen
Paper from responsible sources
FSC® C105338
FSC
www.fsc.org

If you have any concerns about our products,
you can contact us on
ProductSafety@springernature.com

In case Publisher is established outside the EU,
the EU authorized representative is:
Springer Nature Customer Service Center GmbH
Europaplatz 3, 69115 Heidelberg, Germany

Printed by Libri Plureos GmbH
in Hamburg, Germany